Evolution and Victorian Musical Culture

This engaging book explores the dynamic relationship between evolutionary science and musical culture in Victorian Britain, drawing upon a wealth of popular scientific and musical literature to contextualize evolutionary theories of the Darwinian and non-Darwinian Revolutions. Bennett Zon uses musical culture to question the hegemonic role ascribed to Darwin by later thinkers and interrogates the conceptual premise of modern debates in evolutionary musicology. Structured around the Great Chain of Being, chapters are organized by discipline in successively ascending order according to their object of study, from zoology and the study of animal music to theology and the music of God. *Evolution and Victorian Musical Culture* takes a non-Darwinian approach to the interpretation of Victorian scientific and musical interrelationships, debunking the idea that the arts had little influence on contemporary scientific ideas, and by probing the origins of musical interdisciplinarity the volume shows how music helped ideas about evolution to evolve.

BENNETT ZON is Professor of Music at Durham University, Director of Durham's Centre for Nineteenth-Century Studies and Co-Director of the International Network for Music Theology. He is General Editor of *Nineteenth-Century Music Review* and of the book series Music in Nineteenth-Century Britain. His own publications include *The English Plainchant Revival* (1999), *Music and Metaphor in Nineteenth-Century British Musicology* (2000) and *Representing Non-Western Music in Nineteenth-Century Britain* (2007).

Evolution and Victorian Musical Culture

BENNETT ZON
Durham University

CAMBRIDGE
UNIVERSITY PRESS

CAMBRIDGE
UNIVERSITY PRESS

University Printing House, Cambridge CB2 8BS, United Kingdom

One Liberty Plaza, 20th Floor, New York, NY 10006, USA

477 Williamstown Road, Port Melbourne, VIC 3207, Australia

4843/24, 2nd Floor, Ansari Road, Daryaganj, Delhi – 110002, India

79 Anson Road, #06–04/06, Singapore 079906

Cambridge University Press is part of the University of Cambridge.

It furthers the University's mission by disseminating knowledge in the pursuit of education, learning, and research at the highest international levels of excellence.

www.cambridge.org
Information on this title: www.cambridge.org/9781107020443
DOI: 10.1017/9781139103985

First published 2017

Printed in the United Kingdom by TJ International Ltd. Padstow Cornwall

A catalogue record for this publication is available from the British Library.

Library of Congress Cataloging-in-Publication Data
Names: Zon, Bennett.
Title: Evolution and Victorian musical culture / Bennett Zon.
Description: Cambridge ; New York, : Cambridge University Press, 2017. |
Includes bibliographical references and index.
Identifiers: LCCN 2017011387 | ISBN 9781107020443 (hard back : alk. paper)
Subjects: LCSH: Music–Social aspects–England–History–19th century. | Evolution (Biology)–England–History–19th century. | Great Britain–Civilization–19th century. | Great Britain–History–Victoria, 1837–1901.
Classification: LCC ML3917.G7 Z66 2017 | DDC 780.941/09034–dc23
LC record available at https://lccn.loc.gov/2017011387

ISBN 978-1-107-02044-3 Hardback

To Challoner and Gabriel,
my two wonderful children on the cusp of evolution

Contents

Figures

Acknowledgements

I would like to thank the following institutions for their kind permission to reproduce images used in this book: Cambridge University Library; the Ernst Mayr Library, Museum of Comparative Zoology, Harvard University; the Master and Fellows of Trinity College, Cambridge University; the Sherardian Library of Plant Taxonomy, the University of Oxford; the Surrey History Centre; the Wellcome Library, London; and the Zoological Society of London. I am, in addition, deeply indebted to the Institute of Advanced Study, Durham University, for its gracious support through a Christopherson–Knott Fellowship. Many people reviewed draft portions of my work, but I am especially grateful for the assistance of Bernard Lightman, Garry Marvin and Thomas Rutten. Fraser Liddell, Lorenzo Servitje and Flora Wilson were very helpful in providing me with useful information and resources, and lastly Kate Brett and Vicki Cooper (current and previous commissioning editors at Cambridge University Press, respectively) have been immensely attentive. More than anyone else, however, I would like to thank my wife and partner Clare for her intellectual and editorial contribution to this book. Without her, *Evolution and Victorian Musical Culture* would have been simply impossible.

Introduction

Cheesecake. One of the greatest gifts of the culinary gods: 'Decadent and delicious', according to *The Cheesecake Bible* (2015);[1] pure 'seduction', claims the title of *Cheesecake Seduction* (2008);[2] 'excess', declares *The Joy of Cheesecake* (1980).[3] Cheesecake conjures up all sorts of images, but the word excess sums them up. It suggests something surplus to requirement, superfluous, spare, redundant, unwanted, unnecessary, excessive and leftover[4] – something unessential for survival, fundamentally unadaptive and vaguely pointless. In *How the Mind Works* (1997) Stephen Pinker argues that the arts are excessive too; like 'alcohol, drugs, and rich deserts'[5] they represent 'a biologically pointless challenge: figuring out how to get at the pleasure circuits of the brain and deliver little jolts of enjoyment without the inconvenience of wringing bona fide fitness increments from the harsh world'.[6] All the arts may be pointless but one stands out from than the rest; metaphorically speaking, it is cheesecake:

> as far as biological cause and effect are concerned, music is useless. It shows no signs of design for attaining a goal such as long life, grandchildren, or accurate perception and prediction of the world. Compared with language, vision, social reasoning, and physical know-how, music could vanish from our species and the rest of our lifestyle would be virtually unchanged'; 'music is auditory cheesecake, an exquisite confection crafted to tickle the sensitive spots . . . of our mental faculties.[7]

Pinker's hypothesis reflects a common theoretical viewpoint amongst evolutionary psychologists called EEA (environment of evolutionary adaptedness): 'the idea that only those functions that evolved in the distant

[1] Emma Watt, *The Cheesecake Bible*, Homemade Cookbook, n.p.: Emma Watt, 2015, 9.
[2] Catherine Lau, *Cheesecake Seduction*, Singapore: Marshall Cavendish Cuisine, 2010.
[3] Dana Bovberjerg and Jeremy Iggers, *The Joy of Cheesecake*, Hauppauge, NY: Barron, 1980, 3.
[4] https://www.google.co.uk/search?q=excess&rlz=1C5CHFA_enGB748GB748&oq=excess&aqs=chrome..69i57j0l5.1026j0j7&sourceid=chrome&ie=UTF-8.
[5] Joseph Carroll, 'Steven Pinker's Cheesecake for the Mind', *Philosophy and Literature* 22/2 (1998), https://www.google.co.uk/search?q=excess&rlz=1C5CHFA_enGB748GB748&oq=excess&aqs=chrome..69i57j0l5.1026j0j7&sourceid=chrome&ie=UTF-8.
[6] Stephen Pinker, *How the Mind Works*, London: Penguin, 1997/1999, 524.
[7] Pinker, *How the Mind Works*, 528; 534.

evolutionary past have any particular adaptive status'.[8] According to advocates of EEA music appears too recently in the evolutionary history of man to be adaptive (the first instruments appear roughly 40,000 years ago). Cultural archaeologist Steven Mithen offers a compelling – if speculative – alternative. In *The Prehistory of the Mind* (1996) he proposes a theory of 'cognitive fluidity'[9] in which previously separate parts of the brain developed connections when subject to the environmental changes initiated at and by the dawn of culture (around the same time musical instruments and other evidence of culture begin to appear). Mithen gives the impression of rather pitying Pinker: 'Is music no more than auditory cheesecake, as Pinker would have us believe? Is it simply an evolutionary spin-off from language – a lucky break for humankind, providing song and dance as a relief from the tedium of survival and reproduction?'[10] Not at all; for Mithen the Neanderthals evolved 'a music-like communication system that was more complex and more sophisticated than that found in any of the previous species of Homo ... which via iconic gestures, dance, onomatopoeia, vocal imitation and sound synaesthesia, resulted in a further expansion of the brain and changes to its internal structure as additional neural circuits were formed'.[11] Whether or not music is auditory cheesecake one thing is certain: the cheesecake wars continue to rage to this very day without an obvious winner; as Nicholas Bannan suggests, they have been raging ever since Rousseau.[12] *Evolution and Victorian Musical Culture* extends the debate by exploring how and why evolutionary ideas about music shaped Victorian thought and behaviour. It seeks to debunk the common misapprehension that the sciences influence the arts more than the arts influence the sciences, and it aims more broadly to prove that music helped evolution evolve.

Structure and Rationale

Evolution and Victorian Musical Culture is organized around a structure which reflects, I believe, the dominant evolutionary science of Victorian

8 Carroll, 'Steven Pinker's Cheesecake For The Mind'.

9 Steven Mithen, *The Prehistory of the Mind: A Search for the Origins of Art, Religion and Science*, London: Phoenix, 1996/2003, 76, and *passim* 154–246.

10 Steven Mithen, *The Singing Neanderthals: The Origins of Music, Language, Mind and Body*, London: Phoenix, 2005, 11.

11 Mithen, *Singing Neanderthals*, 234.

12 Nicholas Bannan (ed.), *Music, Language and Human Evolution*, Oxford: Oxford University Press, 2012, 4.

musical culture – the Great Chain of Being, a concept ordering nature in a fixed, ascending sequence from the lowest to highest forms of life. Thus each chapter explores a different discipline linked to music in the ascending order of the Great Chain of Being: Chapter 1 (Zoomusicology) focuses on music and animals; Chapter 2 (Ethnomusicology), the music of non-Western savagery; Chapter 3 (Folk Musicology), folk song and the domestic peasant; Chapter 4 (Music Pedagogy), educational music for children; Chapter 5 (Music Biography), biographies of great musical individuals; Chapter 6 (Music History), musical histories of great men; and Chapter 7 (Music Theology), music and concepts of God. An epilogue rounds it off with thoughts on the place of music beyond God. Each chapter includes a brief outline; an introduction to the chapter's key discipline and its interaction with musical culture; and an extended set of case studies. Chapters are intended to be discrete though there is, inevitably, some continuity between them – mostly between Chapters 2 (Ethnomusicology) and 3 (Folk Musicology), and 5 (Music Biography) and 6 (Music History).

As a structure for organizing material the Great Chain of Being has many theoretical benefits but, admittedly, there are a few practical disadvantages as well. Some readers may find fault with my choice of disciplines or genres, for example, omitting as I do fiction and poetry, or psychology, philosophy and art history. As a guiding principle I exclude these (but reference them abundantly) because as disciplines they tend not to focus on any one particular type of being in the Great Chain itself. Some disciplines have, moreover, been studied much more comprehensively than others – fiction and poetry, for example.[13] The disciplinary and genre-related status of biography is also problematical. Birgitte Possing describes biography as '*the interdisciplinary analytic field* known as biographical studies': the field, she argues, 'distinguishes between the *genre*, this being the empirical narratives of individual lives as they can be read from antiquity to the present day, and *the academic discipline*, which this genre comprises'.[14] There is also some variability in the way Victorian

[13] See Phyllis Weliver, *Woman Musicians in Victorian Fiction: Representations of Music, Science and Gender in the Leisured Home*, Aldershot: Ashgate, 2000; Delia da Sousa Correa, *George Eliot, Music and Victorian Culture*, Houndmills and New York: Palgrave Macmillan, 2003; Sophie Fuller and Nicky Losseff (eds.), *The Idea of Music in Victorian Fiction*, Aldershot: Ashgate, 2004; Phyllis Weliver (ed.), *The Figure of Music in Nineteenth-Century British Poetry*, Aldershot: Ashgate, 2005; and Claire Mabilat (ed.), *Orientalism and Representations of Music in Nineteenth-Century British Popular Arts*, Aldershot: Ashgate, 2008.

[14] Birgitte Possing, 'Biography: Historical', www.possing.dk/pdf/historicalbio.pdf, accessed 17/2/16.

culture ordered the Great Chain. Non-Western savages, domestic peasants and children are often grouped together in Victorian thinking, and many believed that man (not God) stood at the apex of the ordered world. I respond to this predicament by problematizing the role of theology within zoology (Chapter 1) and the role of evolutionary thought within theology (Chapter 7). This is intended to give the impression of an inherent circularity in the Great Chain of Being which I believe corresponds to what Arthur O. Lovejoy calls a 'Great Chain of Becoming' (examined in the Epilogue). Daniel Wilson rightly notes that Lovejoy was desperate to preserve God's presence within the created order;[15] indeed, in his seminal book *The Great Chain of Being* (1936) Lovejoy declaims: 'Not only had the originally complete and immutable Chain of Being been converted into a Becoming, in which all genuine possibles are, indeed, destined to realization grade after grade, yet only through a vast slow unfolding in time; but now God himself is placed in, or identified with, this Becoming'.[16]

Lovejoy's reading of the Great Chain of Being provides *Evolution and Victorian Musical Culture* with three underlying concepts: plenitude, continuity and gradation. Plenitude suggests that for every being in the world 'there must be an ultimate reason, self-explanatory and "sufficient"'; continuity demands that 'there are no sudden "leaps" in nature; infinitely various as things are, they form an absolutely smooth sequence, in which no break appears'; and gradation, the Aristotelian idea of arranging 'all animals in a single graded *scala naturae* according to their degree of "perfection"'.[17] I do not observably apply Lovejoy's three evolutionary principles to my study of Victorian musical culture but let them freely permeate the fabric of the book as a whole. Chapter 1 (Zoomusicology), for instance, uses the plenitude of birdsong to explain what it was like to think (and hear) like a Victorian – to rationalize, insofar as possible, the meaning of birdsong for Victorian musical culture. At the same time it reveals how concepts of continuity between humans and animals (what today is called human–animal relations) undergirded that meaning, and how the structurally liminal position of birds within the Great Chain of Being predetermined the contingency of a relationship between men and birds. The same principles operate at the opposite end of the Chain (and book). Chapter 7

[15] Daniel J. Wilson, 'Lovejoy's The Great Chain of Being after Fifty Years', *Journal of the History of Ideas* 48/2 (April to June 1987), 196.

[16] Arthur O. Lovejoy, *The Great Chain of Being: A Study of The History of An Idea*, New York; Harper Torchbooks, 1936/1960, 325–26.

[17] Lovejoy, *Great Chain of Being*, 327; 327; 58.

(Music Theology) uses the plenitude of Christology to explain music's relationship to man and man's relationship to Christ and God's creation. It explores the creative continuity between God and His creation within nature, and the seamless contractually unbroken relationship between human beings and their creator.

Despite the intellectual virtuosity of his *Great Chain of Being* Lovejoy actually spends very little time on the topic of music, but as I show in the following pages their interrelationship is acute in Victorian culture. One factor brings this into high relief, and it is that factor which principally motivated the writing of this book: The Land Without Music. The Land Without Music is a long-standing chauvinism against English musical culture, and much of the Victorian period was spent refuting the accusation and dispelling the myth. What few people realize is that evolutionary thought played an essential role in Britain's musical defence. As I show in Chapter 5 (Music Biography) and 6 (Music History) it was evolutionary language which provided the ammunition in the war against this particular prejudice, especially in the writings of C. Hubert H. Parry and Henry Davey. Music fought that battle armed with the latest science because it was science, not music, which was the universal language of the time. Like Bengali nationalist musicologist Sourindro Mohun Tagore (discussed in Chapter 2, Ethnomusicology) critics of The Land Without Music spoke a language their European musical colonizers could understand – evolution – and they used that language to assert their individual difference, or what Homi Bhabha calls 'a difference that is almost the same, but not quite'.[18] It could go both ways, however – evolutionary ideas could work against Victorian musical culture. Nicholas Temperley, for instance, charts the rise and fall of The Land Without Music in his provocative article 'Xenophilia in British Musical History' (1999), using music histories to measure and reveal four noticeable trends: (1) a period declines into a dark age from the 1700s to 1800s, (2) a mixed period of transition from the 1800s to 1830s, (3) a period of increasingly upward trajectory from the 1830s to 1870s, and (4) an entirely positive trajectory (known as the English Musical Renaissance) from the 1870s. As shown in Chapters 5 and 6 many people have tried to explain historical reasons for The Land Without Music, myself included, but Temperley's reason is amongst the most compelling: 'Social emulation of the upper classes', he suggests, 'was the spur that energized the middle classes to the great achievements of the

[18] Homi K. Bhabha, *The Location of Culture*, London: Routledge, 1994, 86.

Victorian age . . . In music, emulation of the upper classes was disastrous, because they had long since downplayed its value and adopted foreignness as their shibboleth. This is the point of my argument'.[19] Interesting in itself Temperley's argument tells only half the story because behind the musical effect is a sociological cause underpinned by a Great Chain of Being translated into an order of social class. The middle class, in other words, aped the tastes of the upper class – or mimicked its superior colonizer – as an evolutionary mechanism of survival to climb higher on the evolutionary chain. Temperley was right: it was *disastrous* for British musical culture, but conversely it was *good* for the idea of the Great Chain of Being.

Three historical factors lie behind Victorian xenophilia and its brush with the Great Chain of Being: (1) the development of professional disciplinary organization, (2) changes in higher education and university curriculums, and (3) popularization, and these run like a current through *Evolution and Victorian Musical Culture*. The first of these concerns the more positive role science played in helping form the discipline and professional organization of music. The importance of science is evident from the very founding of Britain's first professional musicological society, the Musical Association. William Spottiswoode, one of its first Vice Presidents, records the moment of conception:

> It has been suggested by several leading persons interested both in the theory and practice of Music, that the formation of a Society, similar in the main features of its organisation to existing Learned Societies, would be a great public benefit. Such a Musical Society might comprise among its members the foremost Musicians, theoretical as well as practical, of the day; the principal Patrons of Art; and also those Scientific men whose researches have been directed to the science of Acoustics, and to kindred inquiries. Its periodical meetings might be devoted partly to the reading of Papers upon the history, the principles, and the criticism of Music; partly to the illustration of such Papers by actual performance; and partly to the exhibition and discussion of experimental relating to the theory and construction of musical instruments, or to the principles and combination of musical sounds.[20]

The Musical Association practiced what they preached, and the first issue of the *Journal of the Musical Association* was crammed with scientific articles focusing on experimental work in acoustics, scales, tunings, pitch

[19] Nicholas Temperley, 'Xenophilia in British Musical History', in Bennett Zon (ed.), *Nineteenth-Century British Music Studies*, vol. 1, Aldershot: Ashgate, 1999, 14.

[20] William Spotteswoode, 'Letter, Addressed by William Spottiswoode, Esq., M.A., F.R.S., to Some Leading Members of the Musical and Scientific World, Originated this Association (8 April 1874)', *Proceedings of the Musical Association* 1 (1874–75), iii.

and other areas of music including William Stone's 'On Extending the Compass and Increasing the Tone of Stringed Instruments, with Especial Reference to the Author's and Mr. Meeson's Elliptical Tension-Bars'; Robert Bosanquet's 'Temperament; Or, The Division of the Octave' and 'Temperament; Or, The Division of the Octave (Part II); Sedley Taylor's 'On a Suggested Simplification of the Established Pitch-Notation'; James Baillie-Hamilton's 'On the Application of Wind to String Instruments'; and Alexander John Ellis's 'Illustrations of Just and Tempered Intonation'.[21] Subsequent issues continue this trend but then dip in the 1880s as increasingly historical interests squeezed science out of its pole position. Experimental musical science (which is not the subject of *Evolution and Victorian Musical Culture*) was dominated by the massively influential scientist Hermann von Helmholtz, an enthusiastic Darwinian whose work in practice appears to have influenced Darwin more than Darwin influenced him, especially in the field of optics.[22] Beyond the pages of the Musical Association the experimental literature of Victorian musical science tend, broadly speaking, to reflect Helmholtz's approach, underwriting simpler popular primers like Sedley Taylor's *Sound and Music: A Non-Mathematical Treatise on the Physical Constitution of Musical Sounds and Harmony* (1873), Pietro Blaserna's *The Theory of Sound in its Relation to Music* (1876), Stone's *The Scientific Basis of Music* (1878), John Cook's *Sound and the Physical Basis of Music* (1877) and more comprehensive studies like William Pole's *Philosophy of Music* (1879) to name but a few publications.

A second related cog in the evolutionary machine of Victorian musical culture is higher education and its role in diffusing scientific knowledge about music. Rosemary Golding gives us a good overview of problems associated with higher education. Before the Victorian period musical science was conducted by amateurs largely outside universities but in venues like the

[21] See *Proceedings of the Musical Association* 1 (1874–75): W. H. Stone, 'On Extending the Compass and Increasing the Tone of Stringed Instruments, with Especial Reference to the Author's and Mr. Meeson's Elliptical Tension-Bars', 1–3; R. H. M. Bosanquet, 'Temperament; Or, The Division of the Octave', 4–17, and 'Temperament; Or, The Division of the Octave (Part II)', 112–158; Sedley Taylor, 'On a Suggested Simplification of the Established Pitch-Notation', 18–40; J. Baillie-Hamilton, 'On the Application of Wind to String Instruments', 42–46; and Alexander John Ellis, 'Illustrations of Just and Tempered Intonation', 159–65.

[22] Gary Hatfield, *The Natural and the Normative: Theories of Spatial Perception from Kant to Helmholtz*, Cambridge, MA, and London: MIT Press, 1990, 192.

Royal Institution.[23] Early models propose incorporating musical science into curriculums through acoustics, reflecting the university curriculum of the Middle Ages when music comprised the mathematical component of the quadrivium. But as Golding separately recounts even the term 'musical science' was not consensually understood. For some it meant the experimental science of acoustics; for others the study of musical composition or the wider art of music studied systematically.[24] One of the earliest music courses to appear in Victorian Britain was created at the University of Edinburgh in 1838, and overseen by Edinburgh Chair of Music John Donaldson. True to his contractual duty Donaldson incorporated acoustics, justifying its inclusion as something which, like the liberal arts, would help broaden the intellect.[25] Cambridge, Oxford and London did similarly. Cambridge taught acoustics from the 1830s, as revealed in the dedication to Airy's *On Sound and Atmospheric Vibrations, with the Mathematical Elements of Music* (1868), and by the 1870s it was a fully integral component in curriculum reforms which sought to legitimate musical qualification against more established disciplines – 'to "rebrand" music as a science or liberal art, rather than an art influenced by aesthetic or artistic criteria'.[26] Acoustics was slow to find a home at Oxford but in the 1870s Music Professor Sir Frederick Gore Ouseley agitated for the creation of an acoustics laboratory complementing, but separate from, a music school.[27] Trinity College London followed suit and around the same time registered agreement on the study of acoustics as an essential prerequisite to 'proficiency in the Science of Music properly so called'.[28] Because, however, both higher education and music's professional association considered musical science to be synonymous with methodologically empirical experimental acoustics, evolutionary theories had free reign on the Victorian intellectual imagination. At the same time professional and academic prioritization of experimental science institutionalized a methodological gulf between the theory and practice of evolutionary musical sciences – a gulf which despite all of our advances exists to this very day, if the recent cheesecake wars are anything

[23] Rosemary Golding, *Music and Academia in Victorian Britain*, Farnham: Ashgate, 2013, 4. See also Jamie Kassler, 'The Royal Institution Music Lectures, 1800–1831: A Preliminary Study', *Royal Musical Association Research Chronicle Volume* 29 (1983–85), 1–30.

[24] Golding, *Music and Academia in Victorian Britain*, 4–5, and Bennett Zon, *Music and Metaphor in Nineteenth-Century British Musicology*, Aldershot: Ashgate, 1999, 115–20.

[25] Golding, *Music and Academia in Victorian Britain*, 37.

[26] Golding, *Music and Academia in Victorian Britain*, 106.

[27] Golding, *Music and Academia in Victorian Britain*, 68.

[28] Senate Minutes, Volume IX [University of London shelfmark UA/ST 2/2/9], 56 (21 June 1867), cited in Golding, *Music and Academia in Victorian Britain*, 68.

to go by. Like Darwin and Spencer, Pinker and Mithen battle it out without any real – or at the very least highly speculative – evidence. As the evidence of Victorian publications and university curricula suggests, acoustics tended to be introduced to Victorian students when they were developed enough to comprehend complex mathematical formulations, circularly justifying the teleological propulsion of the Great Chain of Being. As discussed in Chapter 3 (Education), early stages of education were devoted almost entirely to simple historical and technical issues while later ones evolved into more complex intellectual propositions. Henry Keatley Moore and Mrs Curwen embody this common Victorian approach to educational development.

The third factor influencing the role of evolution in Victorian musical culture, both in the United Kingdom and (as shown in Chapter 3) abroad, is also linked to education and professionalization: popularization. Bernard Lightman introduces us to the popularization of Victorian science with a positively tree-like image of Darwinian cultural branching:

> Just two years after the [Great] Exhibition [of 1851] opened, the naturalist and popularizer Henry Gosse predicted in his *Naturalist's Rambles on the Devonshire Coast* (1853) that the marine aquarium would soon be found in many Victorian parlours. Almost overnight the aquarium became a national craze . . . At the same time, fern collecting became a widespread fad . . . followed by an intense curiosity in dinosaurs . . . in the early 1860s the British public became interested in gorillas, in part due to curiosity about the relationship between humans and apes in light of Darwin's theory of evolution and because of the activities of the explorer Paul du Chaillu . . . Patronized by [the anatomist Richard] Owen, du Chaillu's gorillas became entangled in the celebrated debate between Owen and Huxley on the anatomy of human and ape brains. The public's curiosity about apes was lampooned in *Punch* and immortalized in [the writer Charles] Kingsley's *Water Babies* (1862).[29]

To this description add music; like all the arts musical culture played a vital part in this process of popularization. As Lightman explains, a new generation of middle-class consumers had more money to spend on leisure activities, and science was one of those activities. Science 'became fashionable and respectable within a broader spectrum of the populace, not just within the circles of the well to do . . . Scientific knowledge seemed to offer the magical password – the "open sesame" – that unlocked the doors to exhilarating new works in the second half of the century'.[30] In addition to visual and sonic art forms the popularization of science within Victorian musical culture was intimately

[29] Bernard Lightman, *Victorian Popularizers of Science: Designing Nature for New Audiences*, Chicago and London: University of Chicago Press, 2007, 2.

[30] Lightman, *Victorian Popularizers of Science*, 2–3.

bound up with popular literary culture in ways that vernacularize complex scientific ideas[31] through general periodicals, books and all their literary forms, like the extended essay, reportage, poetry, fiction and non-fiction. Ultimately those ideas coalesce in a term coined by Thomas Huxley in *Essays upon Some Controverted Questions* (1892): the 'principle of the scientific Naturalism of the latter half of the nineteenth century'.[32] According to Frank Turner scientific naturalism was 'a historiographic category denoting the secular creeds of the generation of intellectuals' who in the wake of Darwin's *Origin of Species* (1859) 'wrested cultural authority from the old Anglican establishment and installed themselves as a new professional scientific elite',[33] but the literature of popularization reveals many consumers roaming unrepentantly across religious and scientific boundaries.[34] Dawson, Noakes and Topham regard books as secondary to general periodicals in influence. Books are 'date-stamped', to use Margaret Beetham's words,[35] but accordingly it is precisely the ephemeral nature of periodicals which gives them permanent value to historians today. There is, however, the thorny problem of 'literary replication'[36] in the diffusion of ideas through periodicals, with book material often excerpted, extracted, reviewed, advertised and communicated through correspondence and passing comment.[37] While the role of music journalism within Victorian musical culture has received some – though not terribly extensive – treatment[38] the role of the general periodical in disseminating scientific knowledge within Victorian

[31] Katherine Pandora, 'Knowledge Held in Common: Tales of Luther Burbank and Science in the American Vernacular', *Isis* 92/3 (Sept. 2001), 491–92.

[32] Thomas Henry Huxley, *Prologue, Essays Upon Some Controverted Questions*, London: Macmillan, 1892, 35.

[33] Gowan Dawson and Bernard Lightman, 'Introduction', in Gowan Dawson and Bernard Lightman (eds.), *Victorian Scientific Naturalism: Community, Identity, Continuity*, Chicago and London: University of Chicago Press, 2014, 1; 1.

[34] See David Ray Griffin, *Religion and Scientific Naturalism: Overcoming the Conflicts*, Albany, NY: State University of New York Press, 2000.

[35] Margaret Beetham, 'Towards a Theory of the Periodical as a Publishing Genre', in Laurel Brake, Aled Jones and Lionel Madded (eds.), *Investigating Victorian Journalism*, Basingstoke: Macmillan, 1990, 21, 19–32.

[36] James A. Secord, *Victorian Sensation: The Extraordinary Publication, Reception and Secret Authorship of Vestiges of the Natural History of Creation*, Chicago and London: University of Chicago Press, 2000.

[37] Geoffrey Cantor, Gowan Dawson, Graeme Gooday, Richard Noakes, Sally Shuttleworth and Jonathan R. Ropham (eds.), *Science in the Nineteenth-Century Periodical: Reading the Magazine of Nature*, Cambridge: Cambridge University Press, 2004, 3.

[38] See Leanne Langley, 'The English Music Journal in the Early Nineteenth Century', PhD diss., University of North Carolina, Chapel Hill, 1983, and 'The Musical Press in Nineteenth-Century England', *Notes* 46/3, (1990) 583–92, and Nigel Scaife, 'British Music Criticism in a New Era: Studies in Critical Thought, 1894–1945', DPhil diss., University of Oxford, 1994.

musical culture has yet to be explored systematically; nor has its important relationship to books been given adequate attention. Research which focuses on global scientific networks and dissemination – Bernard Lightman's recent *Global Spencerism: The Communication and Appropriation of a British Evolutionist* (2016),[39] for example – paves the way for reconceptualising the interrelationship between national and international networks of popularization, a task begun only implicitly in my previous book, *Representing Non-Western Music in Nineteenth-Century Britain* (2000).

Methodology

The Great Chain of Being may have had an adverse effect on certain aspects of Victorian musical history but it was also used by the Victorian musical thinkers as an ideological weapon against the Land Without Music. To understand that phenomenon, and to show how *Evolution and Victorian Culture* fits into current research, it is essential to locate the Great Chain within the context of two inter-related methodological issues today, the non-Darwinian revolution and interdisciplinarity. Research into the Great Chain of Being (of which this book forms a musicological part) centres around the work of Peter Bowler's transformatively influential *The Non-Darwinian Revolution: Reinterpreting a Historical Myth* (1988). Like all good theorists Bowler makes a simple point: 'there is now a substantial enough body of literature to convince anyone that the parts of Darwin's theory now recognized as important by biologists had comparatively little impact on late nineteenth-century thought'.[40] The Great Chain of Being was an integral and representative part of that non-Darwinian thought. Bowler's notion of a non-Darwinian revolution originates in his counter-reading of important books like Thomas S. Kuhn's *Structure of Scientific Revolutions* (1962). Kuhn defines the introduction of a new theoretical framework as revolutionary, overturning the certitudes of previously ensconced attitudes. Bowler cites the classic example in Darwinian literature as Gertrude Himmelfarb's *Darwin and the Darwinian Revolution* (1959), published to coincide with the 100th anniversary of the *Origin of Species*, and possibly Michael Ruse's broader

[39] Bernard Lightman (ed.), *Global Spencerism: The Communication and Appropriation of a British Evolutionist*, Leiden: Brill, 2015.

[40] Peter J. Bowler, *The Non-Darwinian Revolution: Reinterpreting a Historical Myth*, Baltimore, MD, and London: Johns Hopkins University Press, 1988, ix.

publication, *The Darwinian Revolution: Science Red in Tooth and Claw* (1979), which incorporates the Darwinian revolution amongst others. In these works Darwinian hegemony is sacrosanct, but according to Bowler historically presumptuous:

> My suggestion is that Darwin's theory should be seen not as the central theme in nineteenth-century evolutionism but as a catalyst that helped to bring about the transition to an evolutionary viewpoint within an essentially non-Darwinian conceptual framework. This was the 'Non-Darwinian Revolution'; it was a revolution because it required the rejection of certain key aspects of creationism, but it was non-Darwinian because it succeeds in preserving and modernizing the old teleological view of things.[41]

The non-Darwinian Revolution continued until the 1920s and 1930s when genetics and Darwinian evolutionary theory joined forces, and so Bowler is not wrong when he suggests that for much of the nineteenth century evolutionism centred not on Darwin but on various developmental models, what he calls anti- or pseudo-Darwinism. Bowler counts Lamarckism amongst anti-Darwinian trends. As explained later in this introduction, Darwin's theory of natural selection assumes a branching, adaptive tree-like evolution where variations are random and undirected, so that in theory two identical populations exposed to the same environment could adapt differently to their circumstances. In contrast, anti-Darwinian Lamarckism presumed that all individuals in a new environment would strive to acquire useful adaptations – something called the inheritance of acquired characteristics.[42] The upshot of this distinction for Victorians in both general and – more specifically – musical culture was to deepen the feeling that Darwinian evolutionism included a randomness which Lamarck's did not. Pseudo-Darwinism, on the other hand, provided 'a blend of Darwinian rhetoric with attitudes that can more fruitfully be seen as continuations of the pre-Darwinian view of nature'.[43] Evolutionary thinkers like Thomas Huxley – ironically, Darwin's bulldog – and Ernst Haeckel are amongst Bowler's pseudo-Darwinians. Huxley remained unconvinced about natural selection, in the 1860s, for example, stressing the 'persistence of type' – the fact that some forms remain unchanged over great lengths of geological time. Haeckel is included because he viewed evolution as 'development seen in the form of the growing embryo and in the past history of its type'.[44] Bowler appears confident that differences

[41] Bowler, *Non-Darwinian Revolution*, 5.
[42] Bowler, *Non-Darwinian Revolution*, 9.
[43] Bowler, *Non-Darwinian Revolution*, 47.
[44] Bowler, *Non-Darwinian Revolution*, 83.

between anti- and pseudo-Darwinism can be found, but as he suggests 'they are by no means as clear-cut as one may at first imagine'.[45] In fact, despite his best efforts they remain frequently difficult to disentangle in the evolutionary vortex of Victorian culture. In my own article 'The "Non-Darwinian" Revolution and the Great Chain of Musical Being' (2014) I propose even more refined categories, including A-Darwinism, Pick-'n'-Mix Darwinism and the vaguely irreverent Non-Non-Darwinism. A-Darwinism excludes reference to Darwin altogether; Pick-'n'-Mix Darwinism took what it liked from Darwin without shedding developmental confidence (based largely on Spencer); and Non-Non-Darwinism selectively disregarded Darwin.[46]

The effect of a non-Darwinian revolution on Victorianists would be a change in scholastic emphasis from Darwinian to non-Darwinian orientations, and *Evolution and Victorian Musical Culture* is emblematic of this shift. It is telling, for instance, that in studies of Victorian fiction some of the most important studies of Darwinian influences – Gillian Beer's *Darwin's Plots* (1983, 2nd ed. 2000) and George Levine's *Darwin and the Novelists* (1988) – occur in the heyday of the Darwinian revolution yet were followed by a stream of works recalibrating the relationship of Darwinian and non-Darwinian revolutions, with books like William Greenslade's *Degeneration, Culture, and the Novel, 1880–1940* (1994); Kelly Hurley's *The Gothic Body* (1996); and Angelique Richardson's *Love and Eugenics in the Late Nineteenth Century* (2003), and articles like Jessica Straley's study of recapitulation in Charles Kingsley's *The Water Babies* (1863) 'Of Beasts and Boys: Kingsley, Spencer, and the Theory of Recapitulation' (2007) or Shafquat Towheed's exploration of neo-Lamarckian aesthetics in 'The Creative Evolution of Scientific Paradigms: Vernon Lee and the Debate over the Hereditary Transmission of Acquired Characteristics' (2006).[47] The same pre- and post-non-Darwinian divide holds true of poetry, for example; according to its author *Darwin's Bards: British and American Poetry in the Age of Evolution* (2009) belies a more comprehensive survey of evolutionary influences, including 'theistic evolution,

[45] Bowler, *Non-Darwinian Revolution*, 47.

[46] Bennett Zon, 'The Non-Darwinian Revolution and the Great Chain of Musical Being', in Lightman and Zon, *Evolution and Victorian Culture*, 197.

[47] Cannon Schmitt, 'Evolution and Victorian Fiction', in Lightman and Zon (eds), *Evolution and Victorian Culture*, 24. See also Jessica Straley, 'Of Beasts and Boys: Kingsley, Spencer, and the Theory of Recapitulation', *Victorian Studies* 49/4 (2007), 583–609; and Shafquat Towheed, 'The Creative Evolution of Scientific Paradigms: Vernon Lee and the Debate over the Hereditary Transmission of Acquired Characteristics', *Victorian Studies* 49/1 (Autumn 2006), 33–61.

neo-Lamarckism, orthogenesis (inherent tendencies within the evolutionary process), alongside the Darwinian concepts of natural selection and the tree of life, and various versions of Social Darwinism'.[48] Research in more recent fields replicates Holmes's approach to greater or lesser extents, including studies of Victorian photography, cinema, art, theatre, architecture, exhibition and of course music. Generally speaking, whereas literary scholars like Phyllis Weliver, Delia da Sousa Correa, Sophie Fuller and Nicky Losseff (see footnote 13) have focussed on music in Victorian culture for some time musicologists have been ponderously slow to follow suit. Rachel Mundy is perhaps the most obvious exception, but her groundbreaking work arguably tends to utilize Victorian science to explain broader sweeps of later music and evolutionary history.[49] A paradigm shift from Darwinian to non-Darwinian contexts is evident in Victorian literary studies as much as it is in musicology, however, from Peter Kivy's important article 'Charles Darwin on Music' (1959) to more recent work by Charles M. Brotman, including his article 'Alfred Wallace and the Anthropology of Sound in Victorian Culture' (2001) and later PhD thesis 'The Power of Sound: Evolutionary Naturalism and Music in Anglo-American Victorian Culture' (2005).[50] Brotman and I both work towards a pluralistic non-Darwinian understanding of evolution and Victorian musical culture, but whereas Brotman is inadvertently non-Darwinian (stressing competing music evolutionary ideas and their own evolutionary trees) 'The "Non-Darwinian" Revolution and the Great Chain of Musical Being' and *Evolution and Victorian Culture* are both purposefully non-Darwinian. In my case this undertaking represents a conscious development from purely historical evolutionary interests found in *Music and Metaphor in Nineteenth-Century British Musicology* (2000) and *Representing Non-Western Music in Nineteenth-Century Britain* (2007) to the more

[48] John Holmes, 'The Challenge of Evolution in Victorian Poetry', in Lightman and Zon, *Evolution and Victorian Culture*, 41.

[49] See for example Rachel Mundy, 'Musical Evolution and the Making of Hierarchy', *The World of Music* 48/3 (2006), 13–27; 'Birdsong and the Image of Evolution', *Society and Animals* 17 (2009), 206–23; 'Nature's Music: Birds, Beasts, and Evolutionary Listening in the Twentieth Century', PhD diss., New York University (2010); and 'Evolutionary Categories and Musical Style from Adler to America', *Journal of the American Musicological Society* 67/3 (Fall 2014), 735–768.

[50] See Peter Kivy, 'Charles Darwin on Music', *Journal of the American Musicological Society* 12/1 (Spring 1959), 42–48; Charles M. Brotman, 'Alfred Wallace and the Anthropology of Sound in Victorian Culture', *Endeavour* 25/4 (2001), 144–47, and 'The Power of Sound: Evolutionary Naturalism and Music in Anglo-American Victorian Culture', PhD diss., University of Rochester, 2005.

overtly theorized non-Darwinian position of this book.[51] The change of emphasis is also observable in more overarching and comprehensive studies like *Evolution and Victorian Culture* (2014) and Jonathan Conlin's *Evolution and the Victorians* (2014). *Evolution and Victorian Culture* may focus on the arts, and *Evolution and the Victorians* on scientific, religious and imperial history but collectively they tell the same balanced, full-spectrum evolutionary story: we now have evolutionisms in the same way we have 'musics'[52] or even the 'romanticisms' of a rather earlier and prescient Arthur O. Lovejoy.[53]

The 'discovery' and subsequent promulgation of a non-Darwinian revolution did not happen in intellectual isolation, however; other changes in Victorian and nineteenth-century British music studies were afoot in the 1980s, and one in particular has had a great influence on the research for this book: interdisciplinarity. A good barometer is the pervasive use of interdisciplinarity (or less usually, multidisciplinarity) in Victorian association and journal website publicity. *Victorian Studies*, for example, proclaims that it is

> One of the first interdisciplinary journals in the humanities . . . [and] remains the flagship journal in the field. Publishing essays on topics from literature, social and political history, fine arts, science, philosophy, economics, and law, *Victorian Studies* is the journal of the North American Victorian Studies Association and publishes a special issue annually devoted to the Association conference.[54]

Then there is the self-explanatory *Victorian Review: An Interdisciplinary Journal of Victorian Studies*;[55] Interdisciplinary Nineteenth-Century Studies and their journal *Nineteenth-Century Contexts: An Interdisciplinary Journal*;[56] the Midlands Interdisciplinary Victorian Studies Seminar;[57] the Victorian Interdisciplinary Studies Association of the Western United States;[58] and other societal blurbs, including The North American Victorian Studies Association (encouraging 'a wide variety of theoretical and

51 See Bennett Zon, *Music and Metaphor in Nineteenth-Century British Musicology*, Aldershot: Ashgate, 1999, and *Representing Non-Western Music in Nineteenth-Century Britain*, University of Rochester Press, 2007.

52 Philip V. Bohlman, *World Music: A Very Short Introduction*, Oxford: Oxford University Press, 2002, xv.

53 Arthur O. Lovejoy, 'On the Discrimination of Romanticisms', *Proceedings of the Modern Language Association* 39 (1924), 229–53.

54 www.indiana.edu/~victstu/victorianStudies.shtml, accessed 22/1/16.

55 http://web.uvic.ca/victorianreview/, accessed 22/1/16.

56 http://www3.nd.edu/~incshp/incs3.html, accessed 22/1/16.

57 https://mivssonline.wordpress.com/, accessed 22/1/16.

58 www.visawus.org/, accessed 22/1/16.

disciplinary approaches to the field');[59] The British Association for Victorian Studies ('a multidisciplinary organisation ... including art history, cultural studies, history, literary studies, performance studies and the history of science');[60] the Nineteenth-Century Studies Association ('an interdisciplinary Association interested in exploring all aspects of the long nineteenth century, from science to music, from architecture to religion, from movement to literatures–and beyond');[61] and the Centre for Nineteenth-Century Studies at Durham University ('advancing interdisciplinary research').[62]

Endemic interdisciplinarity does not make practice easy, however, as Linda Shires points out in her thought-provoking article 'Victorian Studies and Cultural Studies: Interdisciplinarity, The Market, and a Call for Critical Realism' (1999): 'Although it houses many critical perspectives, Victorian studies', she asserts, 'is haunted by an over-reliance on the middle-class form which made it famous, the realist novel'.[63] For Matthew Rowlinson interdisciplinary danger comes from a self-destructive, but ultimately palliative, source: 'Successful interdisciplinary research', he suggests, 'estranges the disciplines from their objects';[64] 'disciplines are therefore themselves constituted by a critical tension between method and object; the virtue of interdisciplinary Victorian studies at the moment is to put that tension in evidence'.[65] Musicologist Stephen Banfield highlights the critical tension between method and object in a selective tour of Victorian music studies in 'What Do You Think of Stainer's *Crucifixion*? (2010). Banfield groups interdisciplinarity in amongst research *themes* of social history, cultural history and reception rather than as a research *method*,[66] but virtuous – if problematical – interdisciplinarity is evident as much in Victorian music studies as it is in Victorian studies. My Ashgate book series *Music in Nineteenth-Century Britain* reflects this interdisciplinary panorama in microcosm, with monographs intersecting disciplines through topics like conductors, performers, composers, genres, education,

[59] http://navsa.org/about/general-info/, accessed 22/1/16.

[60] http://bavs.ac.uk/, accessed 22/1/16. [61] www.ncsaweb.net/, accessed 22/1/16.

[62] www.dur.ac.uk/cncs/about/

[63] Linda Shires, 'Victorian Studies and Cultural Studies: Interdisciplinarity, The Market, and a Call for Critical Realism', *Victorian Literature and Culture* 27/2 (1999), 482.

[64] Matthew Rowlinson, 'Interdisciplinarity, Estrangement, and Method in Victorian Studies', *Victorian Review* 33/1 (Spring 2007), 52.

[65] Rowlinson, 'Interdisciplinarity, Estrangement, and Method in Victorian Studies', 54.

[66] Stephen Banfield, 'What Do You Think of Stainer's Crucifixion? Current Victorian Musicology', *Journal of Victorian Culture* 15/1 (2010), 122.

institutions, instruments, religion, provinciality, imperialism, publishers, poetry, fiction, class, historiography and science.[67]

Unsurprisingly all Victorianists were waking up to the clarion call of the interdisciplinary bugle at around the same time that Himmelfarb's Darwinian revolution was evolving into Bowler's non-Darwinian revolution – when disciplinarity was becoming the subject of academic reinvention. The forces of reinvention stem from a similar impulse to locate and examine intellectual plurality where it genuinely exists. It is no surprise that interdisciplinarity appears as an academic presence when Clifford Geertz is imagining 'blurred genres',[68] for example. Writing in 1986 George Marcus and Michael Fischer describe it in terms equally revolutionary to Bowler:

> The present is a time for reassessment of dominant ideas across the human sciences ... extending to law, art, architecture, philosophy, literature, and even the natural sciences. This reassessment is more salient in some disciplines than in others, but its presence is pervasive. It is not just the ideas themselves that are coming under attack but the paradigmatic style in which they have been presented. Particularly in the social sciences, the goal of organizing disciplines by abstract, generalizing frameworks that encompass and guide all efforts at empirical research is being fundamentally challenged.[69]

Ethnomusicologist Richard Middleton echoes this in the field of music more generally: 'To look across the full range of disciplinary perspectives is important. Indeed, the parallelism of the different histories of engagement with "musics and cultures" research, together with their varied dialogues, seems to be integral to its problematic'.[70]

The interdisciplinary problematic is very much a product of the mid- to late twentieth century, though arguably born of the Victorians. Some scholars define the location of an interdisciplinary problematic an 'interdiscipline',[71] a term particularly useful in describing the relationship of

[67] www.ashgate.com/default.aspx?page=1249&series_id=217&calcTitle=1&forthcoming=1.

[68] See Clifford Geertz, 'Blurred Genres', *American Scholar* 49 (1980), 165–79.

[69] George E. Marcus and Michael M. J. Fischer, *Anthropology as Cultural Critique: An Experimental Moment in the Human Sciences*, Chicago and London: University of Chicago Press, 1986, 7.

[70] Richard Middleton, 'Music Studies and the Idea of Culture', in Martin Clayton, Trevor Herbert and Richard Middleton (eds.), *The Cultural Study of Music: A Critical Introduction*, New York and London: Routledge, 2003, 2.

[71] The term 'interdiscipline' has existed since 1930 in array of journals and books (*Oxford English Dictionary*, www.oed.com.ezphost.dur.ac.uk/view/Entry/243540?rskey=cqn4Se&result=1#eid12528631, accessed 20/3/17). While interdisciplinarity has become increasingly topical, however, research into the idea of an 'interdiscipline' has generally lagged behind despite important contributions like *In(ter)discipline: New Languages for Criticism* (2007) or Gabriel

musicology to other disciplines used in this book like zoomusicology, ethnomusicology, and music theology. S. W. Littlejohn defines an *interdiscipline* as 'a field of scholars who identify with various disciplines but share a common interest in a theme that crosses traditional boundaries'.[72] According to Professor of Humanities Computing Willard McCarty,

A true *interdiscipline* is, however, not easily understood, funded or managed in a world already divided along disciplinary lines, despite the standard pieties. Properly so called an interdiscipline is not just another administrative entity with its budget, chair and department members – difficult as this is to carve these days out of existing turf; it isn't an institutionally sactioned [*sic*] kind of poaching. Rather it is an entity that exists in the interstices of the existing fields, dealing with some, many or all of them.[73]

McCarty later develops his observation into a statement on institutionalization as a key factor amongst the professional competencies which underscore in an interdiscipline:

an interdiscipline in my sense is constituted precisely by that unifying perspective on what happens at the intersection of two or more fields. This perspective gives the interdiscipline integrity and basis for its own research agenda, curriculum and publications. Nevertheless, as long as it remains an interdiscipline it depends on continuous activity in the intersecting fields. Preoccupation with what they share puts it in position to foster cross-fertilising exchange among them, as a merchant trader among mutually incomprehending cultures. Thus it serves them, not as a servant his master but collegially – which has radical implications for its institutionalisation.[74]

Solis's 'Thoughts on an Interdiscipline: Music Theory, Analysis, and Social Theory' (2012). *The Oxford Handbook of Interdisciplinarity*, for example, provides few – albeit important – references to the term, and it is missing altogether from both editions of Joe Moran's *Interdisciplinarity*. See Gillian Beer, Malcolm Bowie and Beate Perrey (eds.), *In(ter)discipline: New Languages for Criticism*, London: Legenda, 2007; Gabriel Solis, 'Thoughts on an Interdiscipline: Music Theory, Analysis, and Social Theory', *Ethnomusicology* 56/3 (Fall 2012), 530–54; Robert Frodeman, Julie Thompson Klein and Carl Mitcham (eds.), *The Oxford Handbook of Interdisciplinarity*, Oxford: Oxford University Press, 2010; and Joe Moran, *Interdisciplinarity*, London and New York: Routledge, 2002/2010.

72 S. W. Littlejohn, 'An Overview of Contributions to Human Communication Theory from other Disciplines', in F. E. X. Dance (ed.), *Human Communication Theory: Comparative Essays*, New York: Harper & Row, 1982, 246.

73 Willard McCarty, 'Humanities Computing as Interdiscipline', A Seminar in the Series, 'Is Humanities Computing an Academic Discipline?', held under the auspices of the Institute for Advanced Technology in the Humanities (IATH), at the University of Virginia, Guy Fawkes Day 1999 (Ver. 13 October 1999; rev. 14 October, 15 October, 22 October 1999), www.iath.virginia.edu/hcs/mccarty.html, accessed 22/1/16.

74 Willard McCarty, 'Looking Through an Unknown, Remembered Gate: Millennial Speculations on Humanities Computing', *Interdisciplinary Science Reviews* 26/3 (Autumn 2001), 173–82, www.mccarty.org.uk/essays/McCarty,%20Through%20an%20unknown%20remembered%20gate.pdf, accessed 22/1/16.

Whether or not an interdiscipline is a phantom discipline, or an interdisciplinary problematic, interdisciplinarity is an abiding feature of current Victorian studies in the same way as the non-Darwinian revolution, and *Evolution and Victorian Musical Culture* aims to use music to underscore their interconnectedness. Joe Moran navigates between discipline and interdisciplinarity in roughly the same way that Bowler (re)interprets the Darwinian and non-Darwinian revolutions. For Moran a 'discipline' has two basic connotations: 'it refers to a particular branch of learning or body of knowledge, and to the maintenance of order and control amongst subordinated groups';[75] but

> 'Interdisciplinarity' provides a democratic, dynamic and co-operative alternative to the old-fashioned, inward-looking and cliquish nature of disciplines. And yet this straightforward interpretation begs a number of questions: how exactly does interdisciplinary research aspire to be warm, mutually developing, consultative? Can disciplinary divisions be so easily broken down or transcended? Is it not inevitable that there should be some means of ordering and structuring knowledge?[76]

Moran problematizes the epistemic reality of a discipline by questioning the political nature of its professional institutionalization. On this basis he questions the discipline-ness of a discipline in the same way Bowler questions the Darwinian-ness of the Darwinian revolution – in the same way that Middleton and many others question the musicology-ness of musicology. Recent Victorian and non-Victorian literature on interdisciplinarity is agreed that the idea of a disciplinary core is ultimately evasive – is it conceptual or institutional, hermetic, centrifugal or centripetal? – are not disciplines in a constant state of 'revolution' anyway; that is to say, in a state of 're-volving'? What is the relationship between evolution and revolution, and at what point does evolution become revolution?[77] Interdisciplinarity, like the inter-theoretical connectedness within the non-Darwinian Revolution, is an equally evasive episteme because the non-Darwinian might include the Darwinian in the same way that interdisciplinarity includes disciplinarity. Interdisciplinarity is fundamentally taxonomic, and like all systems of categorization it is an organizational convenience rather than an

[75] Joe Moran, *Interdisciplinarity*, London and New York: Routledge, 2002/2011, 2.

[76] Moran, *Interdisciplinarity*, 3.

[77] See Jouni-Matti Kuukkanen, 'Revolution as Evolution: The Concept of Evolution in Kuhn's Philosophy', in Vasso Kindi and Theodore Arabatzis (eds.), *Kuhn's The Structure of Scientific Revolutions Revisited*, Chicago: University of Chicago Press, 2012, 134–52.

absolute truth,[78] and even more recent branches in its phylogenic tree – transdisciplinarity, cross-disciplinarity and multidisciplinarity, for example – remain ontologically debatable propositions, as Stephen Banfield suggests. Joanne Shattock summarizes the problem for Victorian studies through the words of Martin Hewitt:

> If Victorian Studies is to transcend its current limitations, to thrive as a truly interdisciplinary field based on combination rather than aggregation . . . it must produce for itself a new scholarly integration, in the sense of a shared literature, a common focus of enquiry, a set of broadly accepted modes of enquiry and theoretical assumptions, and ultimately a defence of the utility of "Victorian" as a frame of analysis. The walls of disciplinary particularism will not simply crumble – they need to be deliberately and systematically undermined; a sense of interdisciplinarity will not evolve, it must be built.[79]

According to Shattock, Hewitt summarizes the interdisciplinary problematic when he argues 'for "greater coherence", "discipline" or "rigour" in our field, while acknowledging that interdisciplinarity remains more aspiration than achievement, and that multidisciplinarity has been accepted as the dominant practice'. He admits, too, that 'chasms of disciplinary difference . . . lurk beneath any efforts of accommodation which are required to construct a *shared* enterprise'.[80]

Evolution and the Great Chain of Being

As all this suggests I am not alone in struggling to model my book on the shared methodological enterprise linking Darwinian and non-Darwinian, disciplinary and interdisciplinary, thought. Nor am I alone in trying to model my book on the Great Chain of Being itself; indeed, as I aim to prove, there are many precedents in Victorian musical culture. But to understand them it is essential to understand its own evolutionary history first, and to put the Great Chain in purely historical and scientific context

[78] Julie Thompson Klein, 'A Taxonomy of Interdisciplinarity', in Robert Frodeman, Julie Thompson Klein and Carl Mitcham (eds.), *The Oxford Handbook of Interdisciplinarity*, Oxford: Oxford University Press, 2012, 15.

[79] Joanne Shattock, 'Where Next in Victorian Literary Studies? – Revising the Canon, Extending Cultural Boundaries, and the Challenge of Interdisciplinarity', *Literature Compass* 4/4 (2007), 1290; see Martin Hewitt, 'Victorian Studies: Problems and Prospects', *Journal of Victorian Culture* 6/1 (Spring 2001), 152.

[80] Joanne Shattock, 'Where Next in Victorian Literary Studies?', 1290, citing Hewitt, 'Victorian Studies', 149.

before contextualizing it musically. This is the purpose of the last section of this introduction to *Evolution and Victorian Musical Culture* – to provide a history of the Great Chain of Being and its relationship to Victorian evolutionary thought.

The Great Chain of Being is one of many scientific ideas Victorian culture accepted as 'evolutionary'. Evolution comes from the Latin word 'evolvere', to roll out. It first appears in the early seventeenth century, and frequently implies a process of rolling out something that already exists in some form or other. In *Divine Dialogues* (1667) English philosopher Henry More describes evolution as the process inherent within God's creation of the world: 'the whole evolution of times and ages from everlasting to everlasting is . . . represented to God at once'.[81] In the eighteenth century evolution begins to assume a more recognizably modern form. Natural philosopher John Turberville Needham exemplifies this in *An Account of Some New Microscopical Discoveries* (1745) in which nature is 'ever exerting its Fecundity in a successive Evolution of organized Bodies'.[82] Echoing Needham public intellectual Erasmus Darwin, Charles Darwin's grandfather and a founding father of modern evolutionism, expands this definition across numerous proto-evolutionary works, such as epic poems *The Botanic Garden* (1791) and *The Temple of Nature* (1803), and the scientific treatise *Zoonomia* (1794–96). In addition to poetry *The Botanic Garden* (1791) and *The Temple of Nature* (1803) include meticulously conceptualized annotations revealing the extent of progressive evolutionary thinking. *The Botanic Garden* (1790–91) may wrestle with the creation of the world 'When *Love Divine*, with brooding wings unfurl'd, / Call'd from the rude abyss the living world', but its evolutionary message cannot be clearer: 'From having observed the gradual evolution of the young animal or plant from its egg or seed; and afterwards its successive advances to its more perfect state, or maturity; philosophers of all ages seem to have imagined, that the great world itself had likewise its infancy, and its gradual progress to maturity'.[83] Evolution evolves in Darwin's writings, and in *The Temple of Nature* (1803) man is no longer separately created by God but has evolved naturally from animals.[84] 'Imperious man', muses Darwin,

[81] Henry More, *Divine Dialogues, Containing Sundry Disquisitions & Instructions Concerning the Attributes of Goad and His Providence in the World*, London: James Flesher, 1668, i.§15.

[82] John Turberville Needham, 'Introduction', *An Account of Some New Microscopical Discoveries*, London: F. Needham, 1745, 1.

[83] Erasmus Darwin, *The Botanic Garden*, Part II. The Loves of the Plants, New York: T. & J. Swords, 1790–91/1798, 105.

[84] Patricia Fara, *Erasmus Darwin: Sex, Science, and Serendipity*, Oxford: Oxford University Press, 2012, 239.

'who rules the bestial crowd / Of language, reason, and reflection proud, / With brow erect who scorns this earthly sod, / And styles himself the image of his God; / Arose from rudiments of form and sense, / An embryon point, or microspic ens!'[85] *Zoonomia* methodically defends this thesis, supporting it with 'arguments showing that all vegetables and animals arose from such a small beginning, as a living point or living fibre'.[86]

Erasmus Darwin expresses in England what had already begun to be imagined elsewhere in various forms. In *Philosophie Zoologique* (1809), for example, Jean-Baptiste Lamarck introduces a proto-Darwinian theory of development emphasizing what he called the 'transmutation' of species. Broadly speaking transmutation combines two evolutionary ideas, the inheritance of acquired characteristics and progression from simplicity to complexity. The inheritance of acquired characteristics is a thorny evolutionary idea, suggesting that characteristics are inherited if they are consistently replicated over generations (i.e. the length of a giraffe's neck is due to habit of its ancestors to stretch for food); and progression from simplicity to complexity is intended to characterise the steady progressive force operating on life.[87] Lamarck believed that the simplest life forms began in spontaneous generation sparked into existence by the power of electricity, and in each generation more complex forms arose to succeed its parents in complexity. To illustrate this he defended his theory by using an ancient imagetext (pictures and texts interacting symbiotically)[88] connecting minerals, plants, animals, humans, angels and God in a single unbroken progression: the Great Chain of Being.[89] As a modern evolutionary imagetext the Great Chain of Being can be traced back to the sixteenth century, when ladders and stairways were used to represent a fixed order bridging earth and heaven, as for example in Ramon Llull's *Liber de*

85 Erasmus Darwin, *The Temple of Nature: Or the Origin of Society: A Poem, with Philosophical Notes*, London: J. Johnson, 1803, 28.

86 Erasmus Darwin, *Temple of Nature*, 28, annotation cross-referenced to 'Sect. XXXIX.4.8. on Generation', in Erasmus Darwin, *Zoonomia; or the Laws of Organic Life*, Philadelphia: Edward Earle, 1794–96/1818, 395.

87 Peter J. Bowler, *Evolution: The History of an Idea*, 3rd ed., Berkeley, Los Angeles and London: University of California Press, 2003, 89.

88 W. J. T. Mitchell, *Picture Theory: Essays on Verbal and Visual Representation*, Chicago and London: University of Chicago Press, 1994, 5.

89 Theodore W. Pietsch, *Trees of Life: A Visual History of Evolution*, Baltimore, MD: Johns Hopkins University Press, 2012, 34–38; J. David Archibald, *Aristotle's Ladder, Darwin's Tree: The Evolution of Visual Metaphors for Biological Order*, New York: Columbia University Press, 2014, 59–66; and Mark A. Ragan, 'Trees and Networks Before and After Darwin', *Biology Direct* 4/43 (16 November 2009), 8–9.

ascensu et descensu intellectus (written in 1304 and published in 1512) and Charles de Bovelle's *Physicorum elementorum* (1512) found in Figures I.1 and I.2. From the late 1500s images of the Great Chain of Being had begun to change into the form of dividing keys or tables, such as *Methodi herbariae libri tres* (1592) by Adam Saluziansky à Zaluzian, as illustrated in Figure I.3. From the middle of the eighteenth century the visual vocabulary may have seemed complete but the concepts undergirding it were about to change. In Figures I.4 and I.5 we can see that Charles Bonnet adapts the ladder in his *Traité d'insectologie* (1745, reproduced in volume 1 of *Œuvres d'histoire naturelle et de philosophie*, 1779) and the stairway in *Contemplation de la nature* (1764, volume 4 of *Œuvres d'histoire naturelle et de philosophie*, 1781) but in *Philosophie Zoologique* (1809) his much younger contemporary Lamarck opts for a different image – a tantalizingly tree-like form depicting the origin of animals branching in various directions (Figure I.6).

Lamarck's important innovation was to connect the branches, and although he begins with two separate starting points (both created by spontaneous generation) branches intersect to give the impression of a dynamic evolutionary process, not the static and hierarchically fixed order of the Great Chain of Being. Creatures seem to evolve rather than remaining permanently fixed as they are; the Great Chain of Being was loosening, and so too were the values it represented. Images of the Great Chain of Being reflect a transitional ontology from the 1820s. In *Zoologia specialis* (1829) Carl Edward von Eichwald captures the tree's vicariously theological status in the earliest recognizably tree-like image of an evolutionary tree.[90] Figure I.7 highlights the sunbeams bursting through the storm-clouds (Gerard Manley Hopkins would call this 'shrivelight'), and a tangle of roots clutching beneath the surface of the ground. The evolutionary message is clear: while God may be missing from nature his presence lives on. Charles Darwin's message would be more direct; as Julia Voss says: 'It is in the diagrams that his thoughts on the scientific theories of his day left the clearest traces'.[91] Described by Gillian Beer as 'a condensation of events'[92] and a 'connection between representation and reality',[93] Darwin's trees would eventually embody an evolutionary proposition bereft

[90] Ragan, 'Trees and Networks Before and After Darwin', 8–9.

[91] Julia Voss, *Darwin's Pictures: Views of Evolutionary Theory, 1837–1874*, trans. Lori Lanz, New Haven and London: Yale University Press, 2007/2010, 64.

[92] Gillian Beer, *Darwin's Plots: Evolutionary Narratives in Darwin, George Eliot and Nineteenth-Century Fiction*, 2nd ed., Cambridge: Cambridge University Press, 1983/2000, 33.

[93] Beer, *Darwin's Plots*, 86.

Figure I.1 Charles de Bovelles, *Physicorum elementorum* (1512), Lib. X, Fol. LXXIII, permission granted by Wellcome Library, London

Figure I.2 Ramon Llull, *Liber de ascensu et descensu intellectus* (written in 1304 and published in Valencia in 1512), between two series foliations, 1–49 and 1–62, by permission of the Master and Fellows of Trinity College, Cambridge

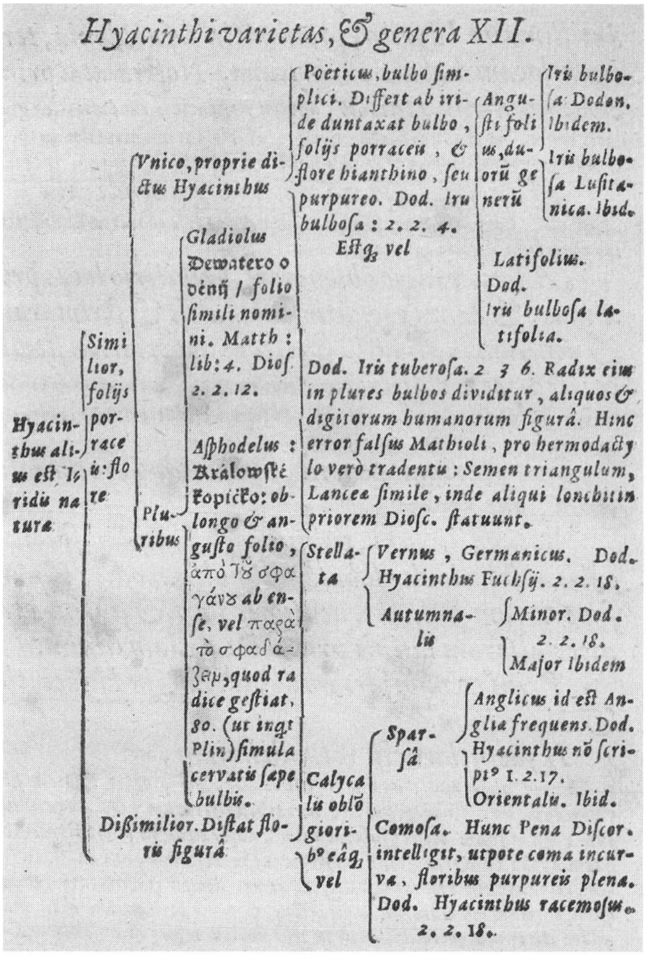

Hyacinthi varietas, & genera XII.

Hyacinthus alius est viridis natura

Similior, folijs porraceis: flore

Vnico, proprie dictus Hyacinthus

Poeticus, bulbo simplici. Differt ab iride duntaxat bulbo, folijs porraceis, & flore hianthino, seu purpureo. Dod. Iris bulbosa: 2. 2. 4.

Estq; vel

Angustifolius, duorū generū

Iris bulbosa Dodon. Ibidem.

Iris bulbosa Lusitanica. Ibid.

Latifolius. Dod. Iris bulbosa latifolia.

Pluribus

Gladiolus Dewatero o dénij / folio simili nomini. Matth: lib: 4. Diosc. 2. 2. 12.

Dod. Iris tuberosa. 2 3 6. Radix eius in plures bulbos dividitur, aliquos & digitorum humanorum figurâ. Hinc error falsus Matbioli, pro hermodactylo verò tradentu: Semen triangulum, Lancea simile, inde aliqui loncbitin priorem Diosc. statuunt.

Asphodelus: Kralowstê topicko: oblongo & angusto folio, ἀπὸ τοῦ σφαγάνου ab ense, vel παρὰ τὸ σφαδάζειν, quod radice gestiat, 80. (ut inquit Plin) simula cervatis sape bulbis.

Stellata

Vernus, Germanicus. Dod. Hyacinthus Fuchsij. 2. 2. 18.

Autumnalis

Minor. Dod. 2. 2. 18.

Major Ibidem

Calycalis oblō giori-b^o eáq; vel

Sparsâ

Anglicus id est Anglia frequens. Dod.

Hyacinthus nō scriptus 1. 2. 17.

Orientalis. Ibid.

Comosa. Hunc Pena Discor. intelligit, utpote coma incurva, floribus purpureis plena. Dod. Hyacinthus racemosus. 2. 2. 18.

Dissimilior. Distat floris figurâ

Figure I.3 Adam Saluziansky à Zaluzian, *Methodi herbariae libri tres* (1592), 'Hyacinthi varietas, & genera XII', p. 130, Sherardian Library of Plant Taxonomy, One of the Bodleian Libraries of the University of Oxford

of God – 'natural selection', a principle without the need for external causation. For Darwin organisms produce more offspring than can survive; all organisms within a species vary; and some variation is inherited by offspring. Since only some offspring can survive, on average survivors will be variants that are better adapted to changing local environment, and since these offspring inherit their parents' favourable variations, organisms of the succeeding generation will on average become better adapted to local environmental conditions.[94] An avid scribbler, Darwin's

[94] Stephen Jay Gould, 'Introduction', in Carl Zimmer (ed.), *Evolution: The Triumph of an Idea*, London: William Heinemann, 2002, xii.

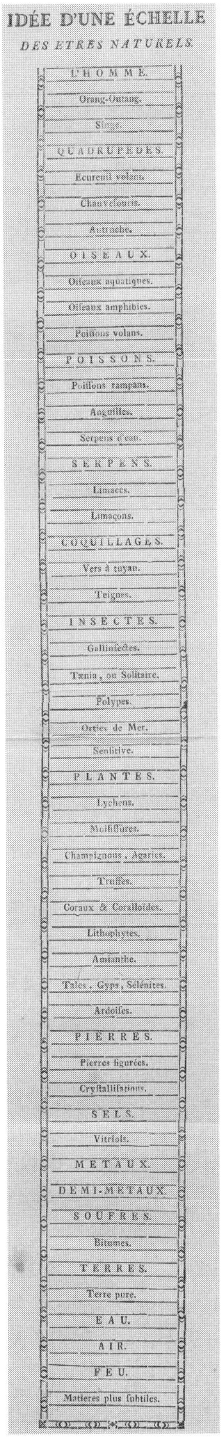

Figure I.4 Charles Bonnet, *Traité d'insectologie* (1745), table in volume 1 of Charles Bonnet, *OEuvres d'histoire et de philosophie* (1779 [1779–83]), between preface and contents, permission granted by Wellcome Library, London

Figure I.5 Charles Bonnet, *Contemplation de la nature* (1764), head ornament to volume 4 of Charles Bonnet, *OEuvres d'histoire et de philosophie* (1781 [1779–83]), p. 1, permission granted, Cambridge University Library

arborial odyssey begins with an unprepossessing image in his first notebook 'On the Transmutation of Species' (1837–38, Figure I.8), and culminates (though does not end) in the *Origin of Species* (1859, Figure I.9). The first substantive phylogenic tree is headed 'I think' – as if Darwin were unsure – but the evolutionary message is clear: all biological diversity stems from a single common ancestor.[95] The famous diagram in the *Origin of Species* – its only illustration – is much less unassured, offering a confident representation of the principle of divergence and the effect evolution (through natural selection) has on the descendants of a common ancestor.[96]

Although Darwin claimed that natural selection 'works solely by and for the good of each being, [and] all corporeal and mental endowments will tend to progress toward perfection'[97] the meandering branches of his tree would struggle to uproot Victorian Britain's firmly held Lamarckian belief in progression from simplicity to complexity, though Lamarck's mechanism of progression – the inheritance of acquired characteristics – was

95 Pietsch, *Trees of Life*, 85.

96 Charles Smith, *Charles Darwin and Victorian Visual Culture*, Cambridge: Cambridge University Press, 2006, 6.

97 Charles Darwin, *On the Origin of Species by Means of Natural Selection, or the Preservation of Favoured Races in the Struggle for Life*, London: John Murray, 1859, 489.

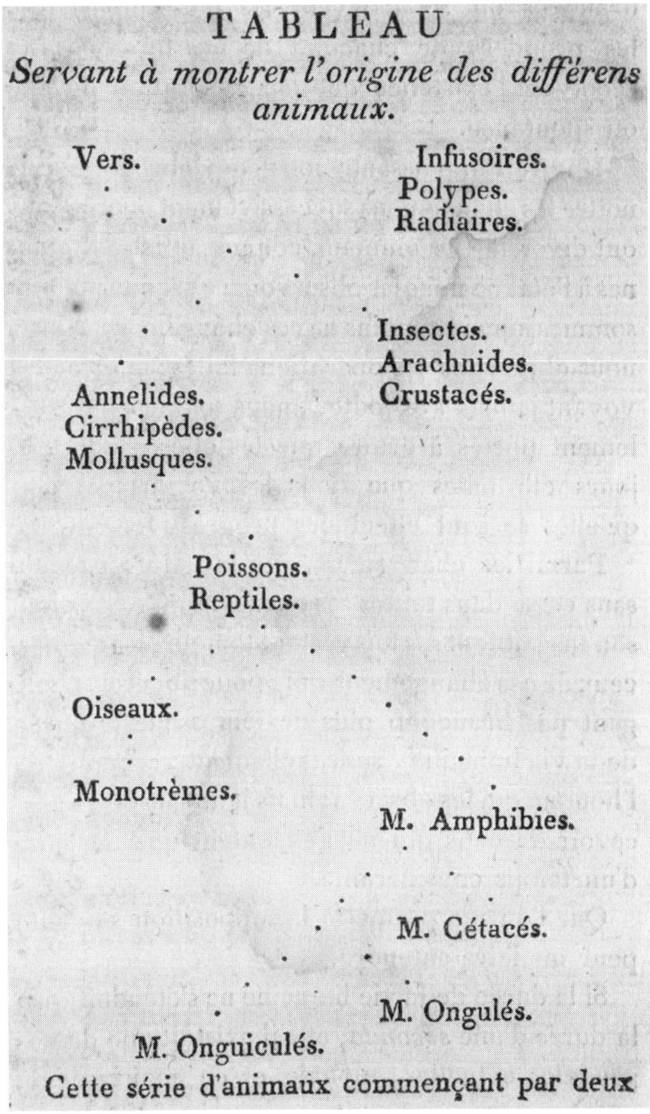

Figure I.6 Jean-Baptiste Lamarck, *Philosophie Zoologique* (1809), volume 2, p. 463

always controversial. In the fabulously popular (and mysteriously anonymous publication) *Vestiges of Natural History of Creation* (1844) Robert Chambers extends Lamarck and describes progression as 'development'. What would become developmentalism led ineluctably from the lowest creatures to man's apogee in creation: 'The whole train of animated beings, from the simplest and oldest up to the highest and most recent, are, then, to be regarded as a series of *advances of the principle of development*, which

Figure I.7 Carl Edward von Eichwald, *Zoologia specialis* (1829), pp. 40–41, from the collections of the Ernst Mayr Library, Museum of Comparative Zoology, Harvard University

have depended upon external physical circumstances, to which the resulting animals are appropriate'.[98] Although *Vestiges* was as sensational as it was roundly criticised by scientists, it nevertheless had the effect of preparing the way for Darwin and other putatively materialistic readings of

[98] Robert Chambers, *Vestiges of the Natural History of Creation*, London: John Churchill, 1844, 203.

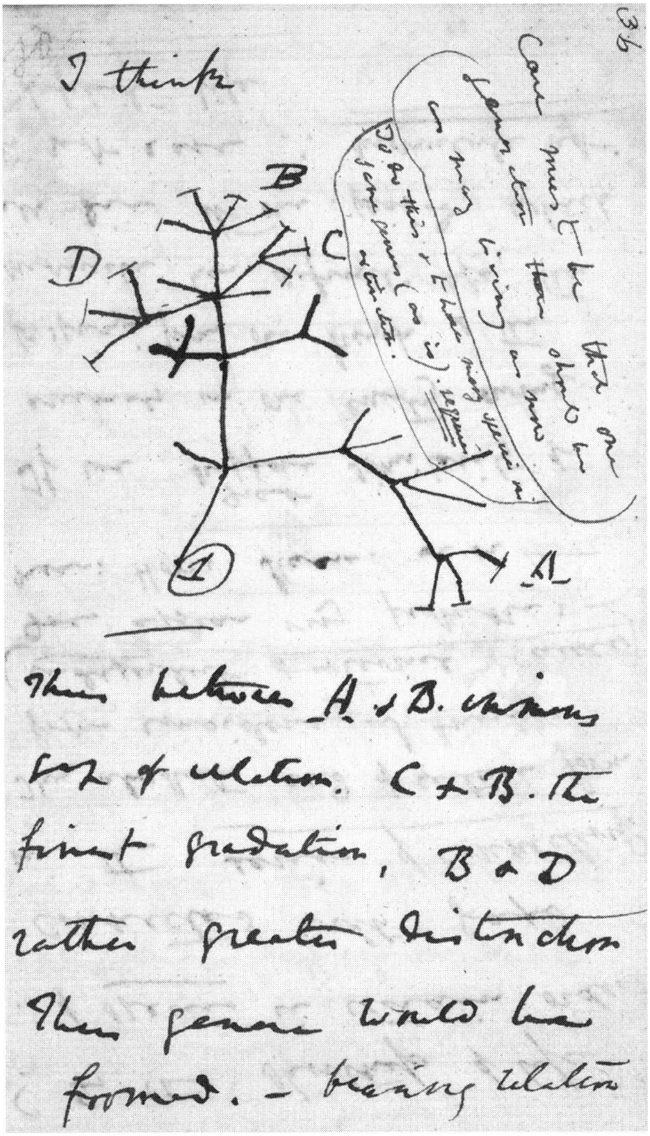

Figure I.8 Charles Darwin, *Notebook B: Transmutation of Species* (1837–38), 36, permission granted, Cambridge University Library

nature. There were many reasons to criticise it: natural theologians and theistic scientists alike despised it for its reduction (or reinterpretation) of God's creative omnipotence, and scientists more generally questioned its inadequate provision of evidence. But there were reasons to like it as well, and the preservation of a Great Chain of Being was one of its saving

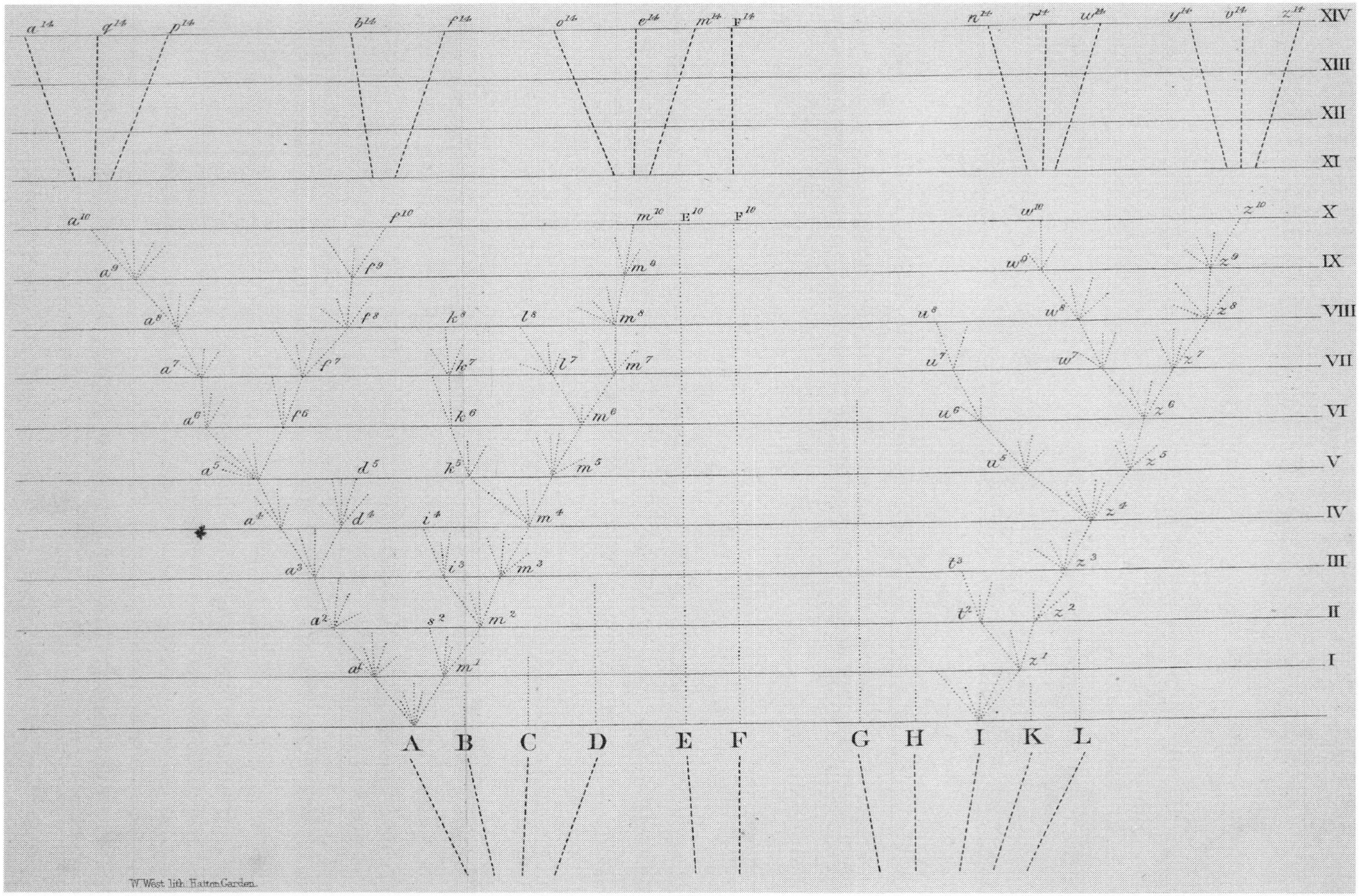

Figure I.9 Charles Darwin, on the *Origin of Species* (1859), pp. 116–117, permission granted, Cambridge University Library

graces. The mutually dependent methodologies of developmentalism and The Great Chain of Being anchored Victorians psychologically at a time of revolutionary change in science when previous theological and scientific certitudes were coming under attack. The geologist Charles Lyell encapsulates the uncertainty in the third edition of his *A Manual of Elementary Geology* (1851) when he describes the absence of elements in a geological column as a 'missing link' in the Great Chain of Being: 'A break in the chain implying no doubt many missing links in the series of geological monuments which we may some day be able to supply'.[99] Although today we tend to think of finding the 'missing link' as a quest to locate a mythic interim species between ancient or primitive and modern man, amongst Victorians the concept catalysed uncertainty in the previously unbroken Great Chain of Being. Other evolutionary paradigms came to the rescue, particularly the sister concept of recapitulation.

Like the Great Chain of Being, recapitulation has an ancient pedigree going back to ancient Greece, but it burgeoned scientifically under influence of eighteenth- and nineteenth-century German Idealism.[100] Recapitulation teaches that embryos do not simply grow larger and more specialized in the womb; they pass through successive stages in the evolution of their animal series. According to J. F. Meckel, 'The development of the individual organism obeys the same laws as the development of the whole animal series; that is to say, the higher animal in its gradual development essentially passes through the permanent organic stages that lie below it'.[101] Recapitulationism evolved little in subsequent generations. Darwin's earlier contemporary Friedrich Tiedemann, for example, claims that 'Every animal, until it reaches its own structure, passes through the organization of one or more of the animal classes standing under it, which is to say, every animal begins its metamorphosis with the simplest animal organization'.[102] German

[99] Charles Lyell, *A Manual of Elementary Geology*, 3rd ed. and rev., London: John Murray, 1851, 220.

[100] Stephen Jay Gould, *Ontogeny and Phylogeny*, Cambridge, MA, and London: Belknap Press of Harvard University Press, 1977, 6.

[101] J. F. Meckel, *System der Vergleichenden Anatomie*, 7 Bde., Halle: Rengersche Buchhandlung, 1821, 514, cited in Ernst Mayr, 'Recapitulation Reinterpreted: 'The Somatic Program', *The Quarterly Review of Biology*, 69/2 (June 1994), 223.

[102] Friedrich Tiedemann, *Zoologie, zu seinen Vorlesungen Entworfen*, 3 vols., Landshut: Weber, 1808–14, Vol. 1 (1808), 56–57, cited in Robert J. Richards, *The Meaning of Evolution: The Morphological Construction and Ideological Reconstruction of Darwin's Theory*, Chicago and London: University of Chicago Press, 1992, 43.

recapitulationism reaches its zenith in the work of Ernst Haeckel, author of the hugely popular *Generelle Morphologie des Organismen* (1866). Haeckel encapsulates recapitulation in the famous scientific axiom 'Ontogenesis is a brief and rapid recapitulation of phylogenesis,'[103] and as before the individual human embryo continues to pass in its development through all stages of human evolution from the lowest to highest life forms – from protozoa to invertebrates, and vertebrates to mammals. Like his predecessors Haeckel illustrates recapitulation through trees, but contrasting with Darwin's rambling evolutionary tangle of branches Haeckel's tree embodies the recapitulation inherent within the Great Chain of Being. As illustrated in Figure I.10 a magisterial oak rises up from the lowest to highest forms, crowned by man, each successively higher position recapitulating the lower stages of life. Haeckel used (and evidently abused) images to great (and sometimes personally damaging) effect.[104] Figures I.11 show the trimesters of increasingly advanced forms of life culminating in man

German recapitulation played into the hands of Victorians desperately clinging to the certitudes of developmentalism and the Great Chain of Being, and in Britain the idea held wide currency amongst thinkers across a spectrum of sciences. The anthropologist Robert Knox maintains that

> the individual [human embryo] is in fact ... passing through forms which represent the permanent forms of other adult beings belonging to the organic world, not human but bestial, of which some are extinct, some belong to the existing world, while others may represent forms which once existed but are now extinct; or finally forms which are some day to appear, running through their destined course, then to perish as [did] their predecessors. The fully developed, or grown-up, brute forms of birds and fishes, or reptiles and mammals, are represented in the organic structures of the human embryo.[105]

Similarly, Baden Powell, in his *Essays on the Spirit of the Inductive Philosophy, the Unity of Worlds, and the Philosophy of Creation* (1855), claims that 'the relations existing between the *permanent* organisation of the lower classes of animals, and the *transitory* embryonic steps

[103] Ernst Haeckel, *Generelle Morphologie des Organismen*, 2 vols., Berlin: Georg Reimer, 1866, cited in Ernst Haeckel, *Riddle of the Universe at the Close of the Nineteenth Century*, trans. Joseph McCabe, New York and London: Harper and Brother Publishers, 1900, 81.

[104] See Robert J. Richards, *The Tragic Sense of Life: Ernst Haeckel and the Struggle over Evolutionary Thought*, Chicago and London: University of Chicago Press, 2008.

[105] Robert Knox, *The Races of Men*, London: H. Renshaw, 1850, 29.

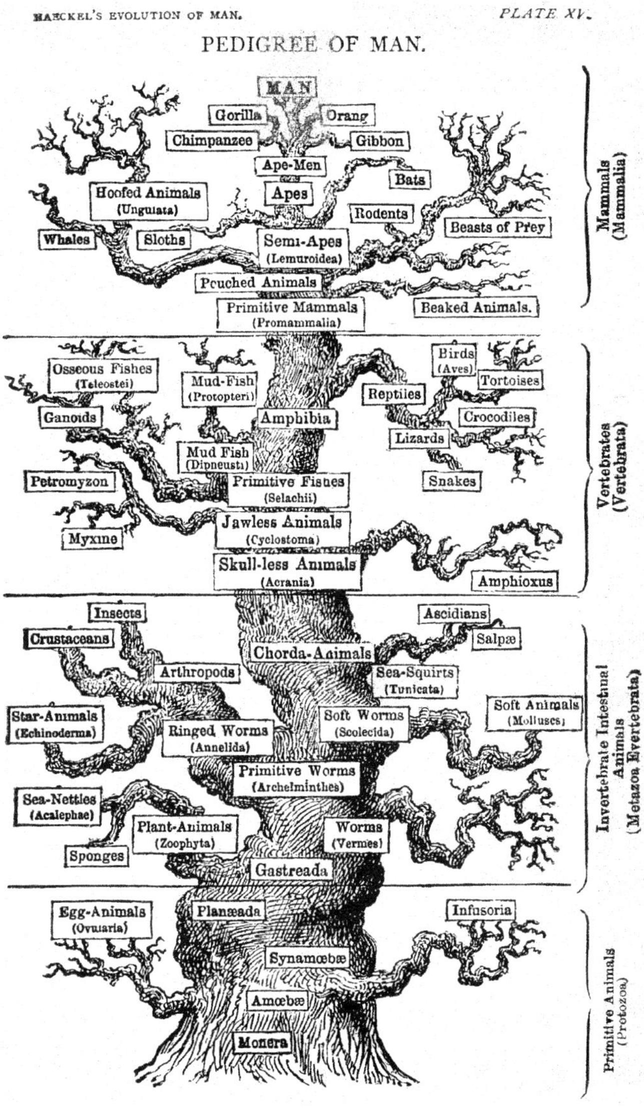

Figure I.10 Ernst Haeckel, *The Evolution of Man* (1879), volume 2, pp. 188–189, originally from *Anthropogenie* (1874)

through which the higher pass and evince the extent to which the resemblances expressed by the term "Unity of Organisation" may be traced between the higher and lower organised animals'.[106] Recapitulation

[106] Baden Powell, *Essays on the Spirit of the Inductive Philosophy, the Unity of Worlds, and Philosophy of Creation*. London: Longman, Brown, Green, and Longmans, 1855, 372–73.

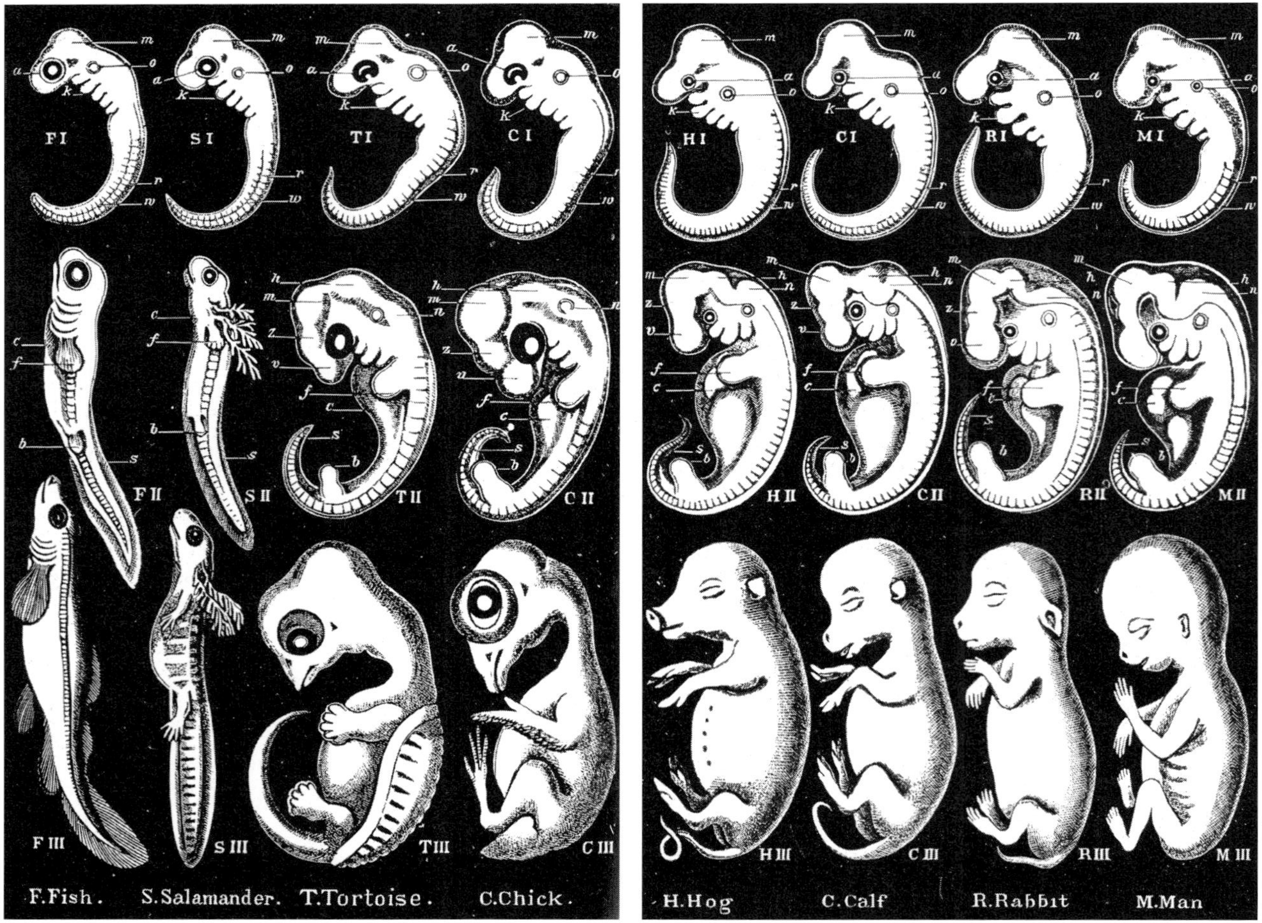

Figure I.11 Ernst Haeckel, *The Evolution of Man* (1879), volume 1, pp. 362–363, originally from *Anthropogenie* (1874)

had its opponents, however, in both Europe and Britain. In what Stephen Jay Gould refers to as 'probably the most important words in the history of embryology'[107] Karl Ernst von Baer insists that 'the embryo of a higher animal is never like [the adult of] a lower animal, but only like its embryo'[108] and Robert Chambers agrees: 'in the reproduction of the higher animals, [it has been said that] the new being passes through stages in which it is successively fish-like and reptile-like. But the resemblance is not to the adult fish or the adult reptile, but to the fish and reptile at a certain point in their foetal progress'.[109]

Recapitulation achieves its ideological apogee in the works of neo-Lamarckian polymath Herbert Spencer, and it is arguably through his systematic, if sometimes philosophically impenetrable, application that the idea became widespread, especially in nineteenth-century British musical culture. There are in fact three basic, if at times contradictory, evolutionary principles operating in what became known as Spencer's 'synthetic philosophy' (interestingly, the source of George Eliot's parable of Casaubon's 'key to all mythologies' in *Middlemarch*):[110] (1) all organic and inorganic matter was subject to the same combination of evolutionary forces – recapitulation, (2) progression from simplicity to complexity, and (3) survival of the fittest. Recapitulation informs practically every one of Spencer's intellectual formulations; in *Principles of Sociology* (3 vols., 1874–96), for example, it appears encapsulated in the following way: 'In the general course of organic evolution from low types to high, there have been passed through by insensible modifications all the stages above described; but now, in the individual evolution of an organism of high type, these stages are greatly abridged, and an organ is produced by a comparatively direct process'.[111] Complementing

[107] Gould, *Ontogeny and Phylogeny*, 56.

[108] Karl Ernst von Baer, *Uber die Entwickelungsgeschichte der Thiere: Beobachtung und Reflexion*, Scholium 5: *Ueber der Verhältniss der Formen, die das Individuum in den verschiedenen Stufen seiner Entwicklung annimmt*, Königsberg: Bornträger, 1828, 224, cited in Alec L. Panchen, *Classification, Evolution, and the Nature of Biology*, Cambridge: Cambridge University Press, 1992, 20.

[109] Chambers, *Vestiges*, 212.

[110] Martin N. Raitiere, *The Complicity of Friends, How George Eliot, G. H. Lewes, John Hughlings-Jackson Encoded Herbert Spencer's Secret*, Lewisburg, PA: Bucknell University Press, 2012, 149.

[111] Herbert Spencer, *Principles of Sociology*, 3 vols. (1874–96), A System of Synthetic Philosophy 6–8, New York: D. Appleton and Co., vol. 1, 1883, 501.

recapitulation is progression from the simple to complex – progress in the broadest sense; in 'Progress: Its Law and Cause' the concept acquires germinal status:

> Now, we propose in the first place to show, that this law of organic progress is the law of all progress. Whether it be in the development of the Earth, in the development of Life upon its surface, the development of Society, of Government, of Manufactures, of Commerce, or Language, Literature, Science, Art, this same evolution of the simple into the complex, through a process of continuous differentiation, holds throughout. From the earliest traceable cosmical changes down to the latest results of civilization, we shall find that the transformation of the homogeneous into the heterogeneous, is that in which all Progress essentially consists.[112]

Lastly and perhaps most importantly is 'survival of the fittest', one of Victorian Britain's most incontestably influential memes. Coined by Spencer in *Principles of Biology* (1864), 'survival of the fittest' came to be used as a byword for evolutionism.[113] Darwin adopted the phrase in *The Variation of Animals and Plants Under Domestication* (1868), the fifth edition of the *Origin of Species* (1869), *The Descent of Man* (1871) and many later publications to distinguish it from natural selection. A 'reciprocal adaptation' is 'social Darwinism', a term not invented by Spencer but associated closely with him from the 1870s.[114] Social Darwinism advocates a laissez-faire policy towards the state giving the laws of evolution unrestrained influence over society. It is easy, today, to

[112] Herbert Spencer, 'Progress: Its Law and Cause', *Humboldt Library of Popular Science Literature* 17/1 (1882), 234.

[113] Although there is consensus about Spencer's authorship of the phrase, its source is disputed. David Crook and James Rogers believe it came from 'Theory of Population, Deduced from the General Law of Animal Fertility', *Westminster Review* 57/112 (April 1852), 468–501. See David Crook, *Darwinism, War and History*, Cambridge: Cambridge University Press, 1994, 216, and James Allan Rogers, 'Darwinism and Social Darwinism', *Journal of the History of Ideas*, 33/2 (April 1972), 266. Current scholarship suggests that it first appeared in *Principles of Biology* (1864, part 3, chapter 12, vol. 1). See Diane B. Paul, 'The Selection of "Survival of the Fittest"', *Journal of the History of Biology* 21/3 (Sept. 1988), 412–14; John Offer, *Herbert Spencer: Critical Assessments*, London and New York: Routledge, 2000, 3; and Peter J. Bowler, 'Herbert Spencer and Lamarckism', in Mark Francis and Matthew W. Taylor (eds.), *Herbert Spencer: Legacies*, Abingdon and New York: Routledge, 2015, 213.

[114] Bernard Lightman and Bennett Zon, 'Introduction', in Bernard Lightman and Bennett Zon (eds.), *Evolution and Victorian Culture*, Cambridge: Cambridge University Press, 2014, 4.

imagine how damaging and divisive such a seemingly heartless social policy can be:

Blind to the fact that under the natural order of things, society is constantly excreting its unhealthy, imbecile, slow, vacillating, faithless members, these unthinking, though well-meaning, men advocate an interference which not only stops the purifying process but even increases the vitiation – absolutely encourages the multiplication of the reckless and incompetent by offering them an unfailing provision, and discourages the multiplication of the competent and provident by heightening the prospective difficult of maintaining a family.[115]

Spencer had competition, and not just from Darwin but from other pre- and non-Darwinian evolutionary theories like monogenism and polygenism. Monogenism asserts that all people are descended from Adam and Eve; polygenism, from multiple origins. According to George Stocking monogenism invites a set of anthropological assumptions: the unity of man, the recentness of man's appearance on earth (famously 4004 BC according to Bishop Ussher's calculation of biblical genealogies), a sharp distinction between man and animals, and the degeneration of non-Christian savages.[116] Although ostensibly innocuous monogenism produced a profoundly racist trope, inverting the Enlightenment principle of progress to account for the presence of 'degenerated' peoples like non-Christians, heathens, savages and other primitive cultures. Monogenism did not preclude evolutionary differentiation and diversity, however; one its great practitioners, James Cowles Prichard, believed that all humans were originally black and that differentiation was a result of civilization.[117] Polygenism also appears benign on the anthropological surface but it contains equally racist implications as well. Whereas monogenism claims that environment can influence difference in physical appearance, polygenism rejects the idea that environment can influence differences in physical appearance, and argues that differences between peoples can only have resulted by descent from different ancestors – a theory which came to be known as 'racial difference'.[118]

[115] Herbert Spencer, *Social Statics; Or the Conditions Essential to Human Happiness Specified and the First of Them Developed*, London: J. Chapman, 1851, 323–24.

[116] George W. Stocking, Jr, *Victorian Anthropology*, New York: Free Press, 1987, 44.

[117] Hanna Franziska Augstein, introduction, in Hanna Franziska Augstein (ed.), *Race: The Origins of the Idea, 1760–1850*, Bristol: Thoemmes Press, 1996, xxiv.

[118] Norris Saakwa-Mante, 'Western Medicine and Racial Constitutions: Surgeon John Atkins, 'Theory of Polygenism and Sleepy Distemper in the 1730s', in Ernst Waltraud and Bernard Harris (eds.), *Race, Science and Medicine, 1700–1960*, London and New York: Routledge, 1999, 30.

Polygenism, in other words, created a theory of innate racial inequality. Although Graham Richards identifies polygenism as the source of Victorian Britain's 'subhumanity question' – racial prejudice against non-whites, and in particular blacks[119] – monogenism and polygenism proved to be mutually compatible when it came to Victorian racism inflected by a hierarchy of the Great Chain of Being which 'apotheosized white European (implicitly British) Man'.[120]

Fuelled by monogenism, polygenism and other popular evolutionary models anthropological racism continued unabated until the 1870s, when an alternative non-Darwinian theory emerged, 'survivals'. 'Survivals' was a term coined by anthropologist E. B. Tylor to denote those vestigial remnants within culture which have remained doggedly unaffected by evolutionary processes but which survived nonetheless. These include processes, customs and opinions which remain as 'proofs and examples of an older condition of culture out of which a newer has been evolved'.[121] Survivals inverted Spencer's concept of survival of the fittest because the Tylorian survivor was non-adaptive. Savages, for example, were deemed to be survivals, yet culture itself was evolutionary. In fact for Tylor culture was synonymous with civilization: 'Culture or Civilization', he maintains, 'taken in its wide, ethnographic sense, is that complex whole which includes knowledge, belief, art, morals, law, custom, and any other capabilities and habits acquired by man as a member of society'.[122] Tylor's evolutionary thinking predisposes him to the accusation of contradiction because survivals are the primitive exception that proves the evolutionary rule of civilization, and as the foregoing chapters aim to prove, in a culture dominated by the principle of developmental progress towards civilization atavistic tendencies and vestigial practices had no place in the Great Chain of Being. Yes, progress can involve stages – Spencer and Freeman[123] were after all deeply influenced by Comte's cyclical thinking.[124] But for proto-modern

[119] Graham Richards, *'Race', Racism and Psychology: Towards a Reflexive History*, London and New York: Routledge, 1997, 7.

[120] Lightman and Zon, 'Introduction', in Lightman and Zon, *Evolution and Victorian Culture*, 5.

[121] E. B. Tylor, *Primitive Culture: Researches into the Development of Mythology, Philosophy, Religion, Language, Art and Custom*, 2 vols., New York: Henry Holt and Co., 1871/1889, vol. 1, 16.

[122] Edward B. Tylor, *Primitive Culture*, vol. 1, 3.

[123] Edward A. Freeman, *The Chief Periods of European History: Six Lectures read in the University of Oxford in Trinity Term 1885*, London: Macmillan, 1896, 176.

[124] Peter J. Bowler, *The Invention of Progress: The Victorians and the Past*, Oxford: Basil Blackwell, 1989, 9.

Tylor, as for educators today, all minds are inherently equal, and the only thing hindering progress is a lack of education: 'active at once in aiding progress, and in removing hindrance, the science of culture is essentially a reformer's science'.[125]

[125] Tylor, *Primitive Culture*, vol. 2, 453.

1 | Zoomusicology

Introduction

The history of science and religion is often portrayed as an ideological battleground, littered with the terminology of warfare. Victorian tracts like John William Draper's *History of the Conflict between Religion and Science* (1875) and Andrew Dickson White's *A History of the Warfare of Science with Theology in Christendom* (1896) do nothing to dispel that impression, and indeed the language persists into the twentieth century with J. Y. Simpson's classic *Landmarks in the Struggle between Science and Religion* (1925). Even as late as 1997 Stephen Jay Gould could reiterate this opinion, describing science and religion unflinchingly as 'nonoverlapping magisteria'.[1] In the broadest sense this chapter uses Victorian zoomusicology to dispute the imagery of conflict and offer a more nuanced appreciation of the complex and often complementary relationship between science and religion. By exploring attitudes towards animal music at the height of the non-Darwinian revolution and using birdsong as a case study, I argue that despite their differences, Victorian scientists and theologians occasionally agreed in their views, if for different reasons. I believe that scientists and theologians heard much the same thing when they listened to birdsong – they heard the sound of 'redemption'. For science, redemption meant identifying similarities between human music and birdsong which collapsed distinctions between man and nature, and creating a more perfect understanding of man's relation to his world; for theology, redemption meant learning how birdsong reflected man's pivotal, liminal position between the animal and spiritual world, and creating a more perfect understanding of his relation to God.

This chapter divides into two unequal parts. The shorter introductory part 'Zoology' provides preliminary background information in two sections, 'Zoological Science' and 'Zoological Culture'. The first section introduces basic issues in Victorian zoology, including hierarchical concepts of type, the

[1] Stephen Jay Gould, 'Nonoverlapping Magisteria', *Natural History* 106 (March 1997), 16–22.

Great Chain of Being and Linnaean taxonomy; the second section explores how these zoological approaches were promulgated through popular literature, museums and the zoo. The second part, 'Zoomusicology', includes two more substantial sections, an introduction to animal music and the chapter's focal point, a case study on birdsong. The first section asks whether Victorians thought all animals made music (and if so how) and provides a close examination of Victorian zoomusicology through anecdotal literary forms, anecdotal types of experimentation (often related to the zoo) and academic writing. The case study explores birdsong variously through theology, Darwinian and non-Darwinian science and the aesthetics of musical unity. It ends with an examination of the ideologically conflicted non-Darwinian ornithologist Charles A. Witchell, arguably the first – if ethically contested – ornithologist to apply evolutionary theory systematically. A conclusion returns to questions raised at the opening of the case study, speculating over the ways scientific and theological ideas interpret the redemptive properties of Victorian birdsong, and how conversely ornithological ideas influenced changes in contemporary science and theology.

Zoology

Zoological Science

Zoology is a branch of biology, which is itself a branch of natural history. Natural history divides between the study of organic (living) and inorganic (not living) matter; organic matter between biology (plants) and zoology (animals); and zoology between animals from the lowest invertebrates (protozoa) to highest vertebrates (bimana – man). According to eminent Victorian zoologist Alfred Newton, zoology means

> a knowledge of animals – a knowledge of where and by what means they live, what food they eat, how they grow and breed, what changes they undergo at various stages of their existence, what peculiar habits they possess, what they look like and how they are framed, a knowledge of their mutual resemblances and differences – in short a knowledge of all about them.[2]

Victorians would gain that all-encompassing zoological knowledge through a variety of approaches, and as later sections of this chapter show, writers on both science and music would consider zoology a perfectly plausible

[2] Alfred Newton, *Zoology*, London: Society for Promoting Christian Knowledge; New York: Pott, Young & Co., 1874, 5.

scientific means to a theological end, even devout Darwinians like Newton.[3] Newton speaks for a multitude of Victorian thinkers when he claims that once we acquire that knowledge, 'we may begin to reason upon such facts and even begin to consider whether what we term our classification may not be a kind of index to the way in which it has pleased God to bring His creatures into being'.[4] Other Victorians omitted God from their classification altogether; equally Darwinian popular zoologist Henry Alleyne Nicholson published *A Manual of Zoology* in seven editions between 1870 and 1887 without so much as a word on God.

If zoology could ignore, debunk or deny a designing God, it could not escape the fact of a meticulously organized cosmos. Victorians loved organization, and because organization meant control and control meant power, they organized animals in their own image. As Harriet Ritvo suggests, Victorian zoological nomenclature 'worked to create and reinforce a range of parallel patterns of human hierarchy'[5] – patterns Victorians eventually formalized into moral hierarchies mapped, for example, onto the ornithological character of songbirds in John Mossop's telling 'Scale of Merit Among British Singing Birds'.[6] While songbirds mirrored the moral values of human hierarchy, Victorian zoology also began to change the way it interpreted the natural order. Whereas earlier in the eighteenth and early parts of the nineteenth centuries nature was regularly ordered in a seamless vertical ascent up the Great Chain of Being, by the Victorian period a competing system of organization was revolutionizing zoology, and this too would effect change in the way writers on animal music like Darwin and Sully appeared to democratize the moral character and musicality of their subjects. Developed by Carl Linnaeus (1707–78) modern taxonomy produced a system of branching classification which categorized variation unhierarchically while preserving the hierarchical nature of teleology. For Linnaeus, there were three kingdoms (animal, vegetable and mineral); kingdoms divided into classes; classes into genera; genera into species;

[3] For an excellent, albeit scientifically orientated, introduction to Newton's significance, see Tim R. Birkhead, 'Alfred Newtown's Contribution to Ornithology: A Conservative Quest for the Facts Rather than Grand Theories', *Ibis* 154 (2012), 887–905.

[4] Newton, *Zoology*, 5–6.

[5] Harriet Ritvo, 'Zoological Nomenclature and the Empire of Victorian Science', in Bernard Lightman (ed.), *Victorian Science in Context*, Chicago and London: University of Chicago Press, 1997, 337.

[6] See Daines Barrington, 'Experiments and Observations on the Singing of Birds, by the Hon., Vice Pres. R. S. In a Letter to Mathew Maty, M. D. Sec. R. S.', *Philosophical Transactions* 63 (1773–74 [1773]), 282, in The Rev. John Mossop, *Ornithologia Britannica: A Synopsis of British Land-Birds, in Verse and Prose*, London: Jackson and Walford, 1841, 68.

and species into what are now called varieties (plants) and subspecies (zoology). Animals and plants were classified binomially (with two names – genus followed by species). Thus, a dog is *canis familiaris*, a wolf *canis lupus*. The practical disadvantages of this system soon outweighed the theoretical advantages, however, as Victorian zoology began to recognize its limitations, especially Linnaeus's method of basing classification on purely external, visible signs of distinction. Naturalist William Swainson expresses in words what many thought:

> One arguing from the flight of bats, looks on it as that animal which constitutes the true passage from quadrupeds and birds. Another, looking to its general aspect, is disposed to place it among mice, fortified by the general name given by the French to the whole tribe of *chauve souris*. A third, chiefly influenced by the peculiarity of its teeth, arranges it in the same group as monkeys, and each, acting on his respective inferences, fashions his system accordingly ... This is the true cause of the number and the mutability of zoological systems.[7]

The implications of these criticisms for animal music are manifold because Victorian zoomusicologists began delving deeper than the aesthetic surface of animal music into its linguistic contexts and reproductive function. Criticisms of Linnaeus did animal music a great service by giving reason to channel discussion into new scientific territory. Swainson's reason for complaint was widespread amongst zoologists, and there were other concerns as well. Thomas Rymer Jones observes an inconsistency between the hierarchical position of creatures in their respective habitats and their disputable position in the Great Chain of Being. Jones describes the complexion of Linnaean classification in a critically uninhibited mid-century assessment:

> In reviewing the multitudinous races composing the animal kingdom, it has of course been necessary to describe consecutively the different classes and order in which they are grouped by naturalists, as though they formed but one extended line, and thus in a somewhat arbitrary manner to assign to each a place in the lengthy procession. To suppose that such is their natural arrangement would, however, be to fall into a very serious error. Although widely separated in our pages, the Tiger-beetle and the Tiger are, in their respective spheres of action, pretty much of equal rank, and we are inclined to think that in its own element the Cuttle-fish holds as high a place among the Mollusca as the Lion does among quadrupeds. The 'vast chain of being', therefore, composed of numerous successive

[7] William Swainson, *A Preliminary Discourse on the Study of Natural History*, London: Longman, Ress, Orme, Brown, Green, and Longon, 1834, 153.

links, exists only in the imagination of the poet, and the young naturalist would be grievously misled by the adoption of such an idea.[8]

The imagination of the poet and naturalist – or poet and zoomusiocologist – is not entirely as separate as Jones would have us believe, however, because like many pre-Darwinians, Jones himself subscribes to an increasingly debated taxonomy of 'type'. Driven partially by idealist philosophy and partially by natural theological principles of intelligent design, the concept of type created a theologically emollient scientific idea which continued to be used to justify scientifically man's theological dominion over nature. Like different musical genres such as the symphony, concerto or sonata, zoological type produced 'one form (usually the indigenous, most "perfect," or best known) and used it as a model. It could be, and was, applied on several taxonomic levels; one could use a single species to characterize (be the type for) a genus, a genus for a family, and so on'.[9] But type also produced intellectual friction: while opposed to the hierarchical nature of the Great Chain of Being, Victorian 'typists' like Jones clung to a theological conception of man's creational apogee:

Every animal is perfect in its kind, and to add to, or to take from, its attributes would deteriorate its usefulness in creation. By the perfect or imperfect structure of an animal we simply mean the degree in which it approximates to Man, the type and pattern of zoological perfection, just as in estimating the value of money, we take the highest coin of the realm as a standard of comparison. Man, the paragon of animals, is the union of what is most perfect and beautiful in them all.[10]

To prove his point Jones updates Paley's famous parable of the watch (nature is like a clock in its integrated mechanical complexity; a clock must have a maker; and God designed nature like a maker designed the clock) and converts his timepiece from a modest instrument found on a beach into the pulsating mechanism of an entire city: 'It is only by such a view as this, that we can at all understand the intricate dependencies whereby so many creatures are combined in one vast system, carrying out harmoniously the laws imposed upon them by their Great Creator'.[11] As this suggests for Jones and like-minded zoologists, type continued to serve a

[8] Thomas Rymer Jones, *The Animal Creation: A Popular Introduction to Zoology*, London: Society for Promoting Christian Knowledge, 1865, 570.

[9] Paul Lawrence Farber, 'The Type-Concept in Zoology during the First Half of the Nineteenth Century', *Journal of the History of Biology* 9/1 (Spring 1976), 94.

[10] Jones, *Animal Creation*, 5. [11] Jones, *Animal Creation*, 570.

dual – if conflicted – purpose. In the evolutionary ferment of the 1860s, it preserved man's premier position in the complex, yet highly (some would say divinely) ordered unity of creation. Animal music – and especially ornithology – would play a leading part in defending this unity, but not for long. Once Darwinian zoomusicologists began contemplating a Godless creation, animal music was used to topple man from his pedestal. Like Darwinian science at large, the study of animal music aimed to prove that man was no different to the animals; in fact, from a musical standpoint, they were, at least in principle, almost entirely alike.

Zoological Culture

Victorian 'typists' practised culturally what they preached scientifically and ordered animals and the quality of their musicality below man in the Great Chain of Being. As Harriet Ritvo suggests, 'animals never exemplified the best human types. But the sense of human dignity that barred animals from realizing, even figuratively, the highest human possibilities made them particularly appropriate representatives of the less admired ranks and propensities'.[12] In pre- and post-Darwinian culture, domesticated animals like sheep and cattle were often compared to the working classes (and vice versa), and savage, wild animals to foreigners. Primates were frequently compared to the upper classes, however, and 'lovingly' anthropomorphized in literature and visual imagery, as well as in zoos and other forms of public and private exhibition.[13] Chimps wore clothing of varying levels of class distinction and sophistication; orangutans read books (apparently); and apes were given Christian names. According to 'typists' chimps could bang a drum or beat a stick, but primates never made music. Eventually that limitation would collapse, however, as Darwinian zoomusicology emphasized greater continuity between their mental and behavioural qualities and capacities. Darwin himself argued in *The Descent of Man* (1871) that 'there is no fundamental difference between man and the higher mammals in their mental faculties',[14] and in *The Expression of the Emotions in Man and Animals* (1872), man and higher mammals share feelings and physical expression embodied in the production, creation and reception of music.

[12] Harriet Ritvo, *The Animal Estate: The English and Other Creatures in the Victorian Age*, Cambridge, MA, and London: Harvard University Press, 1987, 16.

[13] See Helen Cowie, *Exhibiting Animals in Nineteenth-Century Britain: Empathy, Education, Entertainment*, Houndmills: Palgrave Macmillan, 2014.

[14] Charles Darwin, *The Descent of Man, and Selection in Relation to Sex*, 2 vols., London: John Murray, 1871, vol. 1, 35.

Religiously minded scientific popularizers reinforced animal–human asymmetry by amalgamating and sometimes counterpoising seemingly divergent evolutionary and theological perspectives.[15] Alfred Russel Wallace's friend and co-spiritualist Arabella Buckley did this by combining Darwinian evolution and the Great Chain of Being. *Life and Her Children* (1880) begins with the simplest forms of life, tiny slime animals, and hierarchizes animals from those with simple weapons of attack and defence (e.g. sponges) to prickly skinned animals (starfish), shell-inhabiting animals, worms and jointed-foot animals (crabs, centipedes, spiders and six-legged air-breathing insects).[16] Its sequel *Winners in Life's Race* (1882) hierarchizes at a higher level, including amongst other things ancient fish (*pterygotus*, a nine-foot-long sea scorpion), bony fish, backboned animals developing from water-breathing forms to air-breathing forms, cold-blooded air breathers (*iguanadon*), reptiles, birds (toothed, swimming and standing, running to flying, mound-laying to nest building) and mammalia. Buckley's imagetexts sweep unflinchingly from the science of Darwinism to the theology of the Great Chain of Being. One page bears the running header 'creation by evolution' (Darwinism), the next page 'creation by development' (the Great Chain of Being). A methodological bridge spans both pages: to early naturalists,

> the fact that a fish's fin, a bird's wing, a horse's leg, a man's arm and hand, and the flipper of a whale were somewhat akin, had no other meaning than that they seemed to have been formed from the same plan; and when it became certain that different kinds of animals had appeared from time to time upon the earth, the naturalists of fifty years ago could have no grander conception than that new creatures were separately made . . . But a higher explanation was soon to be found . . . [in] Charles Darwin . . . With the great thought before us that every bone, every hair, every small peculiarity, every tine of colour, has its meaning, and has, or has had, its use in the life of each animal or those that have gone before it, a lifelong study even can never weary us in thus tracing out the working of Nature's laws, which are but the expression to us of the mind of the great Creator.[17]

[15] See Bernard Lightman, *Victorian Popularizers of Science: Designing Nature for New Audiences*, Chicago and London: University of Chicago Press, 2007; especially his work on female scientists, 146–63 and 238–53, and the section 'Visualizing a Theology of Nature', 187–96.

[16] Lightman, *Victorian Popularizers of Science*, 246.

[17] Arabella B. Buckley, *Winners in Life's Race or the Great Backboned Family*, London: Edward Stanford, 1883, 345–46.

One should also add music to Buckley's list of Darwinian 'everything', but in *The Fairyland of Science* (1880) she prevaricates: 'Why then do we not hear all sounds as music',[18] she muses. In fact, like many 'typists' Buckley pays lip service to Darwin; for her, nature produces only sounds, not music – even birds (sweet, though, as their voices are). Buckley is not alone amongst zoological popularizers in the way that she openly reconciles scientific and religious models. Many place the natural order in descending evolutionary sequence (from mammals to fish) to foreground the implicit superiority of man. The opening of A *Popular Natural History for Boys and Girls* (1894) offers a characteristic evolutionary imagetext shown in Figure 1.1: on the left, a bold Darwinian – if cuddly – statement of evolution – 'A Young Chimpanzee' hunches in a chair; cutely dressed in a crisply tailored short-sleeved shirt; right hand supporting the chin in a gesture of thought and contemplation; left hand raised and clinging to a rope, as if striving higher; satellite-dish ears are listening receptively, receiving signals; the eyes stare straight at the reader in slightly unsettling evolutionary recognition. On the right, the comfortingly 'Good evolutionary News' –

> The Creation has a long history, lost in the depths of ages marked by constant change and continuous life. The millions of forms that we know to-day are but as nothing in comparison with those which have vanished forever in the onward march. Onward unrestingly has come the great army of the living, each form rising to suit its environment, and failing as its environment was modified; Nature everywhere testifying to the infinity of its Authors; never a living form transmitting its unaltered likeness to distant futurity, and yet nothing in the end which was not also in the beginning.[19]

Victorian zoological culture transmitted conflated scientific and religious mores not only through the imagetexts of books and periodicals[20] but through the richly embodied use of exhibition spaces like museums and zoos – spaces which would in turn have a direct bearing on zoomusicological culture, particularly in relation to ornithology and the meaning it ascribes to music. Like taxonomies that reinforced patterns of natural order in natural history museums, the evolutionary sequence of exhibition

[18] Arabella B. Buckley, *The Fairy-Land of Science*, London: Edward Stanford, 1880, 140.

[19] W. J. Gordon, *Popular Natural History for Boys and Girls*, London: The Religious Tract Society, 1894, 15–16.

[20] Bernard Lightman, 'The Popularization of Evolution and Victorian Culture', in Bernard Lightman and Bennett Zon (eds.), *Evolution and Victorian Culture*, Cambridge: Cambridge University Press, 2014, 287.

POPULAR NATURAL HISTORY

FOR BOYS AND GIRLS

CHAPTER I

MAMMALS

THERE are over two million species of animals on this earth, and if we were merely to print their names, we should have to mass eight thousand of them on every page of this book. The numbers are, in fact, so appalling and incomprehensible that it is simply beyond human powers to realise the immensity and variety of the mighty Creation amid which we live. The wonderful is not necessarily the unknown; in the known we have the wonderful with us.

The Creation has a long history, lost in the depths of ages marked by constant change and continuous life. The millions of forms that we know to-day are but as nothing in comparison with those which have vanished for ever in the onward march. Onward unrestingly has come the great army of the living, each form rising to suit its environment, and failing as its environment was modified; Nature

Figure 1.1 W. J. Gordon, A *Popular Natural History for Boys and Girls* (1894), 15–16

cases 'presented a scientist's view of nature's underlying structure . . . The visitor was expected to absorb the lessons that were offered'.[21] Public museums repeated this lesson: genera were typically separated, but species remained together. Birds, for example, were separated from mammals, but different species of birds, like songbirds, remained expectedly together: 'The route through the museum itself becomes pedagogical: the orders, classes, and families of nature appear to reveal themselves spontaneously to the diligent museumgoer'.[22] By 1855 the London Zoo emblemized that pedagogical approach: 'The principal object [of the London Zoo] will be to present as many types of form as possible, with the view of illustrating the generic variations of the Animal Kingdom';[23] 'To furnish every possible link in the grand procession of organised life, is the aim of the science of zoology',[24] opines a contemporary reviewer of its guidebooks.

21 Rachel Poliquin, *The Breathless Zoo: Taxidermy and the Cultures of Longing*, University Park, PA: Pennsylvania State University Press, 2012, 125.

22 Poliquin, *Breathless Zoo*, 126–27.

23 No author, *Reports of the Council and Auditors of the Zoological Society of London* (30 April, 1855), London: Taylor & Francis, 1855, 14.

24 No author, review of Joseph Wolf, *Zoological Sketches, Made for the Zoological Society of London, from Animals in Their Vivarium in the Regent's Park*, Edited, with Notes, by D. W.

From the 1840s, the physical space of the London Zoo was structured around an 'interrelated, graduated zoological series – an animated representation of the standard vertebrate categories',[25] and by the 1850s, the evolutionary pathway of the Great Chain of Being was fixed around the pragmatic necessity of zoo layout.[26] Fumbling his or her way around a fold-out map (in Figure 1.2), a visitor to the London Zoo in 1851 could expect to follow just this same layout. Entering precisely at the cages most commonly associated with music – the aviaries – the map takes visitors sequentially around sixty-eight sites divided between two unequal halves by a public drive. The Southern half includes (in order) the Western Aviary (rare birds) and then proceeding to paddocks with a range of animals, Carnivores (lions, tigers and so on), the Bear Pit and Pond, the Lawn (waterfowl, seabirds), the Eastern Aviary (exotic birds), the Falcon Aviary, the bison hut, seal pond, more birds (eagles, parrots), small carnivores (ocelots, lynx, etc.), otters, armadillos and tortoises, the Monkey House, wombats, the dove aviary and birds (eagles, owls). The Northern half includes the Reptile House, cattle, hummingbirds, elephants, zebras, rhinos, tapirs and wild hogs, small quadrupeds, hippos, giraffes, ostriches and kangaroos and antelopes.[27]

Following the larger taxonomic groupings emphasized in the guide book in bold lettering, visitors progress through the Southern gardens from Birds (lower forms) to large Carnivores (bear to lions and tigers); back to Birds (higher forms – falcons) to the Seal Pond (sea mammals); to Birds (even higher forms – eagles) and to the Monkey House (Apes, chimps, etc.); and in the Northern gardens from Reptiles (and amphibians), to hummingbirds, large Mammals and unusual birds and exotic smaller mammals. In both halves, visitors trace a zoological path – a 'grand procession' – up the Great Chain of Being: the Southern half progresses from Birds to lower and higher Mammals, climaxing with a tour of the Monkey House; and the Northern half progresses from Reptiles to lower

Mitchell B.A., F.L.S., Secretary to the Society, London, 1856; D. W. Mitchell, *A Popular Guide to the Gardens of the Zoological Society of London*, London, 1855; and Philip Henry Gosse, *The Aquarium: An Unveiling of the Wonders of the Deep Sea*, London, 1854, *Quarterly Review* 98 (1855), 220.

[25] Harriet Ritvo, 'The Order of Nature: Constructing the Collections of Victorian Zoos', in R. J. Hoage and William A. Deiss, *New Worlds, New Animals: From Menagerie to Zoological Park in the Nineteenth Century*, Baltimore and London: Johns Hopkins University Press, 1996, 46.

[26] Ritvo, 'Order of Nature', 47.

[27] D. W. Mitchell, *A Popular Guide to the Gardens of the Zoological Society of London*, London: Printed for the Author, 1852, opening map and i–iii.

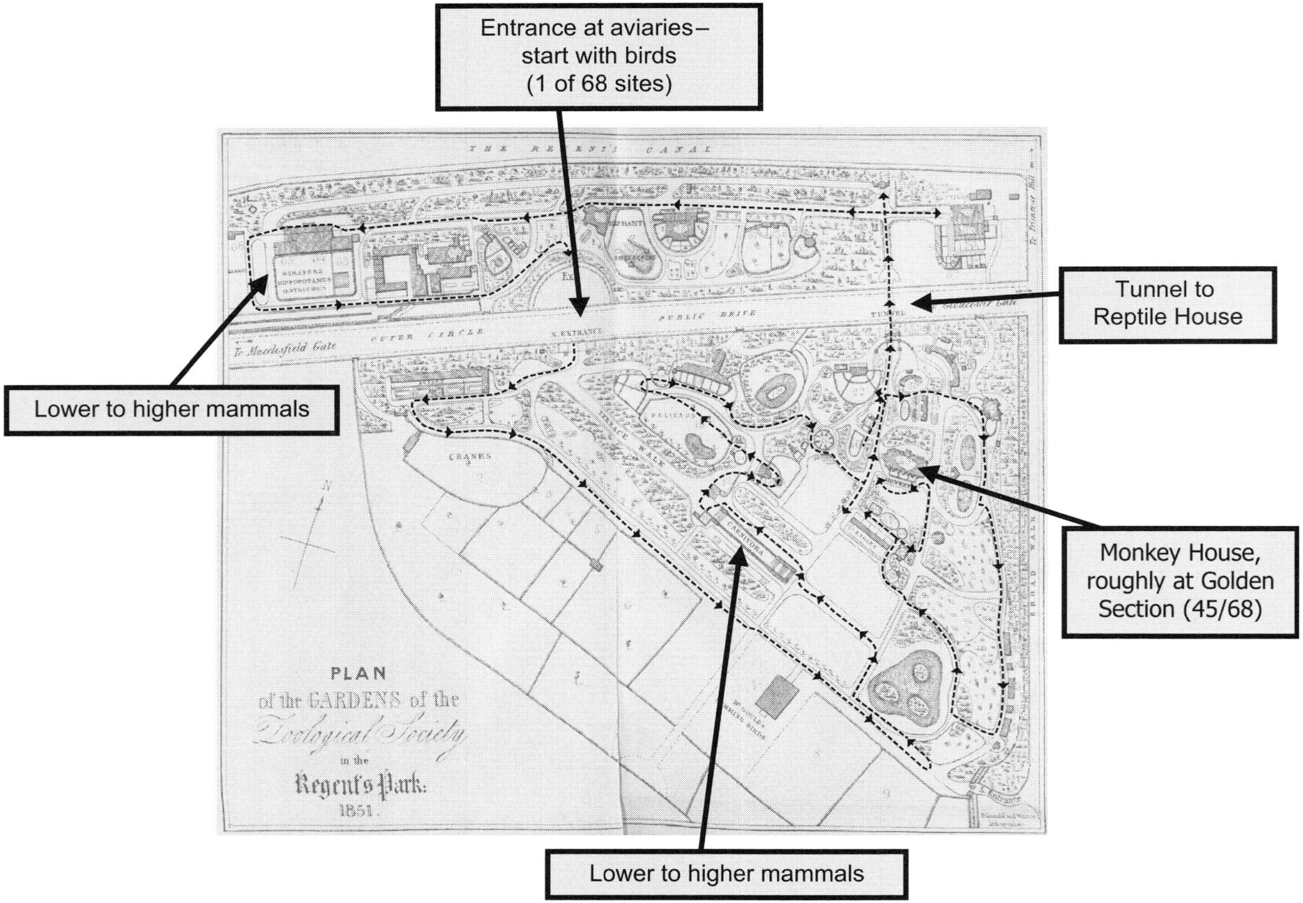

Figure 1.2 Plan of the Gardens of the Zoological Society in the Regent's Park: 1851, D. W. Mitchell, *A Popular Guide to the Gardens of the Zoological Society of London* (1852), i-ii, copyright Zoological Society of London

and higher Mammals, climaxing with elephants (followed by a brief coda of lower Mammals). There are other processional markers as well: to appreciate the full scale and subterranean depth of our evolutionary past, visitors pass through an underground tunnel to view the Reptile House; and if the aesthetic principles of the Fibonacci Series apply to the London Zoo at the time, then the Monkey House (45 on the list of sites) is sited at a key marker of the Golden Section (42/68 sites). Significantly, the entire experience begins and effectively ends at the aviaries (the exit is opposite the entrance), with their sound – or depending upon your point of view, music – strategically impressed upon the memory.

Zoomusicology

Animal Music

While the design of any public exhibition space embodies its cultural context,[28] the grand procession of the London Zoo raises three sets of interrelated questions with implications for Victorian zoomusicology: first and most importantly, is there any particular reason why visitors enter the zoo at Birds rather than lower down the Great Chain of Being – for example, at Reptiles? Is the location arbitrary or intentional; can it be attributed to limitations in the zoo's landscape; or might there be a theological, scientific or aesthetic explanation? What, therefore, did visitors think they heard when they entered the zoo listening to birdsong? Second, what does the position of birds in the Great Chain of Being tell us about the Victorians, and what therefore does the location of the aviaries tell us about Victorian zoomusicological attitudes more broadly? And third and lastly, how did ornithology help or hinder the introduction of Darwinian ideas in a widely non-Darwinian zoomusicological culture?

[28] For Victorian examples, see Sadiah Qureshi, 'Dramas of Development: Exhibitions and Evolution in Victorian Britain', and Carl Yanni, 'Development and Display: Progressive Evolution in British Victorian Architecture and Architectural Theory', in Bernard Lightman and Bennett Zon (eds.), *Evolution and Victorian Culture*, Cambridge: Cambridge University Press, 2014, 261–85 and 227–60, as well as Stephen T. Asma, *Stuffed Animals and Pickled Heads: The Culture of Evolution of Natural History Museums*, Oxford: Oxford University Press, 2003; Sadiah Qureshi, *Peoples on Parade: Exhibitions, Empire, and Anthropology in Nineteenth-Century Britain*, Chicago and London: Chicago University Press, 2011; and Carla Yanni, *Nature's Museums: Victorian Science the Architecture of Display*, New York: Princeton Architectural Press, 2005.

The answers are expectedly complex because, firstly, many Victorians were uncertain whether *any* animal sound qualified as music. Though some believed birds produced music, they were divided on the rest of the animal kingdom's musicality. What made animal sounds musical had in some respects less to do with what sounds animals actually made than the possibility (and for some people the fear) that humans and animals lower down the Great Chain of Being might actually share the same mental capacity to create and respond to music. Indeed for many Victorians, the mere suggestion that humans and animals shared mental faculties had theologically disconcerting implications, collapsing evolutionary differences between humans and animals.[29] This belief did not, however, preclude the everyday naturalist – especially contributors to popular newspapers and scientific literature – from proving that animals showed great powers of response to music, nor from using their findings to reinforce hierarchical certitude in the Great Chain of Being. In popular literature these findings usually emphasize animal 'habits'. Writers survey domesticated and wild animal behaviour ('habits') often through a series of loosely veiled morality tales vaguely resembling experimental observation. The naturalist author of 'Animal Love of Music' (1857) describes his motivation:

The sensibility of animals to music will hardly be questioned in the present day, when the manners and habits of all animated nature are so thoroughly observed and studied. We no longer doubt the dictum of the poet, who sings, 'music hath charms to soothe the savage breast'; and therefore, it is not so much in corroboration of his assertion, as in the illustration of a fact so interesting and pleasing in itself, that we are about to bring to the notice of the reader some few instances of animal love of music, which are too well authenticated to admit of a doubt, and some which are the records of our personal observation and experience.[30]

Some anecdotes are as chatty and conversational as they are informative. The same naturalist recounts:

We once put the question to the keeper of a menagerie: 'Does your band of music have any beneficial effect upon the animals under your care?' 'To be sure it does', said he; 'they like it, and there it does them good. If you were to come in and look at them in a morning when the band is away perambulating the town, you would

[29] Teresa Magnum, 'Dog Years, Human Fears', in Nigel Rothfels (ed.), *Representing Animals*, Bloomington and Indianapolis: Indiana University Press, 2002, 43.

[30] No author, 'Animal Love of Music', *The Leisure Hour: A Family Journal of Instruction and Recreation* 272 (12 March 1857), 165.

see many of them, the more savage beasts in particular, dull and moping, and either sitting or lying in their dens. We who travel with wild beasts cannot of course give them the room they ought to have; and being cribbed up in show-boxes, they degenerate for want of exercise, do what we will with them. But it would be very much worse, I reckon, if it were not for the music. When they hear the band strike up, they rouse themselves, and begin taking what exercise they can: the beasts of prey walking backwards and forwards, and the others by repeating the movements natural to them when at liberty; the birds will begin to chaffer and plume themselves at the sound; and even the snakes at times will uncoil and rear up, and convince the people, who sometimes seem to doubt the fact, that they are living creatures'.[31]

Generally, musical anecdotes follow a generic plot line and outcome: while in a natural environment, a man or men play instrumental music and/or sing (as individuals or in variously numbered groups); animals accidentally overhear the music (seldom if ever specified); and their peculiar response is observed, recorded and interpreted by the astonished onlooker. These range from factually innocuous accounts of the kind described above to more theologically loaded descriptions revealing man's true dominion over nature:

On one Sunday evening five chorister boys were walking on the banks of the River Mersey. After some time, being tired, they sat down on the grass to rest. At one end of the field where they say was a growth of thick wood. They commenced to sing an anthem, but had not been singing long when they observed a hare come out of the words and come directly toward them. She passed by with swiftness, but immediately turned about, stood still and listened to their song till it was completed, when she started back to the woods with the greatest fleetness.[32]

A related genre is the anecdotal experiment, reprising the anecdotal form (focussing on animal reaction) with one significant difference in content: music is played or sung intentionally (experimentally) for the purposes of 'scientifically' observing animal response. Although scientific in its aims, the anecdotal experiment was, however, only nominally systematic. Judging from the number and diversity of reviews it received, C. J. Cornish's *Life at the Zoo* (1897)[33] provides possibly the most popular

[31] No author, 'Animal Love of Music', 165.

[32] James M. Tracey, 'The Power of Music Over Animals', *The Musical Standard* 47/32 (11 August 1894), 106.

[33] C. J. Cornish, *Life at the Zoo: Notes and Traditions of the Regent's Park Gardens*, new ed., London: Seeley & Co. Ltd, 1897, 128–54.

example of the genre.[34] Through a sequence of three chapters titled 'Orpheus at the Zoo', Cornish walks the course of the (London) zoo with the companionship of a musician playing the violin in chapters 1 and 2, and then violin, flute and piccolo in chapter 3. Together, they gauge animal responses with predictably high levels of amazement, beginning with the lowest creatures and rising up through the Great Chain of Being.

Although Cornish fulfils his popular readership's expectation of being rewarded with the musical anecdote's signature narrative gimmick (the unpredictable surprise musical response), his scientific impulse is not to be underestimated. The *Quarterly Review* describes Cornish as breaking 'entirely new ground',[35] and although reliant on tried and true literary forms, there are good reasons to take him seriously scientifically. Firstly, he is unusual in describing the type of music (seldom the actual music) being played, including single notes played low, gentle and melodious to high, sustained, loud, shrill and piercing; scales; juxtaposing soft harmonious and loud discordant chords; slow crescendos; tremolos; soft minor chords; serenading, followed by bursts of discordant chords; playing in different positions around the cage; 'The Keel Row'; 'The Shepherd's Call' from *William Tell*; bagpipe-like violin drones; and alternations between instruments (violin, flute, piccolo). Secondly, he describes the animals involved, including (in the first chapter) the tarantula, scorpion, cobra, talking mynah bird, Axis deer, crocodile, Robben Islands black snake, Madame Paulus's python, Indian cobra and rattlesnake; (in the second chapter) polar bear, grizzly bear, lion, European wolf, jackal, fox, Himalayan sheep, Indian wild boar, wild swine, African elephant, Indian bison, gayal, zebra, African wild ass, ape, silvery gibbon, orangutan, Tcheli monkey, Capuchin monkey, Malbrook monkey, baboon, ostrich, ibex, wild ass, zebras, tapir, tiger and cheetah. And lastly, though not systematically, he does try to establish psychological correspondences between musical sounds and animal reactions: loud, shrill sounds provoke anxious, aggressive responses; soft music elicits attentive postures; tremolos and drones cause physical expansion. Clearly influenced by Darwin's work on emotion (and his illustrative use of photography), Cornish is also significant for probably being the first naturalist to include photographs of animals actually responding to music.

[34] See, for example: no author, 'The Effect of Music on Animals', *The Musical Standard* 9/231 (4 June 1898), 361, which also cites reviews in the *Quarterly Review* (April 1898) and the *Review of Reviews* (May 1898), and Gustav Hirsch, 'Music and Animals', *The Spectator* (17 October 1891), 531.

[35] No author, 'Animals at Work and Play: Their Activities and Emotions', *The Quarterly Review* 187/374 (April 1898), 475.

In whatever anecdotal form, observing music soothe the savage breast was often part of the same anthropomorphizing agenda which saw animals dressed as humans. Animals could neither dress themselves nor (debatably, apart from birds) create music, but if dressed, they could wear clothes, and if exposed to music, they could respond affectedly. Victorian anthropomorphism asserted a common prejudice: animal inability reaffirmed the power structure underlying the assertion of human ability. Although domestic animals were 'friends' and wild animals 'respected' (if feared), the presence of musical absence (response without creation) reinforced evolutionary asymmetry. James Fennell describes it in these terms:

> the highest class of animals [mammals], presents us, as we might naturally expect, with many highly susceptible to the influence of music. Animals being, from their birth, accustomed to the inartificial sounds that occur in their native haunts, are not affected by them in any peculiar manner, but, like the uncivilized races of our own species, remained unmoved by them. Music, however, exercises extraordinary effects upon them, and fully confirms the remark of Racine, that 'Nature has given ears sensible of harmony even to brutes'. We shall find that music subdues the rude disposition of some; arouses the ferocity of others; renders some so docile and tame, that they may be approached without hesitation; while it makes others suspicious and frightened. It must be conceded, however, that that strain of music which may affect one species of animal in some particular way, will not, perhaps, be attended with similar results in all other species of the same class; and in the cases of domestic animals the effects seem to vary upon different individuals of the same species. M. Le Cat remarks, that 'if mammals did not possess a cochlea to the ear, one would not discover in them that sensibility for harmony which they manifest. We should rather find them, in this respect, as stupid as fish, which are destitute of the cochlea'.[36]

Fennell, author of *A Natural History of British and Foreign Quadrupeds* (1841), recapitulates a multitude of Victorian conventions towards the musicality of animals: (1) there is a difference between natural ('inartificial') and artificial sounds (sounds of nature and music), and as animals climb the Great Chain of Being, they become more susceptible to musical stimulus; (2) although animals are variously responsive to music, their reactions are seldom uniform – they are behaviourally inconsistent within and across species; (3) the physical structure of animals determines their corresponding ability to respond to music; and (4) uncivilized races are musically animal, not human. In other words, animals (and savages) prove

[36] James H. Fennell, 'On the Effects of Music on Man and Animals', *The Mirror of Literature, Amusement, and Instruction* 1081 (16 October 1841), 245.

that music is a product of civilization (itself a product of progress and development over time), and as such, music conforms to the highest example of its 'type'. Animals' response-only condition proves that music is the highest class of the species sound – it is *the* specimen type. In Fennell's evolutionary world, music strategically reaffirms conviction in the Great Chain of Being because animals' experience of music remains fixed at a primary level of reception. Beyond the cognitive understanding of animals, music remained something securely human. Animals could be taught how to respond to music through 'ear training', yet they remained categorically 'dumb':

> That animals may be taught to associate certain actions with certain tunes, as a trick, is an undoubted fact, which is proved by the performing of horses and elephants in scores of circuses. Some of them, too, have shown a very inconvenient 'ear for music', in their recognition of the air to which they were accustomed to stand on their hind legs or turn somersaults, long after the necessity of exhibiting those feats had passed away, causing considerable embarrassment by their repetition in wholly inappropriate spheres of action. But this, again, is obviously only reflex action, not a mental emotion.[37]

Inevitably, however, not everyone accepted these popular conventions as conclusive, least of all professional scientists with strong evolutionary leanings. Criticism of popular viewpoints is at the root of James Sully's staunch Darwinian line taken in 'Animal Music (1879)': 'the new doctrine of Evolution', Sully declares,

> which regards all species as connected steps in one complex movement of organic development, has naturally tended to raise the intellectual and moral status of animals by suggesting that in them are to be found the germs of mental qualities previously supposed to be man's exclusive possession.
>
> Among the attributes which science is thus extending to the lower animals is the artistic impulse. Man can no longer boast of being the sole artist . . . With respect to music, it must of course always have been a matter of observation that the lower animals share in our love of song. The first human musicians doubtless noticed the similarity of their rude art to bird-song.[38]

Sully argues from a classic Darwinian perspective when collapsing human–animal relations in regard to music; indeed, according to Rachel Mundy, 'Darwin problematized the human–nonhuman . . . and his writings on

[37] No author, 'The Influence of Music on the Lower Animals', *All the Year Round* 30/735 (30 December 1882), 538.

[38] James Sully, 'Animal Music', *The Cornhill Magazine* (November 1879), 605.

music neatly mapped this "demotion" of the human onto the sonic and onto music'.[39] It is worth noting that Sully's remarks reflect evolutionary comparisons adumbrated in the second edition of *The Descent of Man* (1874). Darwin foregrounds his observations by using the second and third illustrations to highlight inherited similarities in the ear of a human and orangutan – a bold claim not helped by his own apparent lack of confidence:

> it may be seen how different the pointed outline of the ear is at this period (fetal) from its adult condition, when it bears a close general resemblance to that of man. It is evident that the folding over of the tip of such an ear, unless it changed greatly during its further development, would give rise to a point projecting inwards. On the whole, it still seems to me probable that the points in question are in some cases, both in man and apes, vestiges of a former condition.[40]

Like Sully after him, Darwin attributes musical properties (as a function of sexual selection) to practically the whole of the animal kingdom, from various insects, spiders, fish, frogs, toads, tortoises, alligators and gibbons to dogs, seals and singing mice:[41] 'The capacity and love for singing or music', he claims, 'though not a sexual character in man, must not here be passed over. Although the sounds emitted by animals of all kinds serve many purposes, a strong case can be made out, that the vocal organs were primarily used and perfected in relation to the propagation of the species'.[42] Sully and other followers of Darwin rejoice in this positively orchestral proposition for two reasons: firstly, because Darwin maintains that *all* animals make music; and secondly because all animals *make* music. Unlike Fennell, Darwin and Sully believe that animals not only respond to music, they create it as well, and if they *all create* it, humans and animals must share musical ancestry. In other words, for Darwinists music proves that animals and man share common origins; music unites, rather than separates, man and animal.

This is certainly the chosen message musical Darwin wished to convey:

> A critic has asked how the ears of man, and he ought to have added of other animals, could have been adapted by selection so as to distinguish musical notes. But this question shews some confusion on the subject; a noise is the sensation resulting from the co-existence of several aërial 'simple vibrations' of various

[39] Rachel Mundy, 'Nature's Music: Birds, Beasts, and Evolutionary Listening in the Twentieth Century', PhD diss., New York University, 2010, 31.

[40] Charles Darwin, *The Descent of Man, and Selection in Relation to Sex*, 2nd ed. rev. and augmented, London: John Murray, 1871/1874, 16–17.

[41] Darwin, *Descent of Man*, 566–73. [42] Darwin, *Descent of Man*, 566.

> periods, each of which intermits so frequently that its separate existence cannot be perceived. It is only in the want of continuity of such vibrations, and in their want of harmony *inter se*, that a noise differs from a musical note. Thus an ear to be capable of discriminating noises – and the high importance of this power to all animals is admitted by every one – must be sensitive to musical notes. We have evidence of this capacity even low down in the animal scale [crustaceans, gnats, dogs and seals].[43]

Although conflicted over certain other aspects of Darwinian musical origins (particularly sexual selection), musicologist W. J. Treutler exemplifies Darwin's universalism when he considers the experience of music a common attribute of animals with ears: 'For our purpose we may define *music* as a succession of sounds so combined and modulated as to please, not only *the* ear – our ear – but *some* ear; that is the fundamental idea – the object of music is the gratification of the sense of hearing'.[44] Treutler even refers his audience to Cornish to substantiate his opinions:

> Mr. Cornish has made a number of very interesting experiments on the effects of music on the animals in the Zoological Gardens, which show how acute is the perception of pure musical tones, harmonies, and discords in animals, and how animals are variously affected by various kinds of music and in many cases show a distinct appreciation of certain instruments and a dislike to others. But for the details of these observations I must refer you to Mr. Cornish's charming book on 'Life at the Zoo'. It would therefore appear that in animals not only are the sense of hearing and the cerebral auditory centre considerably developed and organized, but that also they are able to some extent to intellectually comprehend and interpret music itself.[45]

Birdsong

Treutler's reference to Cornish is not just empty gesturing towards Darwinism; it is selective. Unlike Cornish, who experiments for the sheer joy of scientific discovery, Treutler has a wider agenda. Although he wants to prove that animals 'comprehend and interpret music itself', he also wants to distinguish that ability from the higher spiritual faculty which gives man his unique aptitude for cognizing music. Proving that forces Treutler to make some pretty fanciful – if not uncommon – claims over the existence

43 Darwin, *Descent of Man*, 68–89.

44 W. J. Treutler, 'Music in Relation to Man and Animals', *Proceedings of the Musical Association* 25 (1998–99), 72

45 Treutler, 'Music in Relation to Man and Animals', 82.

of the animal soul: 'there is a necessary something more than the external organ of hearing and the cerebral auditory centre, however highly these may be developed and exercised in the individual, and this "something more" is a sensitive, feeling, impressible, and highly organized *soul*'.[46] Treutler's spiritual aspirations reveal a conservative evolutionary temperament, for only man combines 'the material ear of the body, and on the other hand, the spiritual ear of the soul'.[47] For many Victorian evolutionists the animal soul was not an incompatible proposition; as Teresa Magnum suggests, 'Darwin's book is merely a milestone among the nineteenth-century books, magazines, and medical studies that debate animals' ability to feel, to reason, and even to possess immortal souls'.[48] Indeed for many naturalists, there was one particular creature with convincingly spiritual credentials: the songbird.[49] Rivalling man in its musicality, the songbird was unquestionably the highest 'type' of musical animal – a liminal creature perched at a physical and metaphysical threshold between science and theology, reality and ideal, material and spirit, immanence and transcendence, being and becoming, nature and God.

[46] W. J. Treutler, 'Music in Relation to Man and Animals', 83.

[47] W. J. Treutler, 'Music in Relation to Man and Animals', 83.

[48] Teresa Magnum, 'Animal Angst: Victorians Memorialize Their Pets', in Deborah Denenholz Morse and Martina A. Danahay (eds.), *Victorian Animal Dreams: Representations of Animals in Victorian Literature and Culture*, Aldershot: Ashgate, 2007, 22.

[49] Despite the importance of birdsong in Victorian poetic and literary culture, there appears to be little if any substantial research specifically aiming to connect birdsong and theology. Good examples include J. R. Watson, 'Echoes of That Voice: Romantic Echoes in Victorian Poetic Birdsong', in Andrew Radford and Mark Sandy (eds.), *Romantic Echoes in the Victorian Era*, Aldershot: Ashgate, 2008, 155–74, but surprisingly it is barely mentioned in Matthew Bevis (ed.), *The Oxford Handbook of Victorian Poetry*, Oxford: Oxford University Press, 2013. Historical research on Victorian birdsong tends to focus almost exclusively on ornithological science or birdwatching, as in Scott Eidensaul, *Of a Feather: A Brief History of American Birding*, New York: Harcourt, Inc., 2007; Mark V. Barrow, *A Passion for Birds: American Ornithology After Audubon*, Princeton, NJ: Princeton University Press, 1998; Peter Bircham, *A History of Ornithology*, London: HarperCollins Publishers Ltd, 2007; Tim Birkhead, *The Wisdom of Birds: An Illustrated History of Ornithology*, London: Bloomsbury, 2008; or Tim Birkhead, Jo Wimpenny and Bob Montgomerie, *Ten Thousand Birds*, Princeton and Oxford: Princeton University Press, 2013. There are substantial musicological works linking theology, philosophy and birdsong, unfortunately outside the nineteenth century, including Elizabeth Eva Leach, *Sung Birds: Music, Nature, and Poetry in the Later Middle Ages*, Ithaca and London: Cornell University Press, 2007, and David Kraft, *Birdsong in the Music of Messiaen*, CreateSpace Independent Publishing Platform, 2013, as well as more popular theological writings like John Stott, *The Birds: Our Teachers: Essays in Orni-Theology*, Abingdon: Candle Books, 2007. For an introduction to the metaphysical issues relating to nineteenth-century birdsong, see Matthew Head, 'Birdsong and the Origins of Music', *Journal of the Royal Musical Association* 122/1 (1997), 1–23.

The songbird sang in the ear of the Victorian imagination like no other creature, featuring prominently as a scientific and theological figure in children's and adult literature, poetry, popular natural history, biblical natural history studies, studies of the zoo, field books and a vast array of academic ornithological and zoological treatises. Its allure can be explained partly because of its moral and ultimately theological potency, partly because of birdsong's proximity to sounds humans understood to be musically expressive, and partly because birdsong was implicated in musical and therefore human origins. The songbird's allure guaranteed its attraction in wide-ranging intellectual debates between religion and science, the Great Chain of Being and Darwinism, absolutism and relativity, and individual and universal notions underlying Victorian issues over taxonomic type. Like all creatures, the image of songbirds (or songsters as they were commonly known) was frequently projected through the lens of a Great Chain of Being which mirrored physical and qualitative attributes, so popular literary convention (e.g. animal 'habit' anecdotes) characteristically dictated a close correlation between musical position and moral compass.

Polymath the Rev. John Mossop reflects this correlation when he appreciatively reproduces Daines Barrington's famous 'Scale of Merit Among British Singing Birds' in his *Ornithologia Britannica* (1841)[50] – in effect a Great Chain of Songster Being. Earlier Vice President of the Royal Society, Barrington graded the musical character of songbirds on a scale up to 20 on topics of mellowness, sprightliness, plaintiveness, compass and execution (Figure 1.3). First place in the Victorian league table of birdsong musical merit is invariably held by the nightingale, making it the indisputably highest type of 'songster' – the 'most enchanting and most celebrated of the warblers'.[51] For the anonymous author of *English Singing Birds: Their Songs, Homes, and Habits*, 'The strength of the note of the nightingale is such that it can be heard fully half-a-mile, but its compass, flexibility, prodigious variety, and harmony of voice have every made it the most admired of all songsters'.[52] In second place, the skylark 'is the most universal, and one of the most delightful, of the

[50] See Daines Barrington, 'Experiments and Observations on the Singing of Birds, by the Hon., Vice Pres. R. S. In a Letter to Mathew Maty, M. D. Sec. R. S.', *Philosophical Transactions* 63 (1773–74 [1773]), 282, in The Rev. John Mossop, *Ornithologia Britannica: A Synopsis of British Land-Birds, in Verse and Prose*, London: Jackson and Walford, 1841, 68.

[51] Mossop, *Ornithologia Britannica*, 11.

[52] No author, *English Singing Birds: Their Songs, Homes, and Habits*, London: W. B. Horner, [n.d.], 15.

SCALE OF MERIT

AMONG BRITISH SINGING BIRDS.

BY MR. D. BARRINGTON.

This Table is designed to exhibit the comparative merits of British Singing Birds; in which 20 is supposed to be the point of absolute perfection.	Mellowness of Tone.	Sprightly Notes.	Plaintive Notes.	Compass.	Execution.
Nightingale, ...	19	14	19	19	19
Sky Lark, ...	4	19	4	18	18
Wood Lark, ...	18	4	17	12	8
Tit Lark, ...	12	12	12	12	12
Linnet, ...	12	16	12	16	18
Goldfinch, ...	4	19	4	12	12
Chaffinch, ...	4	12	4	8	8
Greenfinch, ...	4	4	4	4	6
Hedge Sparrow, ...	6		6	4	4
Siskin, or Aberdevine, .	2	4		4	4
Red Pole, ...		4		4	4
Thrush, ...	4	4	4	4	4
Blackbird, ...	4	4		2	2
Redbreast, ...	6	16	12	12	12
Wren, ...		12		4	4
Reed Sparrow, ...		4		2	2
Black Cap, or Norfolk Mock Nightingale,	14	12	12	14	14

Figure 1.3 The Rev. John Mossop, 'Scale of Merit Among British Singing Birds', *Ornithologia Britannica: A Synopsis of British Land-Birds, in Verse and Prose* (1841), 68

British songsters'.[53] On the bottom branches and in the derisory depths are the blackbird and siskin. The blackbird may have a 'restless and violent temper', but 'if he does wrong, he does what he can to repay us in the Spring, in the beautiful still evenings, with his pensive and mellow song'.[54] The underperforming siskin merely 'resembles that of the Canary, with which it mixes, but it is not so powerful'.[55]

[53] Mossop, *Ornithologia Britannica*, 101.

[54] Rev. W. E. Evans, *The Songs of the Birds; Or, Analogies of Animal and Spiritual Life*, London: Francis & John Rivington, 1845, 122.

[55] Mossop, *Ornithologia Britannica*, 91.

While music and morality are often correlated in the Great Chain of Songster Being – as in the Victorian musical culture more broadly[56] – in much popular zoology, this relationship is projected through an intentionally redemptive theological lens. Thus, for ornithological popularizer H. G. Adams, the nightingale's song conjures up images of heaven, sung when

> the mind is most free from worldly cares, and the heart is most alive to impressions of a gentle, and a soothing, and a holy character; into these impressions there enters, too, a feeling of solemn awe, at a time when we seem to stand face to face with the great Creator of the universe, and to hold high commune with the mightiest spirits of the past ... we walk amid a host of attendant spirits ... to music of an unearthly and seraphic character.[57]

The author of English Singing Birds asks more broadly:

> If God can call forth such notes of song from so small a creature on earth, what must be the music of heaven? What the song of redeemed creation – of those saints, who have not only been saved, but changed into the image of Christ, when they shall sing not the praises of creation, but the glories of Him who hath loved them, washed them from their sins in His own blood, and made them kings and priests unto God?[58]

The nightingale has close competition in the redemption stakes; second-place skylark is also redemptive but for different reasons. It disregards the things of the world (its nest is modestly constructed), yet 'His whole heart is in his song, and it is only at heaven's-gate he loves to pour it out ... [yet] if by faith in the blessed Lord Jesus Christ, you have been delivered from it all [guilt, ruin and death], redeemed to God, then I know your hearts are in heaven'.[59] For the Rev. W. E. Evans, the skylark lives to teach redemption by example: 'Oh! Let thy soul thus daily make / Towards heaven its upward way; / 'Twill then at last rejoicing break / Its prison-house of clay: / Towards heaven soar / To fall no more; / And with the angel throng, / Borne up on its eternal wing, / Before the Lamb for ever sing / Its deathless song'.[60] The skylark is also constant and humble; its 'artless nest' may be lowly, but humility is a virtue: 'Emblem meet of the true Christian spirit', we are told, 'which though it soar so high, returns to

[56] See, for example, Charles M. McGuire, *Music and Victorian Philanthropy: The Tonic Sol-fa Movement*, Cambridge: Cambridge University Press, 2009.

[57] H. G. Adams, *Favourite Song Birds*, London: Groombridge and Sons, 1887, 2.

[58] *English Singing Birds*, 17.

[59] *English Singing Birds*, 28–29.

[60] Rev. W. E. Evans, *The Songs of the Birds; Or, Analogies of Animal and Spiritual Life*, London: Francis & John Rivington, 1845, 50.

the enjoyment of its pure home beneath!'[61] Because songsters serve the purpose of redemption, they (and other non-singing birds) are commonly described in categories of theological type. Note the taxonomic terminology Evans invokes to describe the constancy of the woodlark:

> While the other larks congregate in flocks during the Winter, and fly from field to field, this, for the most part, is stationary; or, if joining the others for a time, soon separates to return and sing in its accustomed haunts. Type of the Christian with his pure and holy home . . . Type, too, of the spirit whose home is with God . . . where all its hopes and all its love are centred![62]

Scottish Evangelist Henry Drummond (1851–97) summarizes this approach in his book *Natural Law in the Spiritual World* (1896), using type taxonomy to define a bird as both 'the' bird (the individual specimen) and 'a' bird (its theological specimen type). In this case, the relationship between a bird and the bird is explained Christologically to characterize a religious (redemptive) relationship between man and nature:

> What goes on then in the animal kingdom is this – the Bird-Life seizes upon the bird-germ and builds it up into a bird, the image of itself . . . The visible bird is simply an incarnation of the invisible Bird-Life . . . As the Bird-Life builds up a bird, the image of itself, so the Christ-Life-builds up a Christ, the image of Himself, in the inward nature of man.[63]

Thus, birds become paradigms, or literally preachers to the young, when the Rev. A. N. Mackray asserts that 'the ministry of birds sets forth in many ways the ministry of children. And just as the least in the bird-kingdom are its best minstrels, so does God out of the mouth of His little ones perfect his praise'.[64] For Mackray and many other religiously minded Victorians, Christian *ministry* and ornithological *minstrelsy* are integrally related through redemption in a permanent state of relational – or relative – theological unity: 'The time of the singing of birds', he asserts, 'is their love season, when they strain their wonderful little throats to tell out their confidences and affections to their mates . . . The sin of many, it is true, / 'Broke the fair music that all creatures made / To their great Lord, whose love their motion swayed', / and a sinful, sorrowful world might well be a song-less one; but 'where sin abounded,

[61] *Songs of the Birds*, 44; 45. [62] *Songs of the Birds*, 37.

[63] Henry Drummond, *Natural Law in the Spiritual World*, London: Hodder and Stoughton, 1896, 292–93.

[64] Rev. A. N. Mackray, *Bird-Preachers: Friendly Talks with Little Folks about Bible-Birds*, London: The Religious Tract Society, [n.d., 1880s?], 115.

grace did much more abound'. The song of Redemption is the first and the last song of the Bible'.[65] Songsters may have reminded the Victorians that redemption is the alpha and omega of theological type, but because of their privileged liminal position in the Great Chain of Being (between nature and heaven), they also reminded them of the intelligent design of God's creation; thus, in the design of created nature, they occupy a liminal position between lower animal forms and man: 'The Almighty has ordained that the objects and operations of nature shall arrest our attention. He designed them to lead our minds to himself, and to produce definite effects upon our hearts. We are not sent into the world, teeming as it does with proofs of its Divine original, to block up our senses to all around us'.[66] Whether intelligent or not, design is a key component in Victorian ornithological appreciation because the design of birdsong is intimately bound up with what evolutionists increasingly understand to be musical expression and its relationship to language.

According to Darwin, the most distinguishing feature of music is that it is arguably not language.[67] As early as 1838 Darwin had begun to ponder the relationship between birdsong and language, and reading Barrington, he inadvertently honed in on what he considered to be one of music's ostensibly non-linguistic characteristics – its beauty. Could the recognition of beauty (or an aesthetic faculty of some sort) affect the function of choosing a mate in a way that language could not? 'Daines Barrington says cock birds attract females by song do they by beauty, analogy of man if so . . .'[68] By 1844 Darwin believed that birdsong, aesthetics and language were connected somehow, but that music's aesthetic design served an observable reproductive function which perhaps paradoxically language did not:

Besides this natural means of selection, by which those individuals are preserved . . . which are best adapted to the place they fill in nature, there is a second agency at work in most bisexual [what we call today heterosexual] animals tending to produce the same effect, namely the struggle of the males for the

[65] Mackray, *Bird-Preachers*, 118–19.
[66] No author, *Our Song Birds*, London: The Religious Tract Society, [n.d., 1872?], 8.
[67] David Rothenberg, *Why Birds Sing: One Man's Quest to Solve an Everyday Mystery*, London: Allen Lane, 2005, 35.
[68] Charles Darwin, *Darwin's Notebooks of Transmutation of Species*, ed. Gavin De Beer, Part II, Second Notebook (February–July 1838), London: Bulletin of the British Museum (Natural History) Historical Series 2/3 (May 1960), 109, Darwin Online, http://darwin-online.org.uk/content/frameset?keywords=the%20differ%20of%20men%20chiefly%20races&pageseq=33&itemID=F1574b&viewtype=text, accessed 13/5/13.

females. These struggles are generally decided by the law of battle; but in the case of birds, apparently, by the charms of their song, by their beauty or their power of courtship.[69]

Darwin did not substantially develop his theory in the *Origin of Species* (1859)[70] but used it as an opportunity to ruminate over the evolutionary aesthetics of music. He arrived at two conclusions: first, unlike language, birdsong is not melody unless it – like all melody – is totally complete; and second, the production of music is akin to a force of instinct, and like instinct, its action must be entirely complete to satisfy its purpose:

When once acquired, they [songs] often remain constant throughout life. Several other points of resemblance between instincts and habits could be pointed out. As in repeating a well-known song, so in instincts, one action follows another by a sort of rhythm; if a person be interrupted in a song, or in repeating anything by rote, he is generally forced to go back to recover the habitual train of thought.[71]

Resuming these discussions in *The Descent of Man* (1871), and having revisited Barrington on the acquired nature of language and birdsong, Darwin came to the following further conclusions: (a) songs are acquired, like language; (b) the action of singing is instinctual, like speaking; and (c) the aesthetic reception of song relies upon and influences sexual selection:

The sounds uttered by birds offer in several respects the nearest analogy to language, for all the members of the same species utter the same instinctive cries expressive of their emotions; and all the kinds that have the power of singing exert this power instinctively; but the actual song, and even the call-notes, are learnt from their parents or foster-parents. These sounds, as Daines Barrington has proved, 'are no more innate than language is in man.' The first attempts to sing 'may be compared to the imperfect endeavour in a child to babble.' The young males continue practising, or, as the bird-catchers say, recording, for ten or eleven months. Their first essays show hardly a rudiment of the future song; but as they grow older we can perceive what they are aiming at; and at last they are said 'to sing their song round' . . . I have given the foregoing details to shew that an instinctive tendency to acquire an art is not a peculiarity confined to man.[72]

[69] Charles Darwin, 'Essay of 1844', in Francis Darwin (ed.), *The Foundations of the Origin of Species: Two Essays Written in 1842 and 1844 by Charles Darwin*, Cambridge: Cambridge University Press, 1909, 92–93.

[70] Peter Kivy, 'Charles Darwin on Music', *Journal of the American Musicological Society* 12/1 (Spring 1959), 43.

[71] Charles Darwin, *On the Origin of Species by Means of Natural Selection, or the Preservation of Favoured Races in the Struggle for Life*, London: John Murray, 1859, 208.

[72] Charles Darwin, *The Descent of Man, and Selection in Relation to Sex*, 2 vols., London: John Murray, 1871/2nd ed. rev. and augmented, 1874, vol. 1, 55–56.

Darwin's argument culminates in *The Expression of the Emotions in Man and Animals* (1872) through the voice of his son-in-law Richard Buckley Litchfield, who postulates a theory of integrated musical relativity. Musical expression is musical (not linguistic) because all of its parts are internally relational (integrated) and therefore not divisible (able to be divided without destroying its total effect):

any law which is found to hold as to the expression of the emotions by simple sounds must apply to the more developed mode of expression in song, which may be taken as the primary type of all music . . . But this leaves unexplained the more subtle and more specific effect which we call the *musical* expression of the song – the delight given by its melody, or even by the separate sounds which make up the melody. This is an effect indefinable in language – one which, so far as I am aware, no one has been able to analyse . . . For it is certain that the *melodic* effect of a series of sounds does not depend in the least on their loudness or softness, or on their *absolute* pitch. A tune is always the same tune, whether it is sung loudly or softly, by a child or a man; whether it is played on a flute or on a trombone. The purely musical effect of any sound depends on its place in what is technically called a 'scale;' the same sound producing absolutely different effects on the ear, according as it is heard in connection with one or another series of sounds. It is on this *relative* association of the sounds that all the essentially characteristic effects which are summed up in the phrase '*musical* expression,' depend.[73]

The idea that musical composition needs to be complete and relationally unified to qualify aesthetically as music (instead of just natural sound) was widespread in a Victorian musical culture influenced by organicism.[74] Writing in the magisterial *Power of Sound* (1880), one of organicism's greatest exponents, Edmund Gurney, reduces music to its constituent parts, including colons, full stops, divisions of paragraphs, as well as transitions and sections, with each component expressing organic unity within itself and the larger form. The absence of unity (differentiation) undermines its purpose and unsettles the listener:

whether or not unified in some special manner, any paragraph which is to be musically valuable, must satisfy the test that each bit shall necessitate, as it were, and so enter into organic union with the *one next to it*: a test which will be found practically to necessitate reference at each point to parts considerably further

[73] Charles Darwin, *The Expression of the Emotions in Man and Animals*, London: John Murray, 1872, 89–90.

[74] For useful background to British contexts, see Ruth Solie, 'Melody and the Historiography of Music', *Journal of the History of Ideas* 43/2 (April–June 1982), 297–308, and 'The Living Work: Organicism and Musical Analysis', *19th-Century Music* 4/2 (Autumn 1980), 147–56.

removed from it, whether behind or in front, than the points which are absolutely contiguous to it.[75]

Inevitably, the influence of a relative, integrated theory of music entered popular post-Darwinian ornithological literature. *Beeton's Book of Birds* (1870s?), for example, claims that a birdsong is not music unless it is completely uninterrupted: 'A perfect note, according to bird-keeping phraseology, is a perfect and complete *toll-loll-loll-chickweedo:* if this phrase is a syllable short, it counts for nothing'.[76] Ornithologist W. P. Pycraft quotes the naturalist Warde Fowler, remarking that the totality of birdsong can never be captured in any other form than itself, and certainly cannot be adequately represented by the representationally limited capability of transcription; indeed, birdsong 'is a highly-developed product of science and art combined: that you cannot write down on our musical scale without depriving them all that freedom and wildness in which their very life and beauty consist'.[77] Further evidence comes in the forms of Charles Louis Hett's *A Dictionary of Bird Notes* (1898). Hett was perfectly willing to represent more limited bird calls or notes as dislocated mnemonic, onomatopoeic syllables, but not birdsong proper. Translating from 'Note to Bird' and 'Bird to Note' in two separate sections, Hett prints text on right-hand pages only, leaving left-hand pages blank for noting observations. The perceptive twitcher who hears the call 'a-chuck, a-chuck' will instantly identify the call of the Common Snipe, and the lucky eagle-eye who identifies a Common Snipe will hear him call 'a-chuck, a-chuck'. Hett's nightingale, however, calls only 'pur, purr', 'wheet' or 'wit', and disappointingly never sings a song.[78]

Later ornithologists like W. Percival Westell would portray birdsong in complete sentences to maintain the relational, unified integrity of its expression, sometimes in satisfyingly moral character reflecting the bird's pedagogical (and theological) nature. The chaffinch sings 'What is here, what cheer, watch here, so queer, which is it, so quiet and queer – A what is she – A teacha'.[79] The necessity for completeness of musical expression

[75] Edmund Gurney, *The Power of Sound*, London: Smith, Elder & Co., 1880, 204. See also Solie, 'Melody and the Historiography of Music', 297–308.

[76] No author, *Beeton's Book of Birds: Showing How to Rear and Manage Them in Sickness and in Health*, London: Ward, Lock, & Co., [n.d.], 111.

[77] W. P. Pycraft, *The Story of Bird-Life*, London: George Newnes, Ltd, 1900, 90.

[78] Charles Louis Hett, *A Dictionary of Bird Notes, to Which Is Appended a Glossary of Popular Local and Old-Fashioned Synonyms of British Birds*, Brigg: Jacksons', 1989, 7; 83.

[79] W. Percival Westell, *The Young Ornithologist: A Guide to the Haunts, Homes, and Habits of British Birds*, London: Methuen & Co. Ltd, 1911.

in birdsong is explained partially in its relation to the evolution of language. Darwin saw birdsong as proof that language evolved from music; Herbert Spencer, that music evolved from impassioned speech. The world of ornithology was undecided. Darwin encapsulates the argument in *The Descent of Man* (1871):

> Mr. Spencer comes to an exactly opposite conclusion to that at which I have arrived. He concludes that the cadences used in emotional speech afford the foundation from which music has been developed; whilst I conclude that musical notes and rhythm were first acquired by the male or female progenitors of mankind for the sake of charming the opposite sex. Thus musical tones became firmly associated with some of the strongest passions an animal is capable of feeling, and are consequently used instinctively, or through association, when strong emotions are expressed in speech. Mr. Spencer does not offer any satisfactory explanation, nor can I, why high or deep notes should be expressive, both with man and the lower animals, of certain emotions. Mr. Spencer gives also an interesting discussion on the relations between poetry, recitative, and song.[80]

Both Spencer and Darwin had their opponents, however. Richard Wallaschek takes umbrage with Darwin because he accords the bird agency (or consciousness) in the compositional unity of its song: ‘Is it possible’, Wallasheck asks,

> that male-birds of handsomer plumage are preferred by the hen-birds, but may we thence infer that birds understand painting? Similarly some male birds may be preferred by the hens for the quality of their song; can we, therefore, say that they understand music? Is the bird’s song a composition? Certainly not. But by music we always understand a musical composition, or at least its reproduction, that is to say, a consciously designed and constructed work of art. I am aware that there is no difference in kind between the bird’s instinct and the human design; there is only a difference of degree, as between mind and instinct generally, but this degree must nevertheless be attained before we are justified in speaking of any group of sounds as music. Birds have no conscious intention of charming by the display of magnificent hues in such and such a manner, nor is it within their power to choose their colours any more than it is to change their songs, so as to make them correspond with their feelings.[81]

The real sticking point for Wallaschek is one frequently expressed by others, namely whether animals are imbued with a capacity for aesthetic

[80] Darwin, *Descent of Man*, 336.

[81] Richard Wallaschek, *Primitive Music: Inquiry into the Origin and Development of Music, Songs, Instruments, Dances, and Pantomimes of Savage Races*, London and New York: Longmans, Green & Co., 1893, 242–43.

discernment which underscores an ability to design complete musical forms. Writing of a not dissimilar debate taking place in regard to plumage, C. Lloyd Morgan considers much the same question: it is 'a very different thing from saying that, either in the selection of flowers by insects, or in the selections of their mates by birds, a consciously aesthetic motive has been a determining cause'.[82] Darwin's protégé, George Romanes, wonders if standards in aesthetic taste differ in different species of birds in the same way that they do in people.[83] Wallaschek also disputes Darwin's theory of correlative aesthetics – that 'the so-called singing of mammalia . . . is really developed from birds' singing, with human song its highest outcome'[84] – and his presumption that birdsong evolved from more primitive calls, alarms or other similarly primitive utterances. How was it possible, he asks, for alarm calls to transform into love melodies if it were true that the 'feeling which originally accompanied the emission of sound could consequently not only accompany it in further stages of development but could also be transmitted to future generations'?[85] Others (following Spencer) strongly contest the presumption that birds sing only during the mating season. Spencer himself argues that

> like the whistling and humming of tunes by boys and men, the singing of birds results from overflow of energy – an overflow which in both cases ceases under depressing conditions. The relation between courtship and singing, so far as it can be shown to hold, is not a relation of cause and effect, but a relation of concomitance: the two are simultaneous results of the same cause.[86]

For some non-Darwinian ornithologists, resistance to sexual selection forced them to develop some genuinely unpalatable theories to understand birdsong. The first ornithologist to apply evolutionary thought systematically in his work was Charles A. Witchell, author of books such as *The Evolution of Bird-Song* (1896), the field book *Cries and Call-Notes of Wild Birds* (1899) and important scientific journal articles like 'The Evolution of Bird-Song' (1890). Witchell is, fundamentally, a sceptical if not unappreciative Darwinist – perhaps what Bowler might call a pseudo-Darwinian. Although he accepts the possibility that 'the notes of a bird, primarily used as a mere call or for some other purpose, might have been improved into a melodious love-song',[87] he also believes that Darwin overestimated the

[82] C. Lloyd Morgan, *Animal Life and Intelligence*, Boston: Ginn & Co., 1891, 413.
[83] George J. Romanes, *Animal Intelligence*, New York: Appleton & Co., 1888, 282.
[84] Wallaschek, *Primitive Music*, 246. [85] Wallaschek, *Primitive Music*, 248.
[86] Herbert Spencer, 'The Origin of Music', *Mind* 15/60 (October 1890), 452–53.
[87] Darwin, *Descent of Man*, 67.

emotional range of primitive creatures when claiming that the voice was 'habitually employed as a serviceable aid under certain conditions, including pleasure, pain, rage, etc'.[88] In fact, for Witchell, primitive utterance is based solely around a type of cry, and from that cry evolves the call-notes of danger, combat or alarm which through various repetitions form the simplest song and ultimately more complex melodic forms.

Heredity, in addition to mimicry, might play a vital part in that process, but proving it is problematical. Certain birds like chicks and other non-singing birds appear to mimic the calls of their parents almost from the point of hatching, whereas songbirds tend to take longer. It is significant that Witchell applies heredity, theoretically at least, to the lowest or most primitive forms of calls amongst the lowest forms of birds, and continues to do so even when applying heredity to more advanced birdsong. The young blackbird, for instance, is more likely to use hereditarily received units to produce a song than its more mature counterpart. Moreover, the young blackbird is more likely to be reliant upon imitating other songbirds, like the thrush, than the older ones: 'The first songs of the young thrush and the young blackbird – heard faintly sung in September, or later – are much alike, and seem to be aimless repetitions of whistled tones and slurs . . . traces of mimicry are soon observable'.[89] Mimicry is also observable where physical similarities exist between species, amongst different types of finches, for example, like the greenfinch, goldfinch, siskin and redpoll. Perhaps unsurprisingly, therefore, Witchell arrives at the conclusion that songbirds have a greater capacity for imitation than non-singing birds, the implication being that they are more advanced:

> We should not allow our regrettable ignorance of bird-song to lead us to conclude that because we understand hardly anything about it, the birds themselves can perceive no more meaning in it. Their fidelity to their songs when with their own species, even though dwelling within hearing and being sometimes the companions of other birds [in a cage or aviary], and the sudden change of voice consequent upon a total change of surroundings, are facts not to be lightly passed over; they indicate that the language of the wild bird is to it as important as is the language of a wild man to him.[90]

This is a largely classic Darwinian stance aiming to prove (a) that birdsong evolved from more primitive melodic units and (b) that there is a

[88] Darwin, *Expression of the Emotions*, 84.

[89] Charles A. Witchell, *The Evolution of Bird-Song*, London: Adam and Charles Black, 1896, 110–11.

[90] Witchell, *Evolution of Bird-Song*, 178.

correlation between primitive animals and primitive man. But Witchell goes further than Darwin in specifying links between specific birdsong, like that of the parrot, and the music of ancient and primitive cultures, and more generally, connections between birdsong, hawkers and railwaymen[91] – in other words, connections, as Harriet Ritvo suggests, mirroring hierarchies of human evolutionary position. Witchell's theory of birdsong is also missing one key element in the Darwinian evolutionary toolbox – sexual selection. For Witchell, birdsong has evolved from the call, and although sexual selection plays a part, it is only one of many factors, including heredity, mimicry, environment and spontaneous variation. Witchell is also more inherently musical than Darwin and examines the music of birdsong forensically, whereas Darwin tended to ponder it more conceptually. Both *The Evolution of Bird-Song* and *Cries and Call-Notes of Wild Birds* are packed with musically notated examples, and these show two traits: firstly, they are used to provide prolific evidence substantiating Witchell's theory that song originates in calls, and secondly (and particularly in *Cries and Call-Notes of Wild Birds*), they are designed as a totality to reveal a strong sense of relativity between (a) units of call within songs, (b) units of call within songs sung in different settings, (c) units of call within songs sung by related species, (d) units of call within songs sung by related species in different settings and (e) units of call within songs sung by unrelated species in different settings.

Why is Witchell so obsessed with substantiating relativity between calls, birds and man? As it turns out, Witchell was not just an ornithologist, but a eugenicist, and like all good (bad) eugenicists, he thought that proving relativity between science and his ideological beliefs (themselves unrepentantly theological) would aid his cause. According to its inventor Francis Galton (first cousin of Charles Darwin), 'eugenics' is the study of what is

> good in stock, hereditarily endowed with noble qualities ... [and] equally applicable to men, brutes, and plants ... We greatly want a brief word to express the science of improving stock, which is no means confined to judicious mating, but which, especially in the case of man, takes cognisance of all influences that tend in however remote a degree to give to the more suitable races or strains of blood a better chance of prevailing speedily over the less suitable than they otherwise would have had. The word *eugenics* would sufficiently express the idea.[92]

[91] Witchell, *Evolution of Bird-Song*, 233.

[92] Francis Galton, *Inquiries into Human Faculty and Its Development*, London: J. M. Dent & Co., and New York: E. P. Dutton & Co., 1883, 17.

In its more fully developed form, the idea defines 'the study of agencies under social control that may improve or impair the racial qualities of future generations either physically or mentally'.[93] Books like Witchell's *The Cultivation of Man According to the Teachings of Common-Sense: Being a Treatise on the Application to Man of the Laws that He Applies to All Other Creatures* (1904) and *The Natural History of Sin* (1905) say it all:

> the cultivation of the human race should proceed on natural lines; that the stock from which the future members of the community are to be born should be carefully selected; that the strong should not be childless; and that defective children should be eliminated very early in life. But the argument is claimed to be founded on natural facts; and, if so, it may well be a sound one – in which case many of our currently accepted notions must be wrong.[94]

The science of ornithology taught Witchell a harsh (and now thankfully discredited) lesson in the theology of eugenics – that God is in nature but that imperfection is not necessarily natural: as he says, 'the ways of Nature are the ways of God, and he who studies the secrets of the one learns the laws of the other . . . The facts with which he has to deal will be physical or mental ones; but the artificially produced deformity must not be mistaken for a natural development'.[95]

Conclusion

Rather than teaching Witchell redemption, songbirds ultimately taught him what today we would consider to be a moral distortion – the refracted illusion of human *authority* over nature. Songbirds taught him that he could make God-like decisions because he was the highest type of highest being in the Great Victorian Chain of Being. Ornithology taught Witchell a different definition of humanity – one that highlights the precarious moral certainties of the time and the unresolved, unredemptive side of a non-Darwinian revolution in transition. As reviews would make clear, Witchell's brand of ornithology would not be shared entirely,[96] despite having

[93] Francis Galton, 'Eugenics: Its Definition, Scope and Aims', *Nature* 70 (1904), 82.

[94] Charles A. Witchell, *The Cultivation of Man According to the Teachings of Common-Sense: Being a Treatise on the Application to Man of the Laws That He Applies to All Other Creatures*, London: W. Steward & Co., 1904, xiv.

[95] Witchell, *Cultivation of Man*, 54.

[96] See J. A. A., 'Witchell's "Evolution of Bird-Song." The Evolution of Bird-Song with Observations on the Influence of Mimicry and Imitation by Charles A. Witchell' (review), *The Auk* 13/3 (July 1896), 250. J. A. A. complains that although fascinating and well researched,

the forensic knowledge of birdsong melodies neither Darwin nor other ornithologists possessed. How did ornithology help or hinder the introduction of Darwinian ideas in a widely non-Darwinian zoomusicological culture? As this chapter suggests, increasingly empirical ornithology concentrated the Victorian theoretical mind. Unlike Witchell, Darwin and other non-Darwinian ornithologists idealized music, and in some way or other, their ignorance released them to treat music conceptually rather than empirically – to imagine rather than describe, to divine rather than merely discover the characteristic properties of birdsong. For Darwin, music is a type of human equalizer: birdsong redeems because it humanizes. Birdsong – like all music – teaches Darwinians that all nature is related from the depths of its origins. Birdsong teaches Spencerians that culture redeems man from nature – that man can rise above his condition and overcome his predicament in nature. For both Darwinians and Spencerians, birdsong is, in some form or other, salvific because it ethicizes – in admittedly different ways – man's relation to nature and nature's relationship to the cosmos. For ornitheologians, birdsong also humanizes. Birdsong – like all music – teaches ornitheologians that all nature is related from the depths of God's creation. For ornitheologians, birdsong is, in some form or other, redemptive because it moralizes man's relation to nature and nature's relationship to God's creation.

What does the position of birds in ornithology's Great Chain of Songster Being tell us about the Victorians, and what therefore does the location of the London Zoo's aviaries tell us about Victorian zoomusicological attitudes more broadly? As this chapter has aimed to prove, no matter how it was interpreted, Victorian birdsong was never the preserve of any one scientific or theological discourse; its liminality is not only metaphysical but disciplinary. Like much of the non-Darwinian revolution, Victorian zoomusicology plays havoc with the idea of a classic war between science and religion: on the scientific side, relativity and evolution; on the theological, absolutism and the Great Chain of Being. But in many respects it is this relationship – between science and theology, man and God, the material and the spiritual – which summarizes the liminal position of Victorian birdsong. Birdsong was caught in ideological crossfire: on the one (scientific) hand, birdsong is only music because melody is

Witchell's ideas are 'either too far-fetched or superfluous' (250); and W. Warde Fowler, 'The Notes of Birds. The Evolution of Bird-Song. By Charles A. Witchell', *Nature* 54 (30 July 1896), 290–91. Warde Fowler suggests that 'Mr Witchell often exaggerates superficial and accidental resemblances' between songs (291).

only fully music when it completely expresses the relative relationship of its constituent parts; on the other (theological) hand, birdsong is only birdsong because songbirds express a universal truth about man's relationship to his Creator. But both viewpoints rely upon relativity and universality to prove what is effectively the same point – that the form of birdsong has the capacity to fulfil both scientific and theological functions.

And lastly, what did this rich zoomusicological imagination mean for Victorian visitors to the London Zoo? How did they interpret what they heard when they entered? Did they identify with the birds? Did they hear the answers to their scientific or theological questions, or both? What did they think when they proceeded up the Zoo's 'grand procession' – its Great Chain of Being? Where were they positioned within nature, society and the world at large? Were they like any other animal or the very highest and most perfect type of created nature? Were they locked into the liminality of their own condition at a physical and metaphysical threshold? And more specifically, is there any reason, therefore, why designers of the London Zoo might locate the aviaries at the entrance – at a metaphorical threshold between the world of humans and the world of animals? The answers, I believe, are found in the minds of the Victorians themselves – in the *intellectual* soundscape Victorian zoomusicology created from the imagetext of its books and what might be called the soundtext of its spaces – in a Victorian intellectual imagination so richly interdisciplinary that it could conceive idea as space and space as idea, theology as science and science as theology; not necessarily in conflict, not necessarily in dialogue, but in what might be termed *overlapping* magisteria (OMA). Coining his theory of *non-overlapping* magisteria (NOMA), Stephen Jay Gould claims that 'the net of science covers the empirical realm: what is the university made of (fact) and why does it work this way (theory). The net of religion extends over questions of moral meaning and value'.[97] But if the scientific and theological interchange in Victorian zoomusicology teaches us anything, surely it is something entirely different – it teaches us OMA.

Until hard evidence is found we may never know if the entrance to the London Zoo was designed to place birds at a metaphorical threshold, nor how visitors responded to their experience of the aviaries; unlike the birds themselves, sources on the history of the Zoo are

[97] Stephen Jay Gould, 'Nonoverlapping Magisteria', *Natural History* 106 (March 1997), 16 –22, in Stephen Jay Gould, *Leonardo's Mountain of Clams and the Diet of Worms: Essays on Natural History*, Cambridge MA: Harvard University Press, 1998/2011, 274.

frustratingly silent.[98] What we do know, however, and what we can construct is a sense for Victorian scientific and theological understanding – a kind of human Victorian equivalent to the subject of Tim Birkhead's *Bird Sense: What It's Like to Be a Bird* (2012) – something like Victorian Sense. Birkhead asks: 'what is it like to be kiwi?' Others have asked similar questions about different animals, like bats.[99] But what was it like to be a Victorian? What did zoology teach Victorians about their world? How did it change their understanding of animals and what role did music play in shaping that understanding? Victorian zoomusicology can help answer these questions, but as Tim Birkhead rightly says, 'we can never know *exactly* what it's like to be a bat or indeed a bird'[100] – nor, I would add, a Victorian.

[98] See Takahashi Ito, *London Zoo and the Victorians, 1828–1859*, Woodbridge: The Boydell Press for the Royal Historical Society, 2014. Earlier sources include Lord Zuckerman (ed.), *The Zoological Society of London, 1826–1976 and Beyond*, London: Academic Press for the Zoological Society of London, 1976; Wilfred Blunt, *The Ark in the Park: The Zoo in the Nineteenth Century*, London: Hamish Hamilton, 1976; and Henry Scherren, *The Zoological Society of London: A Sketch of Its Foundation and Development and the Story of Its Farm, Museum, Gardens, Menagerie and Library*, London: Cassell & Co. Ltd, 1904.

[99] Tim Birkhead, *Bird Sense: What It's Like to Be a Bird*, New York: Walker & Co., 2012, loc. 67–77, Kindle.

[100] Birkhead, *Bird Sense*, loc. 77 Kindle.

2 | Ethnomusicology

Introduction

By most accounts modern ethnomusicology begins with Jaap Kunst's *Musicologica: A Study of the Nature of Ethno-musicology, its Problems, Methods and Representative Personalities* (1950). Dissatisfied with its disciplinary predecessor 'Comparative Musicology' Kunst amalgamated ethnography and musicology to form the newly renamed discipline. Austrian ethnomusicologist Guido Adler defines comparative musicology as 'the comparison of the musical works – especially the folk songs – of the various peoples of the earth for ethnographical purposes, and the classification of them according to their various forms'.[1] To compare musical works comparative musicologists – in the broadest sense early ethnomusicologists[2] – used evolutionary theory to help structure their research. Comparing comparison across disciplines, ethnomusicologist Erich von Hornbostel claims that 'Systematization and theory depend on comparison. In this sense all learning is comparative, and comparison is a general and not a special method. Yet one speaks of *comparative anatomy, comparative linguistics* ... Here again

[1] Guido Adler, 'Umfang, Methode und Zeil der Musikwissenschaft', *Vierteljahrsschrift für Musikwissenschaft* 1/14 (1885), cited in Alan P. Merriam, 'Definitions of "Comparative Musicology" and "Ethnomusicology": An Historical-Theoretical Perspective', *Ethnomusicology* 21/2 (May 1977), 191.

[2] There is no consensus about the origins of ethnomusicology as a discipline, and many scholars apply the word to periods well before 1950. Frank Harrison uses it in *Time, Place and Music: An Anthology of Ethnomusicological Observation c. 1550 to c. 1800* (1973); Helen Myers indicates that 'What we now call ethnomusicology began long before that term was invented', citing reference to European folk, North American and Chinese music in Jean-Jacques Rousseau's *Dictionnaire de Musique*, Paris: Chez la Veuve Duchesne, 1768, as well as in travel literature, missionary and civil servant reports (Helen Myers, 'Introduction', in Helen Myers (ed.), *Ethnomusicology: Historical and Regional Studies*, New York and London: W. W. Norton & Co., 1993, 4); and Stephen Blum considers it to have begun in the 1860s (Stephen Blum, 'European Musical Terminology and the Music of Africa', in Bruno Nettl and Philip V. Bohlman (eds.), *Comparative Musicology and Anthropology of Music*, Chicago and London: University of Chicago Press, 1991, 3).

the concept of evolution presented itself: it pointed to new paths and led to new groupings'.[3]

This chapter explores how contemporary evolutionism influenced the way British ethnomusicology conceptualized music, and how that concept was used for ideological purposes both here and abroad. It examines the hegemonic authority of Spencerian evolutionism, its predominance in Victorian ethnomusicological discourse and its political exploitation by early Indian nationalists. An introductory section focuses on key elements of Social Darwinism and their implications for modelling anthropological interpretations of race and nation. A second section examines the role Social Darwinism played in the development of Victorian ethnomusicology, first through the defining comparative methodological works of ethnomusicologist Carl Engel and then through the writings of his great admirer, contemporary Indian ethnomusicologist Sourindro Mohun Tagore. A final section contemplates the effect of encroaching Darwinism and the rise of modern ethnomusicological thinking.

Social Darwinism

Like zoomusicology, classic Darwinian evolutionism tried to use scientifically democratized human–animal relations to equalize humanity in a way unachievable in many other aspects of Victorian intellectual culture. Albeit merely a theory until well into the twentieth century, Darwinian evolutionary science attacked prejudice by subjecting all nature to the same universal laws. Darwin's 'heresy', to use Desmond and Moore's term, was 'to extend human racial relationships to all the branches of creation, and to push the trunk deep into the geological soil'.[4] As exemplified in his famous evolutionary tree and other schematic jottings in his Notebook B (1837–38) Darwin believed firmly in the idea of 'common descent': 'The races of men differ chiefly in colour, form of head & features (hence intellect?) & what kinds of intellect) quantity & kind of hair forms of legs – hence the father of man kind probably possessed a structure in these points'.[5] Darwin's

[3] Erich Moritz von Hornbostel, 'The Problems of Comparative Musicology', trans. Richard Campbell (1905), in Dieter Christensen, Hans-Peter Renecke and Klaus P. Wachsmann (eds.), *Hornbostel Opera Omnia* 1, The Hague: Nijhoff, 1975, 250.

[4] Adrian Desmond and James Moore, *Darwin's Sacred Cause: Race, Slavery and the Quest for Human Origins*, London: Allen Lane, 2009, 112.

[5] Charles Darwin, Second Notebook C, 204, *Darwin's Notebooks on Transmutation of Species*, ed. Sir Gavin de Beer, *Part II, Second Notebook [C], February to July 1938, Bulletin of the British*

musings over common descent attracted considerable hostility in Victorian Britain, not least because it threatened to destabilize the permanence of teleological conviction in Britain's imperial project. It also threatened to destabilize a Victorian musical culture confident – despite sustained Continental detraction – in its own global supremacy. In the same way that non-Darwinian 'typists' refused to believe that all animals could create music, Victorians often believed that non-Western – and particularly tribal and savage – music was not music at all. So when it came to music, asymmetry applied as much to human–animal relations as it did to the relations of civilized and (what were considered to be) uncivilized races. Darwinian attitudes, conversely, were 'heretical', and 'dangerous',[6] smashing the idols of religion and nation in one fell swoop. In a Darwinian world all nations, like all people, were fundamentally equal: a human brotherhood – 'all nations of one blood'.[7] To ethicize evolution Darwin broadened his theory of natural selection to operate not only on individuals but communities: 'I concluded that natural selection might act on the parents & continually preserve those which produced more & more aberrant offspring, having any structure or instincts advantageous to the community'.[8] The *Origin of Species* applies this interpretation to natural selection, and *The Descent of Man* develops altruism into 'community selection'.[9]

Community selection may broadly embody Darwin's 'moral directive'[10] but Darwinian evolutionism manifestly failed to establish consensual definitions of equality; nor did other hybridized, non-Darwinian, types of evolution like Social Darwinism. Social Darwinism accepts four tenets of Darwinian orthodoxy, to which a fifth is crucially added: (1) the universality of evolutionary law on organic nature, (2) competition for food and the consequent struggle for survival, (3) the inheritance of favourable

Museum (Natural History), Historical Series 2/3 (May 1960), 75–118, 107, Darwin Online, http://darwin-online.org.uk/content/frameset?keywords=the%20differ%20of%20men%20chiefly%20races&pageseq=33&itemID=F1574b&viewtype=text, accessed 13/5/13.

6 See Daniel Dennett, *Darwin's Dangerous Idea: Evolution and the Meaning of Life*, New York: Touchstone, 1995.

7 Acts 17:26, New King James Version.

8 Charles Darwin, *Charles Darwin's Natural Selection: Being the Second Part of His Big Species Book written from 1856 to 1858*, ed. R. C. Stauffer, Cambridge: Cambridge University Press, 1975, 510.

9 Robert J. Richards, 'Darwin's Romantic Biology: The Foundation of His Ethics', in Jane Maienschein and Michael Ruse (eds.), *Biology and the Foundation of Ethics*, Cambridge: Cambridge University Press, 1999, 144.

10 Michael Ruse, *Darwinism and Its Discontents*, Cambridge: Cambridge University Press, 2006, 208.

characteristics, (4) the cumulative effect of selection and inheritance in the creation of new species, and (5) the laws of evolution applying equally to the organic and inorganic world, including reason, religion and morality – and as we shall see later, music.[11] Like orthodox Darwinism Social Darwinism was open to conflicting interpretations. Darwin himself wavered over it, concerned over the moral implications of favouring fitness, but towards the end of his life he suggested that natural selection helped the progress of civilization.[12] Its author Herbert Spencer believed Social Darwinism was an important social corrective:

> Blind to the fact that under the natural order of things, society is constantly excreting its unhealthy, imbecile, slow, vacillating, faithless members, these unthinking, though well-meaning, men advocate an interference which not only stops the purifying process but even increases the vitiation – absolutely encourages the multiplication of the reckless and incompetent by offering them an unfailing provision, and discourages the multiplication of the competent and provident by heightening the prospective difficulty of maintaining a family.[13]

As extreme as these words sound it was arguably Spencer, not Darwin, who spoke the language of evolution to the Victorian multitude.[14] Spencer eventually diminishes the importance of individual variation by conversely emphasizing the totalizing social force of natural selection. Fit variations are rewarded with survival; unfit variations vanish unceremoniously. This applies as much to the individual physical human as it does to the corporate, cultural context in which he lives, and music plays its part. Fit music, like Western classical music, is rewarded with survival; unfit music like savage music (if it is music at all) fails to evolve.

Natural inequalities of social Darwinism are replicated in Spencer's anthropology[15] which itself builds upon the theories of Lamarck. As mentioned previously Lamarck developed a theory called 'inheritance of acquired characteristics': an organism acquires a characteristic after birth, and if replicated over successive generations gradually evolves the capacity

[11] Mike Hawkins, *Social Darwinism in European and American Thought 1860–1945*, Cambridge: Cambridge University Press, 1997, 31.

[12] Michael Ruse, *The Darwinian Revolution*, Chicago and London: University of Chicago Press, 1979, 264.

[13] Herbert Spencer, *Social Statics; Or the Conditions Essential to Human Happiness Specified and the First of Them Developed*, London: J. Chapman, 1851, 323–24.

[14] Peter Bowler, *Darwin Deleted: Imagining a World Without Darwin*, Chicago and London: Chicago University Press, 2013, 155.

[15] A. P. Thakur, *The Legacy of Social Darwinism*, New Delhi: Global Vision Publishing House, 2005, 46.

to become inherited. Although superficially Lamarck's theory would seem to give precedence to environment, beneath its cool evolutionary exterior lives an unashamedly teleological certitude. Lamarck espouses 'an inherently progressive trend which carried life up to higher levels of organization;[16] in other words, the Great Chain of Being. In Lamarck's version of the Great Chain two separate but not unrelated mechanisms contribute to evolution: a mechanism of progress and a mechanism of adaptation. These two parallel mechanisms coalesce in Spencer's essay 'The Development Hypothesis' (1853), arguing after Lamarck 'that any existing species – animal or vegetable – when placed under conditions different from its previous ones, immediately begins to undergo certain changes fitting it for the new conditions ... in successive generations these changes continue; until, ultimately, the new conditions become the natural ones'.[17] Soon afterwards Spencer refers to Lamarck in *Principles of Psychology* (1855) using the inheritance of acquired characteristics to elevate discussion from the level of individual to that of nation and race: 'the established truths of Psychology and Physiology unite in indicating ... the view that generalizes the phenomena of habit, of national characteristics, of civilization in its moral aspects'.[18] And not much later Spencer would return to Lamarck for directly musical purposes in his seminal essay 'On the Origin and Function of Music' (1857): 'Music', Spencer contends, 'is a product of civilization'.[19]

Undergirding Spencer's musical Lamarckism is a conflation of individual and society beginning in the mind, and its increasingly graded differences in animals and man. Unlike in animals, the mind of man has

[16] Peter J. Bowler, *Evolution: The History of an Idea*, Berkeley, 3rd ed., Los Angeles and London: University of California Press, 2003, 87.

[17] Herbert Spencer, '*The Development Hypothesis*' (*The Leader*, 20 March 1852), in *Essays: Scientific, Political & Speculative*, London: Williams and Norgate, 1891, http://archive.org/stream/essaysscientific29869gut/pg29869.txt. The Victorian Web notes: 'Joachim Dagg, Abteilung für Entomologie, Institut für Phytopathologie und Pflanzenschutz, Göttingen, has written to point out that "the version Spencer later published differs from the original in *The Leader* of 20 March 1852 in one crucial phrase: 'Those who cavalierly reject the Theory of *Evolution*...' originally read 'Those who cavalierly reject the Theory of *Lamarck*...'" [emphasis added]', www.victorianweb.org/science/science_texts/spencer_dev_hypothesis.html, accessed 14/5/13.

[18] Herbert Spencer, 'Bain on The Emotions and The Will' (*The Medico-Chirurgical Review*, January 1860), *Illustrations of Universal Progress: A Series of Discussions*, New York: D. Appleton and Co., 1865, 315.

[19] Herbert Spencer, 'The Origin and Function of Music', *Fraser's Magazine* (Oct. 1857), in Herbert Spencer, *Literary Style and Music: Including Two Short Essays on Gracefulness and Beauty*, New York: Philosophical Library, 1951, 68.

evolved higher levels of volition, progressing from 'a simple homogeneous mental state, forming the link between feeling and action'[20] to a more complex psychical state in which 'imperfectly coherent' perceptions are organized:

Hence the difference is, that whereas, in the case of the involuntary movement, the psychical states accompanying the impression and the action, are so coherent that the one follows the other instantaneously; in the voluntary movement they are so imperfectly coherent, that the psychical state accompanying the action does not follow instantaneously, but slowly – is partially excited before it is fully excited; and so occupies consciousness for an appreciable time before it actually occurs. And thus the cessation of automatic action and the dawn of volition, are one and the same thing.[21]

Refuting the widely held belief that 'the *ego* is something more than the composition state of consciousness which then exists'[22] Spencer supplants conventional notions of a 'Will' (the separate faculty governing human behaviour) with progressive, directional evolution:

Whether we trace mental progression through the grades of the animal kingdom, through the grades of mankind, or through the stages of individual growth; it is obvious that the advance, alike in cognitions and feelings, is, and must be, from the presentative to the more and more remotely representative. It is undeniable that intelligence ascends from those simple perceptions in which consciousness is occupied in localizing and classifying sensations, to perceptions more and more compound, to simple reasoning, to reasoning more and more complex and abstract – more and more remote from sensation.[23]

For many non-Western (especially savage) cultures, however, that ascendancy stopped dead in its tracks, and in musical terms its ramifications were severe: 'for though some of the lowest savages have their dance-chants, these are of a kind scarcely to be dignified by the title musical: at most they supply but the vaguest rudiment of music properly so called'.[24]

Heredity partially explains this phenomenon, but in the same way non-Darwinian zoomusicologists refused to ascribe musical creativity to animals Spencer implicitly problematizes the innate mental endowments – and

[20] Spencer, 'Bain on The Emotions and The Will', 321.

[21] Herbert Spencer, *The Principles of Psychology*, London: Longman, Brown, Green and Longmans, 1855, §206, http://oll.libertyfund.org/simple.php?id=1394, accessed 15/5/13.

[22] Spencer, *The Principles of Pshychology*, §207.

[23] Spencer, 'Bain on The Emotions and The Will', 324.

[24] Spencer, 'The Origin and Function of Music', 68.

therefore evolutionary capability – of savage peoples. If Spencer could replace 'the *tabula rasa* of the individual with that of the race'[25] how does man actually progress from animal to man and from savage to civilized? The simple but unsatisfactory answer is 'not easily'. While Social Darwinism extruded uncompetitive sectors from the population the causes presuming uncompetitiveness were extreme and irrefragable. For all the ethically questionable socio-political lobbying of *Social Statics* (1851) Spencer's Social Darwinism was largely reactive rather than proactive, accepting the presence of degraded peoples and their music as a matter of universal fact. Like Darwinism itself the evolutionary principle behind Social Darwinism was materialistically mechanistic rather than morally encoded, and so Spencer's undeserving poor, savages, children and similarly undeveloped beings were – like the music they produced – all subject to exactly the same biological laws. Those laws 'preserved human unity while simultaneously insisting on racial hierarchy'.[26] Elaborated in three forensically observed chapters, 'Primitive Man – Physical', 'Primitive Man – Emotional' and 'Primitive Man – Intellectual', *Principles of Sociology* (1874–96) details the anthropological physiognomy of high Victorian racism: (1) physically the uncivilized man is less powerful than his civilized counterpart, yet he is hardier, tolerates pain more easily and reaches maturity earlier;[27] (2) emotionally he is impulsive, unable to conceptualize the future, childish, improvident, proprietary, unaltruistic and malevolent;[28] and (3) intellectually he is unable to conceptualize general facts because he cannot conceptualize beyond particular truths. He is experientially limited and therefore unlikely to develop abstract ideas such as cause, property, uniformity and truth; and his imagination merely receives and repeats without originality.[29] Had it not already been written as 'The Origin and Function of Music' a fourth chapter could easily be added: 'Primitive Man – Musical'.

These considerations unite in Spencer's spectacularly unproven theory of sensory acuity commonly labelled the 'Spencer hypothesis'. Savages surpass civilized people in psycho-physical performance because more

[25] Robert M. Young, *Mind, Brain and Adaptation in the Nineteenth Century: Cerebral Localization and Its Biological Context from Gall to Ferrier*, Oxford: Oxford University Press, 1970, 178, http://human-nature.com/mba/chap5.html, accessed 15/3/13.

[26] George W. Stocking, *Victorian Anthropology*, New York: Free Press, 1987, 225.

[27] See Herbert Spencer, 'The Primitive Man – Physical', *Principles of Sociology*, 3 vols., A System of Synthetic Philosophy, vols. 6–8, New York: D. Appleton and Co., vol. 1, 1883, 44–58.

[28] Herbert Spencer, 'The Primitive Man – Emotional', *Principles of Sociology*, vol. 1, 61.

[29] Herbert Spencer, 'The Primitive Man – Intellectual', *Principles of Sociology*, vol. 1, 86.

of their energy is necessarily devoted to rudimentary aspects of life.[30] Musically and otherwise the Spencer hypothesis is both a blessing and a curse. It bequeaths primitive people a superpower – an 'animal-like "savage" sensory virtuosity'[31] – yet because they cling to the elemental characteristics of nature their culture fails to advance. Savage music, Spencer avers, supplies 'the vaguest rudiment of music properly so called'. By Spencer's time the back-handed compliment enjoyed a long meta-narrative in the annals of Victorian anthropology traceable to what Linnaeus called the 'Wild Man'.[32] Though temporarily redeemed by Rousseau as the 'noble savage',[33] the Wild Man became indistinguishable from other forms of savagery by the beginning of the nineteenth century. By Spencer's time it had embodied well-worn characteristics of spontaneity, naturalness, lust and violence[34] to which other anthropological thinkers added childishness, dependence, aggression, ignorance and impudence.[35] The Spencer hypothesis also tapped into profoundly Victorian moral uncertainties being exacerbated by an increasingly survivalist elucidation of man's condition. Spencer was not alone in considering welfare immoral, for example. He was not opposed to transitory acts of human kindness, nor did he consider improvement socially unredemptive. But the lowly savage could never overcome the Lamarckian immutability of his condition, and neither could his music consequently: 'music has been by slow steps developed in the course of civilization'.[36] Spencer's prejudices were not always shared and as the century wore on they lost some of their doctrinal heat. A later generation of anthropologists, including E. B. Tylor and John Lubbock, would express less immoderate tendencies and with them try to ameliorate the least sociologically palatable elements of the Spencerian hypothesis.

[30] Graham Richards, 'Getting a Result: The Expedition's Psychological Research 1890–1913', in Anita Herle and Sandra Rouse (eds.), *Cambridge and the Torres Strait: Centenary Essays on the 1898 Anthropological Expedition*, Cambridge: Cambridge University Press, 1998, 137.

[31] Richards, 'Getting a Result', 144.

[32] John G. Burke, 'The Wild Man's Pedigree: Scientific Method and Racial Anthropology', in Edward J. Dudley and Maximillian E. Novak (eds.), *The Wild Man Within: An Image in Western Thought from the Renaissance to Romanticism*, Pittsburgh: University of Pittsburgh Press, 1973, 266.

[33] See Ter Ellingson, *The Myth of the Noble Savage*, Berkeley, Los Angeles and London: University of California Press, 2001.

[34] Geoffrey Symcox, 'The Wild Man's Return: The Enclosed Vision of Rousseau's Discourses', in Dudley and Novak, *The Wild Man Within*, 228.

[35] Gustav Jahoda, *Images of Savages: Ancient Roots of Modern Prejudice in Western Culture*, London and New York: Routledge, 1999, 9.

[36] Spencer, 'The Origin and Function of Music', 68.

The racial image of the savage remained intractable, however, not only in particularized individual forms but as a generalized imaginary concept: it remained Other in space and time.[37] Being Other made savages relatively easy to identify in space and time for a Victorian Britain gradually reaching the height of its imperial dominion. The Crystal Palace Exhibition of 1851 typifies mid-Victorian attitudes, with its circulation carefully choreographed to replicate the glories of contemporary anthropological truth. National exhibits led visitors 'along a line of progress from the Tasmanian savage through the "barbaric" civilizations of the East, northwest across the European continent toward an apex in Great Britain'.[38] Contributing nations located material objects in one of four categories: raw materials, machinery, manufactures and fine arts, often locating items from less developed colonies in 'raw materials'. Her Majesty Pomare, Queen of the Society Islands (Tahiti), for example, exhibited eight fine mats, five head-dresses or coronets, bonnets, three pieces of white cloth, and a hinai or Indian vase.[39] In the absence of genuine cultural understanding the Exhibition catalogue accorded these objects purely scientific descriptions.[40] The mats, for example, came from 'a variety of the *Pandanus odoratissimus* of Linnaeus', and the head-dresses were made from a 'plant commonly known in these islands by the name of pia, arrow-root by the English, *Tacca pumalifida* by botanists'.[41] Ethnographic music and musical instruments played their part, of course, including scenes and tableaux recreating native ceremonies. An Arab marriage was 'performed', with native musical instruments and a marriage procession,[42] and ethnographic instruments were displayed from China, the East Indies, India, Trinidad, Turkey and Western Africa.[43] Pitted against instruments designed to illustrate Western

[37] Johannes Fabian, *Time and the Other: How Anthropology Makes Its Objects*, New York: Columbia University Press, 1983, 75.

[38] Stocking, *Victorian Anthropology*, 5.

[39] *Great Exhibition of the Works of Industry of all Nations, 1851. Official Descriptive and Illustrated Catalogue*, 3 vols., vol. III, 'Foreign States', London: Spicer Brothers, 1851, 1428.

[40] Paul Young, 'Mission Impossible: Globalization and the Great Exhibition', in Jeffrey A. Auerbach and Peter H. Hoffenberg (eds.), *Britain, the Empire, and the World at the Great Exhibition of 1851*, Aldershot: Ashgate, 2008, 16.

[41] Great Exhibition, Official Descriptive Catalogue, vol. III, 1428.

[42] Edward Ziter, 'Teaching Nineteenth-Century Orientalist Entertainments', in Diane Long Koeveler and Jeffrey Cass (eds.), *Interrogating Orientalism: Contextual Approaches and Pedagogical Practices*, Columbus: Ohio State University Press, 2006, 236.

[43] Peter Mactaggart and Ann Mactaggart (eds.), *Musical Instruments in the 1851 Exhibition: A Transcription of the Entries of Music Interest from the Official Illustrated Catalogue of the Great Exhibition of the Art and Industry of all Nations, with Additional Material from Contemporary Sources*, Welwyn: Mac & Me Ltd, 1986, 80–81.

'works of industry', however – what Flora Wilson describes as 'signs of progress'[44] – ethnographic instruments struggled to face down conventional anthropological denigrations. Responding to these an admirer of Indian music retorts that although some contend that Indian 'ideas of music, if we may judge from their practices, are barbarous' others, like William Jones, find 'the Hindoo system has been formed on truer principles than our own ... some of their tunes are pleasing, even to a European ear'.[45]

Possibly more economic necessity than anthropological reality,[46] the official utopian illusion of fraternity promulgated an image of universal nationhood not shared in the actual experience of exhibitors. Note the dissimulation in William Whewell's commendation of egalitarian Crystal Palace:

> We had ... the choicest productions of human art in all nations; or, at least, collections which might be considered as representing all nations. Now in nations compared with nations there is a difference; in a nation compared with itself at an earlier time, there is a progress ... in virtue, in morality, in happiness. But there always is, except when very adverse influences roll back the common course of things, a progress in art, and generally in science. In the useful and ornamental arts nations are always going forwards, from stage to stage. Different nations have reached different stages of this progress, and all their difference stages are seen at once ... in the magical glass ... The infancy of nations, their youth, their middle age, and their maturity, all appear, in their simultaneous aspect.[47]

While the Great Exhibition touted national brotherhood, the structural contradiction of Social Darwinism continued to produce examples of racially anomalous cultural practice all over the world, especially in the form of excitingly live, primitive entertainment involving music at international exhibitions and other less aspirationally cosmopolitan settings. Hoping to avert declining ticket sales the Jardin d'Acclimatation, for example, exhibited fourteen Nubians in Paris in 1878 to great acclaim,

[44] Flora Wilson, 'Hearing Things: Musical Objects at the 1851 Great Exhibition', in James Q. Davies and Ellen Lockhart (eds.), *Sound Knowledge: Music and Science in London, 1789–1851*, Chicago and London: Chicago University Press, 2016, 242.

[45] 'Observations on the Art and Manufactures of India', cited in Mactaggart and Mactaggart (eds.), *Musical Instruments in the 1851 Exhibition*, 80.

[46] See Paul Young, *Globalization and the Great Exhibition*, Houndsmill: Palgrave Macmillan, 2009.

[47] William Whewell, 'The General Bearing of the Great Exhibition on the Progress of Art and Science', *Lectures on the Results of the Exhibition of 1851, Delivered Before The Society of Arts, Manufactures, and Commerce*, London: David Bogue, 1852, 13.

and the 1889 Paris Exhibition showcased small settlements of its own colonial peoples, including the Senegalese, Congolese, New Caledonians, Gabonese, Dahomeans, Cochin-Chinese and Kampong-Javanese. While systemically depriving people of their human dignity, native displays subsumed ethnic and national identity into colonial semiological contexts. Natives were not the instantiated human beings of a particular culture (the signifier), but objectivized universal exemplars (the signified). As international exhibitions attest, savage Others came from all nations of the colonized world proving, paradoxically, that evolution can produce 'survival of the unfittest'.[48]

Anthropologically deracinated yet biologically universalized the imperial unfittest held only notional national and musical citizenship with the West. Without national citizenship the Unfittest held tenuous human citizenship as well. Crowning the inverse scientific glory of Social Darwinism was pioneering evolutionist Francis Galton, encountered in the previous chapter as the inventor of 'eugenics' and indirect influence upon the ideologically contentious ornithologist Charles Witchell. President of the Anthropological Institute of Great Britain and Ireland from 1885 to 1889 Galton is the eminently logical – if ethically questionable – personification of institutionally entrenched scientific racism. The Anthropological Conference on Native Races in British Possessions held in June 1886 encapsulates their descent into objectification. Following Galton's opening remarks three Bantus were exhibited and played musical instruments. Using the opportunity to validate the Spencer hypothesis a Bushman was persuaded to test his hand strength by squeezing a pressure gauge called a dynamometer.[49] Ultimately, Galton brought something to evolution singularly lacking in Spencerian evolution: data. Galton was an inveterate counter, measurer and statistician, and confidently cited empirical evidence to substantiate the heredity underpinnings of eugenics. *Hereditary Genius* (1869) and *Inquiries into Human Faculty and its Development* (1883) are full of figures, and in 1884 he measured 9000 people at the International Health Exhibition in South Kensington.[50] Eventually, however, statistics failed to convince and eugenics was eventually discredited – but not before Social Darwinism cast a very long shadow over Victorian and

[48] See Pat Shipman, *The Evolution of Racism: Human Differences and the Use and Abuse of Science*, Cambridge, MA: Harvard University Press, 1994/2002, 107–21.

[49] John C. Kenna, 'Sir Francis Galton's Contribution to Anthropology', *The Journal of the Royal Anthropological Institute of Great Britain and Ireland* 94/2 (July–Dec. 1964), 85.

[50] Paul B. Rich, *Race and Empire in British Politics*, 2nd ed., Cambridge: Cambridge University Press, 1990, 18.

post-Victorian musical culture. It would be well into the 1930s when an ethnomusicologist (Charles Samuel Myers) finally tackled the Spencer hypothesis and the shadow of Social Darwinism began to recede for good.

Ethnomusicology

West Looks East

Long before the shadow of Social Darwinism crept up on Victorian Britain progressive musical thinkers had already begun to attack Western ignorance about non-Western musical cultures. It comes as no surprise, therefore, that the Great Exhibition's Indophile correspondent of 'Observations on the Arts and Manufactures of India' should cite William Jones as a fellow lover of Indian music. Intellectual polymath and founder and president of the Asiatic Society of Bengal Jones wrote the first English treatise on foreign music 'On the Musical Modes of the Hindoos' (1784; rev. 1792).[51] Using reason to discredit common prejudices towards its 'noise and jargon'[52] and 'hideous astonishing sounds'[53] Jones casts Indian music in a convincingly scientific light by analyzing and comparing tuning systems of the veena and violin. All music, whether Indian or not, affects the nervous system similarly 'in a singular manner by combinations of sound'.[54] Jones offers an aesthetic emollient as well. Comparing Hindu modes to the music of ancient Greece and Rome and to the 'affecting strains of the Plain Song'[55] he translates the complexities of Indian music into an interpretable European experience: 'The fancy of Shakspeare [*sic*] and the pencil of Albano might have been finely employed in giving speech and form to this assemblage of new aerial beings, who people the fairyland of Indian imagination ... the *Sanscrit* language is "equal to Italian in softness"'.[56] Jones was unusual, however, in defending the notion of a

[51] See Sir William Jones, 'On the Musical Modes of the Hindoos: Written in 1784, and since much enlarged, By the President [of the Asiatic Society of Bengal]', reprinted in Sourindro Mohun Tagore, *Hindu Music*, Delhi: Low Price Publications, 1882/1994, 125–60.

[52] Charles Burney, *A General History of Music from the Earliest Ages to the Present Period. To which is Prefixed, a Dissertation on the Music of the Ancients*, 4 vols., London: Printed for the Author, 1776–89, vol. 1 (1789), 186.

[53] John Hawkins, 'Preface', in *A General History of the Science and Practice of Music*, London: T. Payne and Sons, 1776.

[54] Hawkins, 'Preface', *A General History*, 127.

[55] Jones, 'On the Musical Modes of the Hindoos', 131.

[56] Jones, 'On the Musical Modes of the Hindoos', 146.

universal brotherhood. James Mill captures the hostility to his approach in the acerbically critical *History of British India* (1817): 'It was unfortunate that a mind so pure, so warm in the pursuit of truth, and so devoted to oriental learning, as that of Sir William Jones, should have adopted the hypothesis of a high state of civilization in the principal countries of Asia . . . The term civilization was by him, as by most men, attached to no fixed and definite assemblage of ideas. With the exception of some of the lowest states of society in which human beings have been found, it was applied to nations in all the stages of social advancement'.[57]

Another early ethnomusicological detractor, N. Augustus Willard, faults Jones – perhaps justifiably – for his predominantly bibliographical approach, and in *A Treatise on the Music of Hindustan* (1834) criticizes him for failing to differentiate between theory and practice: 'This is the reason why even so able and eminent an Orientalist as Sir William Jones has failed. Books alone are insufficient to the purpose – we must endeavour to procure solutions from living professors, of whom there are several, although grossly illiterate'.[58] Though Willard makes a good point he inexplicably glosses over Jones's principal innovation, which was to use a comparative method to subvert the congenital developmental lies at the root of Enlightenment thought. William Dauney recognizes Jones's achievement in his own ethnographic writing on music and puts the importance of his method in greater context: although the modern European musical system

> may possibly be the best which can be adopted . . . this can only be known by an extensive comparison with other systems. The furtherance of such inquires, therefore, may lead to a direct improvement in the cultivation of music, while there can be no doubt that the resources of that art would be immensely enriched by a more complete knowledge of the different styles of melody which prevail in foreign countries, and copious and authentic collections of the airs themselves.[59]

Like Willard, Dauney insists that the comparative method succeeds only when foreign music is studied in its own terms: certain music of the East

[57] James Mill, *The History of British India*, 4th ed., 8 vols., London: James Madden and Co., 1840–48, vol. 8, 155–56.

[58] N. Augustus Willard, *A Treatise on the Music of Hindoostan: Comprising a Detail of the Ancient Theory and Modern Practice*, Calcutta: Baptist Mission Press, 1834, reprinted in Tagore, *Hindu Music*, 3.

[59] William Dauney, 'Observations with a View to an Inquiry into the Music of the East', *Journal of the Royal Asiatic Society*, old series 6 (1841), 1.

must be treated with reference to its capabilities for melody only; and this is the more necessary to be kept in view, as many excellent practical musicians are apt to suppose (although there cannot be a greater mistake) that where a foreign melody will not yield to the application of modern European harmony, it must be defective – that there must be something wrong in the manner in which it is performed, and for this reason intractable airs of this kind are often thrown into a modern shape, in order to be adapted to our major and minor scales, with their diatonic and chromatic intervals.[60]

This forward-looking strain of comparative method points to a methodologically modern future in which foreign – and indeed all – music would eventually be defined by ethnomusicology as 'humanly organized sound'.[61] To arrive at that future Victorian ethnomusicology would first have to contend with the anthropological contradictions posed by Spencer and Social Darwinism. Based on de-ethicized principles of natural selection and competitive fitness Social Darwinism takes an altogether materialistic view of man and music, as we have seen previously: 'That music is a product of civilization is manifest: for though some of the lowest savages have their dance-chants, these are of a kind scarcely to be signified by the title musical: at most they supply but the vaguest rudiment of music properly so called'.[62] The corollary for civilized music is a progressive integration which gradually complexifies music as it advances towards the most fully evolved and elaborated forms:

In music, progressive integration is displayed in still more numerous ways. The simple cadence embracing but a few notes, which in the chants of savages is monotonously repeated, becomes, among civilized races, a long series of different musical phrases combined into one whole; and so complete is the integration, that the melody cannot be broken off in the middle, nor shorn of its final note, without giving us a painful sense of incompleteness. When to the air, a bass, a tenor, and an alto are added; and when to the harmony of different voice-parts there is added an accompaniment; we see exemplified integrations of another order, which grow gradually more elaborate. And the process is carried a stage higher when these complex solos, concerted pieces, choruses, and orchestral effects, are combined into the vast ensemble of a musical drama; of which, be it remembered, the artistic perfection largely consists in the subordination of the particular effects to the total effect.[63]

60 Dauney, 'Observations with a View to an Inquiry into the Music of the East', 3.

61 John Blacking, *How Musical Is Man?*, Seattle and London: University of Washington Press, 1973/2000, 32.

62 Spencer, 'The Origin and Function of Music', 68.

63 Herbert Spencer, *First Principles*, 4th ed., New York: D. Appleton and Co., 1896, 336.

Five aspects of Social Darwinism are reflected in Spencer's disquietingly simple, doctrinal statements on musical evolution: (1) perhaps most importantly music is the exclusive function or by-product of civilization, and like all organisms it is subject to the same universal evolutionary laws as nature itself. By definition savages cannot and do not have music – what they have is a simpler, less developed form of proto-musical expression; (2) because competition exists anthropologically it also exists musically so that the fittest (civilized) peoples enjoy the fittest (civilized) music. In a sense music is by definition *only* fittest (3) if, as Lamarck suggests, it is possible to inherit favourable characteristics anthropologically, the ability to produce music (as fittest) can be acquired through inheritance; (4) the cumulative effect of selection and inheritance creates new musical forms which by necessity reveal evolutionary paradigms at work. Thus there is a progression from simplicity to complexity, and further integration and development from the simplest cadences to the most complex vocal and instrumental forms. Spencer conceives music as a completely advanced, civilized organism, and even if relatively simple (beginning with ordinary cadences) it is distinct from the monotone dance chants of savages. Music – *not* savage chant – is so systemically complete that completeness becomes both its structural strength and weakness. As we found in birdsong one missing note and melody loses its recognizable identity; and lastly (5) music is subject to the same laws that govern the organic and inorganic world, and only becomes distinguishably music when it rises to an advanced evolutionary condition like reason, religion or morality.

Three overarching issues emerge from these five points: (1) when Spencer describes music he discusses it at the level of specific compositional type. Savages may have their generic dance-*chants*, but when chant becomes music (when it is civilized) it acquires the individuality and complexity of a song, chorus, concerto, symphony and opera; (2) these are the musical forms that have been selected by the ineluctable force of evolution. They have evolved because they are the fittest, and circularly, they are the fittest because they have evolved. Unfit, pre-musical forms are not only less differentiated, but because they lack differentiation they also remain inseparable from other proto-civilized artistic forms, like dance and poetry. When savages repeatedly 'have their dance-chants', they express an atavistic form of communication in which neither the arts nor to some extent language have been disaggregated:

In the co-ordinate origin and gradual differentiation of Poetry, Music, and Dancing, we have another series of illustrations. Rhythm in speech, rhythm in sound, and rhythm in motion, were in the beginning, parts of the same thing; and have only in process of time become separate things. Among various existing barbarous tribes we find them still united. The dances of savages are accompanied by some kind of monotonous chant, the clapping of hands, the striking of rude instruments;[64]

and lastly (3) by suggesting that music is strictly a function of civilization Spencer gives the contradictory impression that natural selection operates only upon organisms already at a more advanced stage of development: 'While between the old monotonous dance-chant and a grand opera of our own day, with its endless orchestral complexities and vocal combinations, the contrast in heterogeneity is so extreme that it seems scarcely credible that the one should have been the ancestor of the other'.[65] As such, savage dance-chant is more physical, and less emotional and intellectual than civilized music. It is the living musical embodiment of the Spencer hypothesis – more physical than psychical, more emotive than emotional, more sensory than sensitive.

Until the 1870s the absence of a credible Darwinian counterargument gave the Social Darwinist project free reign over Victorian musical culture, and the ethnomusicologist who most diligently reflected its position was German expatriate Carl Engel (1818–1882). Engel was an organologist with a particular interest in non-Western instruments as well as author of many articles and books on domestic and foreign national music, amongst them *The Music of the Most Ancient Nations* (1864), *An Introduction to the Study of National Music* (1866), *Descriptive Catalogue of the Musical Instruments in the South Kensington Museum* (1870), the musical section for the anthropological travel manual *Notes and Queries* (1874) and a compilation of *Musical Times* articles published as *The Literature of National Music* (1879). Engel adds a vital comparative component to Social Darwinian method – the past: 'As I proceeded', he considers, 'I became more and more convinced that, in order to understand clearly the music of the various modern nations, it was necessary to extend my researches to the music of ancient nations'.[66] Inevitably, Engel draws upon comparative models exemplified in contemporary anthropological practice. Henry Maine, for example, extrapolates a patriarchal society from Semitic peoples of the Bible,[67] and

[64] Spencer, *First Principles*, 864. [65] Spencer, *First Principles*, 869.

[66] Carl Engel, *The Music of the Most Ancient Nations, particularly of the Assyrians, Egyptians, and Hebrews; with Special Reference to Recent Discoveries in Western Asia and in Egypt*, London: John Murray, 1864, v.

[67] Henry Maine, *Ancient Law: Its Connection with the Early History of Society and Its Relation to Modern Ideas*, London: John Murray, 1861/1908, 109.

John Lubbock (referring to Maine) identifies ways of studying the origin of the social state through 'accounts by contemporary observers of civilizations less advanced than their own; [and] the records which particular races have preserved concerning their primitive history, and ancient law'.[68] But Engel's most undoubtedly profound influence comes from England's foremost Victorian anthropologist E. B. Tylor. Tylor sets out his comparative methodological stall in the seminal work *Primitive Music* (1871):

Culture or Civilization, taken in its wide ethnographic sense, is that complex whole which includes knowledge, belief, art, morals, law, custom, and any other capabilities and habits acquired by man as a member of society. The condition of culture among the various societies of mankind, in so far as it is capable of being investigated on general principles, is a subject apt for the study of laws of human thought and action. On the one hand, the uniformity which so largely pervades civilization may be ascribed, in great measure, to the uniform action of uniform causes: while on the other hand its various grades may be regarded as stages of development or evolution, each the outcome of previous history, and about to do its proper part in shaping the history of the future. To the investigation of these two great principles in several departments of ethnography, with especial consideration of the civilization of the lower tribes as related to the civilization of the higher nations, the present volumes are devoted.[69]

Engel learns several things from Tylor. First, through Tylor's anthropologically pivotal theory of 'survivals' he acquires a methodological vocabulary for distinguishing the historical past from the savage present. Tylor defines 'survivals' as 'antiquarian relics'[70] which contain clues to our developmental present: 'the savage state', he claims, 'in some measure represents an early condition of mankind, out of which the higher culture has gradually been developed or evolved'.[71] 'Nothing', Engel proffers,

is more usual than the notion that, in order to trace the art of music from its most primitive state and to observe its gradual development, we must commence our inquiries by penetrating the most remote periods. This, however, is erroneous ... among the most ancient nations known to us – the Assyrians as well as the ancient Egyptians and Hebrews – music had already attained a degree of perfection considerably higher than we meet with in many nations of our own time.[72]

[68] John Lubbock, *The Origin of Civilization and the Primitive Condition of Man*, New York: Appleton and Co., 1870/1879, 2.

[69] Tylor, *Primitive Culture*, vols. 1, 3.

[70] Tylor, *Primitive Culture*, vol. 1, 158.

[71] Tylor, *Primitive Culture*, vols. 1, 7.

[72] Engel, *Music of the Most Ancient Nations*, 8.

According to Engel, therefore, it is possible that historically unfit nations produced fit music in the context of their own historical past. It is not the case that degenerated cultures necessarily produced degenerated music; nor is it the case that contemporary primitives necessarily represent remnants of degenerated nations or peoples. Second and not unrelatedly, Engel accepts the principle that contemporary primitives hold the key to unlocking the musical past: 'A clear idea of the gradual development of the art of music, from its most primitive condition to that degree of perfection in which it at present exists among ourselves, may be best obtained by examining the music of contemporary nations in different stages of civilisation'.[73] Lastly, he adopts an unstintingly unilinear approach to cultural development, believing that the comparison of equally unadvanced musical cultures can help us evaluate the earliest music: 'in order to ascertain how music reveals itself in its earliest infancy, we ought to observe it not only among the natives of the Fuegian Archipelago, or the Esquimaux, but also among the natives of Australia, of New Guinea, and others in a similarly low state of civilisation'.[74]

A Tylorian approach would also help Engel overturn some fairly entrenched Spencerian presuppositions about music. Unlike Spencer Engel rejects the idea that music is the exclusive function or by-product of civilization though he does accept that it evolved from primitive to civilized stages. He also rejects the idea that the supposedly fittest (civilized) people necessarily have attained the fittest (most civilized) music. Music is not by definition the preserve of the fittest alone but the expression of different stages in development. However, the ability to produce music (as fittest) can be acquired through a kind of Lamarckian acquired cultural inheritance. Writing of the perfection of Assyrian music, for instance, Engel postulates that they must have 'derived their music in an already somewhat advanced state from some other nation or nations unknown to us, and that they only further developed what had been transmitted to them'.[75] New musical forms, structures and technologies reveal evolutionary paradigms at work. Although there is no such thing as 'completeness' any musical culture can achieve the relative quality of 'perfection' (even though civilized music is more advanced than primitive music). And lastly, in its early stages music is inseparable from its physio-cultural context, and only becomes discernibly more musical when it rises to an advanced evolutionary

[73] Engel, *Music of the Most Ancient Nations*, 8–9.

[74] Engel, *Music of the Most Ancient Nations*, 8–9.

[75] Engel, *Music of the Most Ancient Nations*, 27.

condition; thus 'The songs of savages in the lowest scale of civilization are generally confined to the compass of few notes', and their motives 'have not unfrequently been derived . . . from the songs of birds'.[76]

East Looks West

Comparison forms an essential ingredient in Engel's semi-Spencerian methodology, used to convince readers that music derives from a common source or set of interrelated sources: 'The similarities [between music of different nations] are often of such a nature that they cannot possibly be explained as accidental coincidences, but must either have originated in a former connexion between the nations, or must have been derived from a common source which perhaps no longer exists'.[77] For Engel and most Victorian ethnomusicologists that common source originates in the East – the font of musical brotherhood.[78] In *The Music of the Most Ancient Nations*, for example, the 'Eastern Origin of Our Own Music' is placed strategically right at the end of the book, forming an invisible, suppositional bridge from an Oriental past to European future. Inevitably, not everyone accepted the Eastern origins of music. Without any real ethnomusicological understanding, Spencer, like so many people, positively loathed Oriental music. For him and like-minded Victorians civilized European music could never originate in the degraded cultures of the East: 'That recitative – beyond which, by the way, the Chinese and Hindoos seem never to have advanced – grew naturally our the modulations and cadences of strong feeling'.[79] However much the Victorians loved to hate Eastern culture most students of music accepted the Eastern origins of Western music – even, paradoxically, students from Eastern countries denigrated by the same comparative anthropological method they would use to argue for national self-determination. West-facing Bengali evolutionist Sourindro Mohun Tagore (1840–1914) provides an arresting example of this phenomenon.

Born in Calcutta, into a 'heterogenic urban center' – arguably a hotbed of colonial mimicry[80] – Tagore grew up in 'the golden age of the

[76] Engel, *Music of the Most Ancient Nations*, 19; 20.

[77] Carl Engel, *An Introduction to the Study of National Music; Comprising Researches into Popular Songs, Tradition, and Customs*, London: Longmans, Green, Reader, and Dyer, 1866, 317.

[78] See Bennett Zon, *Representing Non-Western Music in Nineteenth-Century Britain*, Rochester NY: University of Rochester Press, 2007, especially 145–56.

[79] Spencer, 'Origin and Function of Music', 63.

[80] Charles Capwell, 'Marginality and Musicology in Nineteenth-Century Calcutta: The Case of Sourindro Mohun Tagore', in Bruno Nettl and Philip V. Bohlman (eds.), *Comparative*

moderates'[81] fuelled by the pervasive influence of Western thought. In his time he was an eminent Indian musicologist at the hub of nineteenth- and early twentieth-century Indian musical culture. Descended from an aristocratic family with equally incontestable European and Hindu national sympathies he was educated at the English-style Hindu College, Calcutta,[82] and became a central – and often polarizing – figure in the ideologically riven world of *fin-de-siècle* Indian musical politics. From the 1870s he founded numerous music schools reflecting his own educational upbringing and complemented these with the publication of some of the first general music treatises and instruction books written in Bengali. Sometimes controversial, divisive activities followed with writings on Hindu music and its notation, involvement in the abortive National Anthem project, and lastly, a well-known kowtow to Western musicology *Universal History of Music* (1896). Like other unrepentantly Anglo-Hindu nationalists Tagore would, soon after his death, fall into obscurity as his ideological mixture of European values and Hindu nationalism lost currency to increasingly effulgent Indian nationalist discourse. Charles Capwell situates Tagore's place in music history predicating his views on the marginality of Bengal within the mainstream of Hindustani music,[83] and Martin Clayton locates Tagore at the centre of an Indian music renaissance mirrored on its English counterpart.[84] The following section takes an altogether different approach broadly interpreting Tagore as an unrepentant colonial mimic, using his career to highlight mimicry as a 'paradoxical feature of colonial resistance'[85] or 'discourses of domination [that] may become rhetorics of resistance'.[86]

By the middle of the Victorian period Tagore's India had closely aligned evolutionism with Hindu religion and nationalism. Keshub Chunder Sen (1838–84) proclaims:

Musicology and Anthropology of Music: Essays on the History of Ethnomusicology, Chicago and London: University of Chicago Press, 1991, 230.

81 Leonard A. Gordon, *Bengal: The Nationalist Movement 1876–1940*, New York and London: Columbia University Press, 1974, 15.

82 David Trasoff, 'Tagore, Sir Sourindro Mohun', in *Grove Music Online*, *Oxford Music Online*, www.oxfordmusiconline.com/subscriber/article/grove/music/27381, accessed 15/4/16.

83 Capwell, 'Marginality and Musicology in Nineteenth-Century Calcutta', 228.

84 Martin Clayton, 'Musical Renaissance and its Margins in England and India', 1874–1914, in Martin Clayton and Bennett Zon (eds.), *Music and Orientalism in the British Empire, 1780s–1940s: Portrayal of the East*, Aldershot: Ashgate Publishing Ltd, 2007, 85.

85 Bill Ashcroft, *Post-Colonial Transformation*, London and New York: Routledge, 2001, 3.

86 Richard Henry Brown, 'Cultural Representation and Ideological Domination', *Social Forces* 71/3 (March 1993), 672.

Let me proceed to take up the great idea of the day – I mean Evolution. I am not going to discuss the details of the philosophy of evolution. Your protoplasm, your natural selection, leave to be discussed by men like Huxley and Darwin. But whether there is a progressive evolution going on in the individual life of man is a question in which we are all interested . . . The animal lives in us still, and wars with incipient humanity. Now triumphs the animal, and now the man. If the lower passions repeatedly win, and if man wholly succumbs and yields to them, the man sinks in the brute. But if the war goes on, the ultimate result of this protracted series of struggles will be the evolution of pure humanity.[87]

For Chunder Sen there are four stages of human evolution, from protozoa to invertebrates, and from vertebrates to mammals. These correspond to the four stages through which man has to pass – the inorganic, the carnal, human and divine[88] and again to increasingly developed incarnations of the Vishnu:

The silent Divinity began to speak, and His speech, His word, a continued breathing of force is creation. What a grand metaphor is the Logos! The Hindu, too, like the Christian, believes in the continued evolution of the Logos, and its graduated development through ever-advancing stages of life. The Puranas speak of the different manifestations or incarnations of the Deity in different epochs of the world's history. Lo! the Hindu Avatar rises from the lowest scale of life through the fish, the tortoise, and the hog up to the perfection of humanity. Indian Avatarism is, indeed, a crude representation of the ascending scale of Divine creation. Such precisely is the modern theory of evolution.[89]

For other Hindu nationalists evolution implied not only a spiritual but more overtly cultural set of overtones, especially when considered a mechanism of natural selection and therefore survival. Bankim Chandra Chattopadhyaya (1838–94), 'the poet and prophet of romantic nationalism',[90] disputes the suitability of Christianity as a form of moral philosophy in an age of science, elevating Hinduism as its more scientifically rational replacement.[91] This complements an allegiance to the kind of

[87] Keshub Chunder Sen, 'Philosophy and Madness in Religion', in *Life and Works of Brahmananda Keshav*, Prem Sundar Basu (ed.), Calcutta: Navavidhan Publication Committee, [1877]/1940, 330–31, cited in C. Mackenzie Brown, *Hindu Perspectives on Evolution: Darwin, Dharma, and Design*, London and New York: Routledge, 2012, 112–13.

[88] Mackenzie Brown, *Hindu Perspectives on Evolution*, 113.

[89] Keshub Chunder Sen, 'That Marvelous Mystery – The Trinity', in *Keshub Chunder Sen's Lectures in India*, London, Paris, New York and Melbourne: Cassell and Co., 1882/1904, 12–13.

[90] Tapan Raychaudhuri, 'Europe in India's Xenology: The Nineteenth-Century Record', *Past and Present* 137: The Cultural and Political Construction of Europe (Nov. 1992), 165.

[91] Partha Chatterjee, *Nationalist Thought and the Colonial World: A Derivative Discourse?*, Minneapolis: University of Minnesota Press, 1993, 67.

utilitarianism exemplified in articles like 'Mill, Darwin and Hindu Religion' (1875), subsequently published as 'What Science Has to Say About the [Hindu] Trinity'. Here Bankim models the Hindu conception of the Trinity on an evolutionary pedestal, suggesting that the threefold Divinity overseeing creation, preservation and destruction bore similarities to processes of natural selection.[92]

For Bankim and Keshub spiritual evolutionism breathed life into a Hindu national politics struggling to assert its identity, and for Bankim specifically that struggle for identity would only be achieved by the evolutionary mechanism of colonial mimicry. Responding to an article by Rajnarayan Basu (Bose) entitled 'Then and Now' (1874) in which Bengalis were castigated for imitating the English, Bankim argues that 'One cannot learn except by imitation. Just as children learn to speak by imitating the speech of adults, to act by imitating the actions of adults, so do uncivilised and uneducated people learn by imitating the ways of the civilised and the educated. Thus it is reasonable and rational that the Bengalis should imitate the English'.[93] While Bankim and Kebush's evolutionary mimicry has clear implications for nationalism it is Lajpat Rai (1865–1928) who, utilizing fitness, converts spiritual evolutionism into a more politically aggressive hermeneutical tool. Rai is a paradigmatic Hindu nationalist, averring, in the face of staunch criticism, the validity of Hindu nationality as a concept.[94] For him the origins of Hindu nationality lie nowhere else but in the Aryan Vedic period: 'We the English educated Hindus of the present day, who claimed to have imbibed the new spirit of nationality and patriotism from the West would really do well to study a few chapters of the Vedic literature with care and thought'.[95] Like many nationalists Lajpat Rai's sense of nationalism was born out of concepts of struggle broadly indicative of resistance culture, and so his brand of evolutionism draws naturally on Spencer's notion of fitness: 'We will not advance the cause of nationality by one inch if we decide to preserve an attitude of silent quietude and non-disturbing peace in all matters, religious and social. Such

[92] Tapan Raychaudhuri, *Europe Reconsidered: Perceptions of the West in Nineteenth-Century Bengal*, 2nd ed., Oxford: Oxford University Press, 1988/2002, 145–46.

[93] Bankimchandra Chattopadhay, 'Anukaraṇ', in *Baṅkim Racanābalī*, Jogesh Chandra Bagal (ed.), Vol. 2, Calcutta: Sahitya Samsad, 1965, 201, cited in Chatterjee, *Nationalist Thought and the Colonial World*, 65.

[94] Chetan Bhatt, *Hindu Nationalism: Origins, Ideologies and Modern Myths*, Oxford: Berg, 2001, 51.

[95] Lala Lajpat Rai, *Lala Lajpat Rai: Writing and Speeches*, ed. V. J. Joshi, 2 vols., Delhi: University Publishers, 1966, vol. 1, 40.

an attitude can only mean stagnation and gradual extinction. Struggle, hard struggle is the law of progress. Yes struggle we must, both inter se as well as others'.[96] The implications, for Lajpat Rai, are almost self-explanatory:

We do not require a Herbert Spencer to tell us that the social efficiency of a social organism as such, depends upon the sense of social responsibility amongst the members of such an organism. The greater and the intenser the sense of responsibility amongst the individual members, regarding the safety and welfare of the whole, the greater and the stronger the efficiency of the organism.[97]

Like many Hindu nationalists Tagore adapts Social Darwinism to fit the particular nationalistic purpose of creating the most level evolutionary playing field as possible. Spencer manufactures theoretical equanimity through the laws of natural selection yet sustains a practical inequality between races and nations. Tagore attempts to resolve the disparity by subverting conventional teleologies with a universalizing methodology not unlike Engel's. Like Engel and Tylor Tagore begins with the presumption that music helps ethnology decode ancestral relationships between peoples:

There is such a connection between music and Man's nature that how far there are differences and in what particular subjects there is unity regarding the quality and character of different races, these questions, too, could be determined if one were to know about national music quite thoroughly. Therefore, it is no less essential for ethnologists than for musicologists to investigate through the study of music what connections exist among the various natures of men.[98]

For Engel,

in an ethnological point of view, an acquaintance with national music must be, I should think, interesting. There are sometimes striking resemblances in musical scales, in peculiarly characteristic modulations, or in the construction and employment of the musical instruments, between two different nations, so that the ethnologist may find therein hints, either affording him additional evidence in substantiating a preconceived theory, or perhaps even suggest some entirely new investigation . . . as in their songs the people express generally their innermost feelings and aspirations, national music would even on this account alone be highly interesting to the intelligent mind, because it exhibits the true character of a nation.[99]

[96] Rai, *Lala Lajpat Rai: Writing and Speeches*, vol. 1, 43.

[97] Rai, *Lala Lajpat Rai: Writing and Speeches*, vol. 1, 55.

[98] Tagore, 'ṭhākura', 39–40, cited in Charles Capwell, 'Marginality and Musicology in Nineteenth-Century Calcutta: The Case of Sourindro Mohun Tagore', in Bruno Nettl and Philip V. Bohlman (eds.), *Comparative Musicology and Anthropology of Music: Essays on the History of Ethnomusicology*, Chicago and London: University of Chicago Press, 1991, 234.

[99] Engel, *The Music of the Most Ancient Nations*, 4–5.

Tagore, like Engel, also accepts that that most 'countries have a music of their own, the character of which may be called national,'[100] and that this music projects 'the inward man ... [which] displays the character and temperament of different races, and the relation they bear to one another'.[101] Strikingly, but perhaps unsurprisingly, aristocratic Tagore retains the belief that music is only a function of civilization. Paraphrasing Spencer, Tagore maintains that 'The primitive tones of the human voice are much the same in all countries prior to the progress of art ... [but] civilized music is cultivated on scientific principles in the upper strata of society ... where it has reached a high degree of development'.[102] But Tagore abandons a Spencerian paradigm when it comes to the question of musical origins, convinced that music originates in the biologically equalizing sounds of nature rather than in the human linguistic turn of primeval man. Rooting music in nature rather than language places Tagore firmly within an Engelian/Tylorian ethnographic model in which the 'antiquarian relics' of savage songs betray their origins in animal mimicry. This attribution would seem to place Tagore within an implicitly more Darwinian camp in which music precedes language. In fact music may originate in nature but it begins in pleasure or pain – in the direct ancestor of (Spencerian) impassioned speech. Spencer writes that

> All music is originally vocal. All vocal sounds are produced by the agency of certain muscles. These muscles, in common with those of the body at large, are excited to contraction by pleasurable and painful feelings. And therefore it is that feelings demonstrate themselves in sounds as well as in movements ... We have there, then, a principle underlying all vocal phenomena, including those of vocal music and by consequence those of music in general.[103]

In Tagore's own words:

> Music pervades all nature. It is co-eval with the creation. There is nothing in nature that arouses our attention or affects our feelings so quickly as sound. The murmuring of water, the sighs of the zephyrs, the whispers of the evening breeze, the roar of the storms, the chirpings of the birds, the cries of the animals, the hum of distant multitudes, and the concussion of sonorous bodies, excite in our minds feelings of pleasure, pain, or fear, and contain in them the germs of music.[104]

[100] Sourindro Mohun Tagore, *Universal History of Music Compiled from Divers Sources Together With Various Original Notes on Hindu Music*, Delhi: Low Price Publications, 1896/1999, 11.
[101] Engel, *Music of the Most Ancient Nations*, 12.
[102] Engel, *Music of the Most Ancient Nations*, 11.
[103] Spencer, 'Origin and Function', 48–49.
[104] Tagore, *Universal History of Music*, 1.

The human voice is an essential evolutionary ingredient for Tagore, as for Engel, Tylor and Spencer, not least because it has the power to separate man from animal and mind from emotion. The human voice can also express and reveal favourable environmental adaptation. In Southern climates, for example, where the mouth can remain unrestricted by climate, 'the voice expands, ripens, and attains to perfection',[105] and so in India, where the climate 'is noted for its beauty and mildness, the vocal art has risen to so much fame'.[106] The voice, therefore, becomes an organ of immense evolutionary importance, because it is the mouthpiece of musical expression: 'By the curious structure of the vocal organs, man is capable of making a greater variety of tones than any other animal, and has at his disposal the power of giving expression to every emotion'.[107] That power is extraordinarily important in Spencerian terms because the capacity to express emotion is an advanced developmental acquisition, as Spencer outlines in 'Primitive Man – Emotional'. For Spencer and Tagore emotional capacity is interlinked to other musical functions, like time and rhythm. In his doctrine of survivals Tylor aggregates music, poetry and dance in pre-civilization, and likewise Spencer aggregates loudness, timbre, pitch, intervals, and rate of variation in song in his theory of musical origins.

Tagore thinks similarly, aggregating in nature time, rhythm and harmony in the primeval stirrings of musical utterance. Time 'is born in nature'; harmony 'is an effect inherent in nature'; and a musical scale is 'formed from an observation of the effects of every sound in nature, may be called the *prism* of the art whereby all combinations of sound are divisible into their component parts'.[108] Nature is not merely evolutionary, however; it is also divine: 'With the Hindus, music is of divine origin. In fact, it is considered as divinity itself'.[109] Because Hindu music is divine it also embodies man's deepest human emotions; as Tagore says, 'The order in which some of the Sanskrit writers have enumerated the Tasa chimes in with the theory of evolution'.[110] Emotions saturate Hindu music from its very foundation in eight Rasas, or affections of the mind. Each has a Ragíní (melody type) or number of Ragínís capable of giving them expression. The eight Rasas include Sringára (love), Hásya (mirth), Karuná (tenderness), Raudra (anger), Víra (heroism), Bhayánaka (terror), Vibhatsa

[105] Tagore, *Universal History of Music*, 2.
[106] Tagore, *Universal History of Music*, 2.
[107] Tagore, *Universal History of Music*, 2.
[108] Tagore, *Universal History of Music*, 2–3.
[109] Tagore, *Universal History of Music*, 50.
[110] Tagore, *Universal History of Music*, Addenda, iii.

(disgust), and Adbhuta (surprise). The human body is divided into six chakras (depressions, rings or circles), and of these Visuddhi, the hollow between the frontal sinuses, is the seat of music and home of Sarasvati, the goddess of music. Visuddhi is divided into sixteen folded petals, and location of the vital spirit determines the success or failure of musical expression.[111]

As shown in Figure 2.1 Tagore suggests that the eight Rasas elicit parallelisms with evolutionary theory, and more particularly the Great Chain of Being. Sringa (love) is 'a feeling common to all sentient beings, and lies at the root of the law of procreation'; Vira (heroism) 'is observed in the next higher stages of created beings, such as mice and snakes'; Karuná (tenderness) 'is non-existent in the lower creatures, such as fish, frogs, mice, snakes, &c.'; Raudra (anger) 'is found in the next higher grades of living beings, such as dogs, tigers, &c'; Hasya (mirth) 'is confined to the highest creation, man'; Bhayanaka (terror) 'is that of man in a state of barbarism'; Bibhatsa (disgust) 'is the feeling of man when he has made strides in the path of civilization'; and Adbhuta (surprise) 'is realized by man only when he has reached the summit of civilization'. A ninth Rasa, Santi (quiesence), is the highest development of human feeling and often conflated with Karuná (tenderness). Santi is excluded from the domain of music because 'it is not capable of being reflected by the art'.[112] The evolutionization of Hindu musical emotion occurs elsewhere in Tagore's work, especially in regard to scales. He notes that Sanskrit authors derive the musical scale from 'the cries of birds and beasts', adding that 'the cries of birds have furnished hints for several musical compositions in Europe'.[113] They appear in the following order, progressing, albeit unevenly, up the Great Chain of Being from large bird to large mammal: C, Peacock; C to D, Bull; C to E, Goat; C to F, Jackal; C to G, Kokil (a sweet-voiced Indian bird); C to A, Horse; and C to B, Elephant. The six original ragas are said to omit the seventh degree from their scales because the pentatonic scale was common in Asia at an early period in time. Whereas Western scales include seven pitches, C, D, E, F, G, A and B, Indian uses B selectively. Quoting Engel Tagore likens scales to the developmental position of children: 'Children, in their first attempts to repeat the diatonic scale after it has been sung to them are

[111] Tagore, *Universal History of Music*, Addenda, iii–iv.
[112] Tagore, *Universal History of Music*, Addenda, iii–iv.
[113] Tagore, *Universal History of Music*, 9.

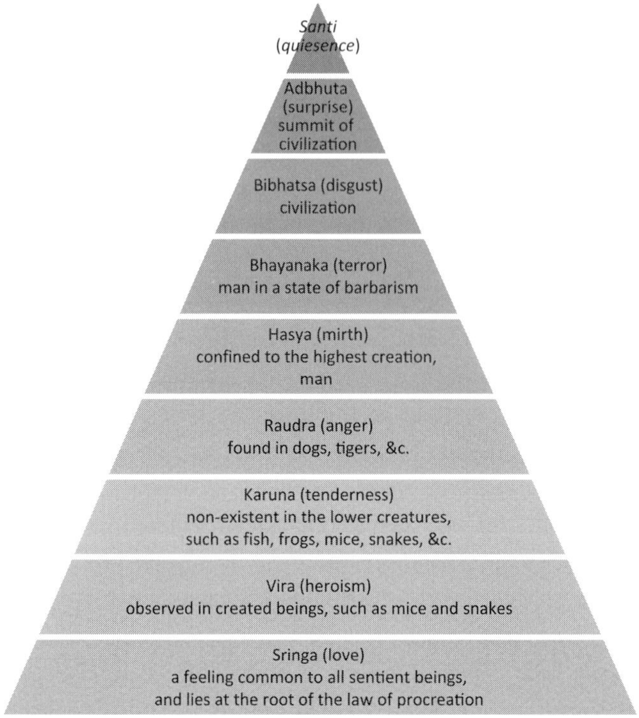

Figure 2.1 Tagore's Great Chain of Musical Being

apt to omit the *fourth* and *seventh*'.[114] *Six Principal Ragas* develops this Engelian precept:

The sounds of animated nature, especially the songs of birds, appear to be another source from which the formation of peculiar scales has been originally derived. The melodious notes of singing birds delight the savage as well as, and perhaps even more than the civilized man who cultivates music as an art. The former is familiar with them from his childhood. He unconsciously receives his musical instruction in the field and forest. The natural instinct for imitation, generally so powerful in the savage, soon leads him to produce intervals similar to those which he hears. He finds his success in the chase facilitated by imitating the sounds of the animals which he desires to capture. His imitation must be so exact as to be deceptive. Thus his ear and his voice are practised and become accustomed to certain minute intervals, which are involuntarily transferred by him to his primitive songs, and from which, in the course of time, a fixed order of intervals is developed, and gradually extended to the octave, constituting a scale.[115]

114 Tagore, *Universal History of Music*, appendix, iii.

115 Engel, *Introduction to the Study of National Music*, 60, cited in Sourindro Mohun Tagore, *Six Principal Ragas with a Brief View of Hindu Music*, Delhi: Low Price Publications, 1877/2006, 12–13.

Although developmentally early the Indian scale is civilized by the advanced system of emotional meaning of the rasas, and from musical meaning larger evolutionary structures begin to take shape. A *Universal History of Music* (1896) maps this phylogenic process onto an historical level. Tagore divides his book into sections according to the five continents, beginning with Asia, and then progressing to Africa, Europe, America and Oceania (Australia). There is an introduction, preliminary information on national music, and a brief essay on the music of savage nations. Each subsection is roughly chronological in relation to Western historical knowledge, thus Asia begins with China, which 'is one of the oldest countries in the world'.[116] The first three sections provide a roughly chronological movement from the Old World of Asia (including India) to Africa and the New World of Europe. The last two sections mirror these conceptually like an ideological palindrome, regressing chronologically from the New (yet frequently savage) World of America to the aboriginal cradle of civilization in Oceania. Contained within Asia, like Russian dolls, Indian musical history is stacked into Hindu, Mahomedan and British periods. Figure 2.2 reveals how Tagore situates historiography along popular Comtean lines.

Here Tagore reflects Engel's use of a Spencerian brush with a Comtean palette in *An Introduction to the Study of National Music* (1866):

> Uncivilized nations consider scarcely anything worthy of note except extraordinary events, which surprise by their novelty or by their immediate and startling effects ['theological'] . . . The conviction that the gradual development of the mental and moral faculties of the people especially deserves consideration, because it affects most deeply the destiny and happiness of man, gains ground only after much experience, and with the increase of knowledge ['metaphysical'] . . . The interest evinced, since the beginning of the present century, by several European nations in the popular songs, folk-lore, and other monuments of the mental condition of man in different parts of the world, is therefore a sign of progress not less delightful than the most important discoveries which have been made through the agency of practical science ['scientific'].[117]

Comte is also an integral component in John Frederick Rowbotham's three-volume *A History of Music* (1885) and C. Hubert H. Parry's *The Evolution of the Art of Music* (1893). Rowbotham divides musical prehistory in three: 'the three stages in the development of Prehistoric Music, [are] the Drum Stage, the Pipe Stage, and the Lyre Stage, which, it seems

[116] Tagore, *Universal History of Music*, 22.
[117] Engel, *Introduction to the Study of National Music*, vii.

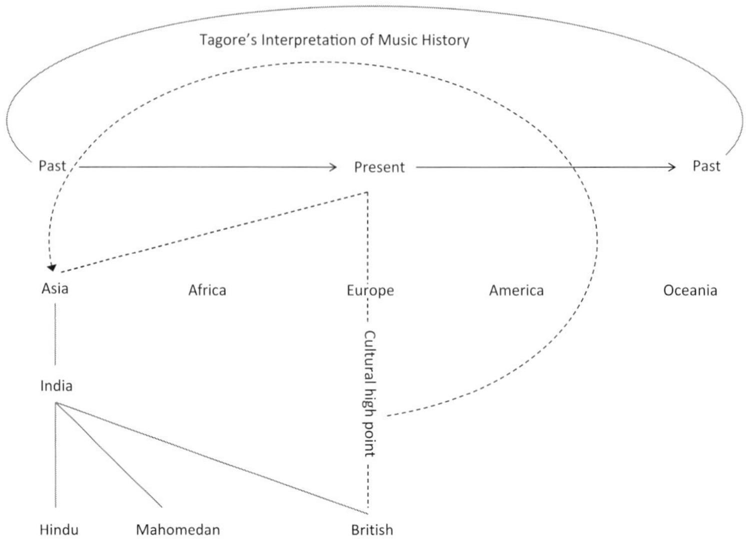

Figure 2.2 Tagore's Interpretation of Music History

to me, are to the Musician what the Theological, Metaphysical, and Positive Stages are to the Comtist, or the Stone, Bronze, and Iron Ages to the archaeologist'.[118] Parry echoes this in *The Evolution of the Art of Music*:

Taking the most comprehensive view of the story of musical evolution, it may be said that in the earlier stages, while the actual resources were being developed and principles of design were being organised, the art passed more and more away from the direct expression of human feeling ['theological']. But after a very important crisis in modern art, when abstract beauty was specially emphasised and cultivated to the highest degree of perfection, the balance swung over in the direction of expression again ['metaphysical']; and in recent times music has aimed at characteristic illustration of things which are interesting and attractive on other grounds than mere beauty of design or of texture ['scientific'].[119]

Because for Tagore India is sheathed within Asia – itself sheathed within the broader sweep of a Western history, again sheathed within a universal history – Hindu music forms a germinal, ontogenic core in

[118] John Fredrick Rowbotham, *A History of Music*, 3 vols., London: Trübner & Co., 1885, vol. 1, 20.

[119] C. Hubert H. Parry, *The Evolution of the Art of Music*, ed. H. C. Colles, New York: Greenwood Press, 1893/1968, 18.

the narration of his uncompromisingly phylogenic story: Hindu music progresses from Hindu to Muslim to British within a history progressing from Asia to Western to Universal. But the music of the Hindu period is invariably 'divinity itself'.[120] And so refracted through a Comtean lens the British period begins in the recapitulationary 'cradle of much of the festive and periodical songs that are sung in connection with Krishna's career'.[121] Yet by the end of the most recent period the British have facilitated 'the revival of Hindu music and its restoration to its pristine glory and purity'.[122] Tagore follows the conflicted Comteanism of many other Hindu nationalists by presenting Hindu culture (and by extension music) as an idealized form gradually less purified – yet increasingly resuscitated – through waves of foreign imperial domination. Jasodhara Bagchi describes Comtean Hindu nationalism as an ambivalent model of order and progress – modelling a progressive vision of industry and commerce against an idealized notion of Hindu order.[123] *Universal History* fits neatly into Bagchi's model: 'The [Hindu] challengers did not want to overthrow the existing Imperialist power structure, but to extend their hegemonic control over the intellectual and material conditions in society and thereby subvert Imperialist domination'.[124]

Self-conflict and determination also mark *Hindu Music* (1875; 2nd ed., 1882), Tagore's compilation of English-language sources on Hindu music from William Jones to the 1880s. Tagore obviously chose authors for their political as much as musical value. Indianist Willard, for example, helpfully decries the Muslim period in Hindu culture as representative of cultural decline:

> The conquest of Hindoostan by the Mahomedan princes forms a most important epoch in the history of its music. From this time we may date the decline of all arts and sciences purely Hindoo, for the Mahomedans were no great patrons to learning, and the more bigoted of them were not only great iconoclasts, but discouragers of the learning of the country.[125]

120 Tagore, *Universal History of Music*, 50. 121 Tagore, *Universal History of Music*, 60.

122 Tagore, *Universal History of Music*, 89.

123 Jasodhara Bagchi, 'Order and Progress: The Reception of Comteian Positivism in Nineteenth-Century Bengal', in G. R. Taneja and Vinod Sena (eds.), *Literature: East and West*, Bombay: Allied Publishers Ltd, 1995, 158.

124 Bagchi, 'Order and Progress', 159.

125 N. Augustus Willard, *A Treatise on the Music of Hindoostan: Comprising a Detail of the Ancient Theory and Modern Practice*, Calcutta: Baptist Mission Press, 1834, in Tagore, *Hindu Music*, 120.

It comes as no surprise as well that Tagore's article for the *Hindoo Patriot* precedes Engel's 'On the Musical Scales of different Nations' excerpted from *Introduction to the Study of National Music*. Tagore held fast to his belief that Hindu music should be written in Bengali, not Western style, claiming sensibly from his standpoint that

> Every nation that has a music of its own has also its own system of notation for writing it. Whether that system be an advanced one or not cannot be correctly expressed in the notation of another nation, however improved and scientific it may be . . . Every civilized nation, that has a music of its own, has also a system of notation adapted to the peculiarities of that music.[126]

In this way Tagore paraphrases Engel paraphrasing Spencer: 'The music of every nation has certain characteristics of its own. The progression of intervals, the modulations, embellishments, rhythmical effects, &c. occurring in the music of extra-European nations are not unfrequently too peculiar to be accurately indicated by means of our musical notation'.[127]

Like Engel, for Tagore the study of national music 'shows the different stages of progress which music has made in different countries'.[128] Like most non-Darwinian revolutionists Tagore exploits the muddle of Victorian evolutionary thought to suit his ideological purposes. His nationalism shows that he does not accept the Spencerian notion of an inherent, natural inequality of individuals, preferring to believe that, as in the case with Hindu national music, evolution – however competitive – has the capacity to restore as much as destroy. For Tagore, as Spencer, the nation begins in the individual mind, but for Tagore that mind is born of the natural, divine world and represents animal as much as man. Though Tagore accepts Spencer's concept of music *as* civilized, his evolutionism refutes the radical distinction Spencer conceives between the animal and human world. Hindu music is and represents the perfectly integrated, inextricably linked natural and spiritual worlds in evolutionary ascent.

India is an analogy for that relationship, not only recapitulating the individual in the nation, the immanent in the transcendent, the simple

[126] Sourindro Mohun Tagore, 'Hindu Music from the "Hindoo Patriot"', in *Tagore, Hindu Music*, Delhi: Low Price Publications, 1882/1994, 366, 380.

[127] Carl Engel, 'Music', in *Notes and Queries on Anthropology, For the Use of Travellers and Residents in Uncivilized Lands [Drawn up by a Committee appointed by the British Association for the Advancement of Science]*, London: Edward Stanford, 1874, 110.

[128] Tagore, *Universal History of Music*, 12.

in the complex, but the animal within man and the man within God. The preservation of that formative evolutionary relationship impinges upon the evolution of national identity: 'National music means a faithful expression of national feelings, and these feelings are best manifested under circumstances that are not controlled by extraneous influences'.[129] 'National music' is not, in other words, a contradiction in terms, but an example of the evolutionary process at work. Hindu music, as national music, is by its nature fittest. By definition all national music is fittest, because nation denotes civilization and civilization denotes music. Though civilized, however, national music is only found amongst 'the less educated classes',[130] and as such represents a lower order of music echoing the racial Otherness of Spencerian Social Darwinism. Inevitably, Tagore's non-Darwinian confusion imbeds only some precepts of Social Darwinism. Playing on the trope of childness, for example, he describes the savage Mexican *Ajacaxtli* as an instrument 'little better than a child's rattle'.[131] But his savage is less savage than Spencer's because man's anthropological spectrum starts with animals, not in a primordial human notionally higher up the Great Chain of Being. Tagore's savage is arguably more like Engel's than Spencer's. He is savage but not entirely undeveloped, and because of that holds the key to the comparative understanding of the evolution of music.

Conclusion

Tagore reinterprets Social Darwinism in a febrile atmosphere of scientific, theological, political, social and technological reinterpretation – the heyday of the non-Darwinian revolution. Darwin's notion of 'common descent' frightened uncomprehending Victorians but Spencer comfortingly repackaged it with his crucial, but ultimately discreditable addition of Social Darwinism. Born in the image of its father Victorian ethnomusicology clung to Spencer but it would not be long before Spencer would come under attack. The first salvo came from anti-Spencerian – though not Anti-Social-Darwinist – Edmund Gurney. No less confused than any other evolutionist of the time Gurney takes a Darwinian view of

[129] Tagore, *Universal History of Music*, 11. [130] Tagore, *Universal History of Music*, 11.
[131] Tagore, *Universal History of Music*, 15.

musical origins but a Social Darwinian view of music. Music may have evolved from music (not speech), but if so savage music *is* therefore music. And if savage music is music the evolution of music must express common descent: 'it is open to us to suppose that our human or semi-human ancestors, having been naturally moved by physiological causes to desert the region of neutral tones in order to be pleasingly emphatic, may have utilised for independent musical purposes the material of increased resonance so discovered and developed'.[132] Gurney maintains that 'the primary and essential function of Music is to create beautiful *objective forms*'[133] but as we know from Spencer's three essays on primitive man (physical, emotional, intellectual) savages lack 'those most complex emotions'; they have 'no conceptions [*sic*] of *general facts*'; and their imagination is '*reminiscent* only – and *constructive* . . . the mind merely receives and repeats – cannot initiate, has no originality'. In fact apart from his disagreement over musical origins Gurney appears to embrace the spirit if not the letter of the Spencer hypothesis:

> To a developed musical ear indefinite transitions without the formative element of fixed degrees are quite devoid of interest; but in any embryonic use of the voice for purposes of pleasure, as among young children or in the grown-up childhood of savage life, the mere change of note, differentiated from other changes of sense-impression, *e.g.* of colour, by the peculiar sense of advance, is very clearly enjoyable. The fundamental point seems to be the *change*: that the change should very commonly be to a related note, as the octave or the fifth, may be, in the earliest stage, a secondary and negative provision, due to the fact that if it were to some unrelated point of pitch, this last would be felt as disagreeably discrepant and unexpected. And much as the more advanced stages of the Ideal Motion may perplex us, this rudimentary pleasure in change seems specially easy to understand in the case of tone, owing to the peculiarly exceptional and intermittent, and therefore peculiarly exciting effect, of this form of sense-stimulation.[134]

Although he admits that 'we cannot judge music with the savage ear till we can remake ourselves into savages,'[135] Gurney reiterates and refines the implacable contradiction of Social Darwinism between the ontogeny of the individual and the phylogeny of the race. Other anti-Spencerian Darwinians would achieve similarly confusing results. Darwinian ethnomusicologist Richard Wallachek also adamantly opposes Spencer's theory

[132] Edmund Gurney, 'On Some Disputed Points in Music', *Fortnightly Review* (July 1876), 107–8.
[133] Edmund Gurney, *The Power of Sound*, London: Smith, Elder, 1880, 490.
[134] Gurney, *The Power of Sound*, 141–42. [135] Gurney, *The Power of Sound*, 492.

of musical origins, and like Gurney espouses a racial interpretation of musical savagery through the comparative method. The subtitle of *Primitive Music* (1893) gives the clue away: *Primitive Music: An Inquiry into the Origin and Development of Music, Songs, Instruments, Dances, and Pantomimes of Savage Races*: 'Now we have seen that even uncivilized races know how to accompany a simple song by ear, while some of the more civilised ones, as the Chinese and other oriental people, do not understand our harmony, although they have every opportunity of hearing our music. Thus the difference between people with and without harmonic music is not *a historical but a racial one*'.[136] *Primitive Music* is testimony not only to the longevity of the comparative method but to the enduring racial inequality it helped sustain into the late Victorian period. Wallaschek contends that although the average child of civilized parents is not necessarily born 'with higher mental equipment than the little savage' nevertheless 'it comes into another world, into a world with a settled tradition. Into this the child gradually grows, and out of it its individuality is formed, while a negro boy will always have the impression that the European world is something strange to him, not made by his equal. His very colour, if not the general behaviour of his white "friends," will soon convince him of this'.[137] The racial Otherness of the comparative method was epitomized in more than Gurney and Wallaschek of course, as ethnomusicology proved again and again that savage races existed in a teleological space and time ontogenically related to the phylogeny of civilization.

The sliding scale of savagery is abundant in J. Donovan's *From Lyre to Muse: A History of the Aboriginal Union of Music and Poetry* (1890) when certain types of death songs originate in the lowliest savage imaginable: 'The most extreme example of this is perhaps to be found in savage death-songs. Even in the most savage of savage races the death-songs sung by dying warriors glow with a spirit of self-glorification'.[138] Ethnomusicology continued to prosecute this paradigm of racial Otherness until the appearance of Charles Samuel Myers (1873–1946), the founding father of British ethnomusicology. Myers studied medicine at Cambridge and as an amateur violinist took an active part in the musical life of the university, eventually becoming president of the University Music Club. In 1895 he

136 Richard Wallaschek, *Primitive Music: An Inquiry into the Origin and Development of Music, Songs, Instruments, Dances, and Pantomimes of Savage Races*, London: Longmans, Green, and Co., 1893.

137 Wallaschek, *Primitive Music*, 280.

138 J. Donovan, *From Lyre to Muse: A History of the Aboriginal Union of Music and Poetry*, London: Kegan Paul, Trench, Trübner, & Co., Ltd, 1890, 69.

was asked by the distinguished Cambridge anthropologist A. C. Haddon to accompany him and others on a multi-disciplinary expedition to Torres Straits (New Guinea) and Sarawak (Borneo). He finished his studies in 1898 and headed to Torres Straits, where he began his ethnomusicological research. In 1899 he returned to England and subsequently held important posts as House Physician at St. Bartholomew's Hospital, London; Cambridge University Demonstrator and later University Lecturer and Reader in Experimental Psychology; Professor at King's College London; President of the Psychology section of the British Association, and one of the founding editors of *The British Journal of Psychology*. Myers's contribution to ethnomusicology stems from his seven months in the Torres Straits where he was responsible for testing on hearing and music and musical instruments, eventually contributing to two of six volumes of the expedition, Volume 2 (*Physiology and Psychology*, 1901 and 1903) and Volume 4 (*Arts and Crafts*, 1912).

In the Torres Straits Myers developed a theory of 'individual differences', the study of what makes people different from one another. He determined that individuality is a function of progress and improvement but elided in Spencerian Social Darwinism the comparative method failed to identify or recognize their respective scientific and ethical differences: 'Every step in evolution, *i.e.*, every change making for increased differentiation of function, increased co-ordination of parts and increased integration of previously independent units, may be accepted as progress. But improvement implies something more – namely, "betterment." Unlike progress, improvement is bereft of any scientific, objective criterion. We have to judge of improvement by subjective, ethical criteria'.[139] By purposefully separating the relationship of improvement from progress Myers tried to disentangle the crucial fifth tenet – the synthetic philosophy – from Social Darwinism, explicitly negating the concept of innate racial difference: 'For my part, I cannot but believe that there are profound racial mental differences which have thus produced, and been maintained by, cultural differences, and that these racial differences will persist long after attempts have been made, as modern civilisation is not attempting, to bring all races under the same social environment'.[140] On the basis of his experiments on sensory acuity on both natives of Torres

[139] Charles Samuel Myers, 'Human Improvability', in *A Psychologist's Point of View: Twelve Semi-Popular Addresses on Various Subjects*, London: William Heinemann (Medical Books), 1933, 1–2.

[140] Myers, 'Human Improvability', 10.

Straits and Britons at home in England Myers arrived at the conclusion that difference can only originate in environment: 'That the relation between the organism and its environment (considered in its broadest sense) is the ultimate cause of variation, bodily and mental . . . [and] That this being admitted, the possibility of the progressive development of all primitive peoples must be conceded, if only the environment can be appropriately changed'.[141] The distinction between progress and improvement is important for Victorian ethnomusicology because it freed Myers to evolve a semi-Darwinian theory of cultural adaptation unencumbered by the socio-political linguistics of Social Darwinism. Accordingly, all music progresses if culturally improved and if culturally improved it is identifiably different: 'improvements do not arise broadly and directly from the fact that most members of that race or people are innately improving, but rather from the fact that the social heritage, the civilisation [*sic*] into which they are born is improving'.[142] Identifiably different music arrives at the apex of a universal evolutionary process, beginning with the discrimination between as noises and tones, and advancing from the awareness of differences in loudness, pitch, duration, character and quality; to the awareness of absolute pitch; the appreciation and use of (small) approximately equal tone-distances; the appreciation and use of (larger) consonant intervals and the development of small intervals in relation thereto; melodic phrasing; rhythmic phrasing; and ultimately arriving at the cultural triumph of musical meaning.[143]

Myers sedulously avoids terminology which compromises the universality of his conception. Harmony, for example, is studiously avoided in his evolutionary summary, and even when it is discussed elsewhere he redefines it to include sounds that are either simultaneous (in Western music) or consecutive (in primitive music).[144] In *Reports of the Cambridge Anthropological Expedition to Torres Straits* (1912) Myers explores the evolution of musical meaning through advances in rhythmic and

[141] Charles Samuel Myers, 'On the Permanence of Racial Mental Differences', in G. Spiller (ed.), *Papers on the Inter-Racial Problems Communicated to the First Universal Races Congress held at The University of London July 26–29, 1911*, London: P. S. King & Son, 1911, 73.

[142] Myers, 'Human Improvability', 16.

[143] Charles Samuel Myers, 'The Beginnings of Music' (1913), in Charles Samuel Myers (ed.), *A Psychologist's Point of View*, London: William Heinemann (Medical Books), 1933, 196.

[144] Bennett Zon, 'More than the Sum of the Life of Its Several Parts: Individual Differences in the Psychology and Ethnomusicology of C. S. Myers', *Nineteenth-Century Contexts* 29/4 (Dec. 2007), 381.

harmonic complexity. The more complex, the more evolved; and by implication the more evolved, the more meaningful:

> The songs of the Miriam or Murray Islanders, which form the subject of this section, are of considerable interest from the standpoint of musical history and development. For they differ among one another not only in complexity of structure but also in date of composition and place of origin. They thus afford an opportunity of tracing the changes in musical expression which may occur in course of time within a primitive community.[145]

To describe this process of increased complexity Myers develops an evolutionary theory called 'accrescence', the first theory of its kind to provide Spencer with any genuinely systematic competition. Spencer claims that evolution consists 'essentially in the addition of matter to matter, *i.e.*, in increasing integration and diversity' but according to Myers '*integration is not the primary aspect of evolution*. What is primary is the "carving" of new parts out of the old. The old may grow in the process; it may and does by *accrescence* add new material to itself. It also combines into larger unity by union with others'.[146] For Myers and later psychologists accrescence produces individual differences and individual difference produces variation in musical traditions. And with an evolutionary premise orientated on variation – rather than permanence – the whole edifice of the Spencer hypothesis, Social Darwinism and the comparative method begins to strain under its own weight. For ethnomusicology – the accrescence of Victorian anthropology – that change would usher in a new period of interpretation, replacing the ethnomusicology of Engel, Tagore and many others. The very concept of nation and race would be revised, and individual differences would put equality on the ideological map. A long-standing victim of scientific prejudice, the evolutionary Other – the Unfit and his music – would emerge from Victorian culture more equal than before, more varied and ultimately more fit.

145 Charles Samuel Myers, 'Music', in Alfred C. Haddon (ed.), *Reports of the Cambridge Anthropological Expedition to Torres Straits*, vol. 4, Arts and Crafts, Cambridge: Cambridge University Press, 1912, 238.

146 Charles Samuel Myers, 'Principles of Development', in Charles Samuel Myers (ed.), *A Psychologist's Point of View*, 25, 26.

3 | Folk Musicology

Introduction

By definition something cannot be defined by what it is not. But in the history of evolutionary thought not everything is as it seems, and what looked to everyone like a Darwinian revolution was in fact a myth; according to Peter Bowler it was a 'non-Darwinian revolution'. Few figures can, like Darwin, be defined by what they are not, but there is another Victorian thinker, introduced in the last chapter, whose impact was arguably as great as Darwin's. At least within his own time E. B. Tylor was the most important anthropologist for thirty years.[1] Tylor's great contribution to anthropology was the anthropological equivalent of natural selection, the doctrine of survivals, and like Darwin, Tylor occupied a hegemonic position in Victorian intellectual culture. But like Darwin's theory of natural selection, the doctrine of survivals was, I would argue, constantly reinterpreted by contemporaries. Just as Bowler observes a 'non-Darwinian revolution' I believe there is also a parallel 'non-Tylorian revolution'. Inevitably there is one substantial difference, however: Tylor may have become the father of modern anthropology but many aspects of his theory are now discreditable; Darwin survived and flourished. This chapter explores how the 'non-Tylorian revolution' affected musical culture through an examination of late Victorian definitions of folk song. It begins by investigating and testing the theoretical consistency of Tylor's doctrine of survivals and its influence over the Folk-Lore Society's definition of folklore; it then traces the movement and dispersion of those definitions in early folk song collections and the Folk-Song Society, and culminates in a close reading and redefinition of Cecil Sharp's non-Tylorian attitudes towards folk song, along with a reappraisal of second folk song revival criticism. A conclusion examines briefly the legacy of Sharp's work in early twentieth-century writing.

[1] James Urry, *Before Social Anthropology: Essays on the History of British Anthropology*, Reading: Harwood Academic Publishers, 1993, 21; John Monaghan and Peter Just, *Social and Cultural Anthropology: A Very Short Introduction*, Oxford: Oxford University Press, 2000, 35.

Folklore as Survival

Tylorian Survival

By almost all accounts one of the most profound influences on the Victorian study of folk culture is that of E. B. Tylor and his theory of survivals. Developed to an apogee in *Primitive Culture* (1871), Tylor's theory bequeathed anthropology a metaphor to represent Britain's commanding geo-political reality. The deep past became an even more foreign country, to paraphrase David Lowenthal,[2] ripe for metaphysical conquest proving through the science of culture the evolutionary superiority of modern (European) man.[3] As observed in the previous chapter, Tylor maps the past with sweeping terminology; survivals are 'processes, customs, opinions, and so forth, which have been carried on by force of habit into a new state of society different from that in which they had their original home, and they thus remain as proofs and examples of an older condition of culture out of which a newer has been evolved'.[4] The peasant using a handloom after the discovery of a 'flying shuttle'; the Midsummer bonfire; the Breton peasant's All Soul's supper; and words onomatopoeiacally imitative of musical instruments like pipe, trumpet or tom tom – these are just a few typical examples of survivals. While modern culture retains vestiges of the past not all vestiges remain irrepressibly submerged, however, and some even become resurgent:

> The serious business of ancient society may be seen to sink into the sport of later generations, and its serious belief to linger on in nursery folk-lore, while superseded habits of old-world life may be modified into new-world forms still powerful for good and evil. Sometimes old thoughts and practices will burst out afresh, to the amazement of a world that thought them long since dead or dying; here survival passes into revival.[5]

Tylor calls survivals 'mines of historic knowledge'[6] but like all mines they attract danger. For Tylor the danger was developmental. Struggling to explain the residue of primitiveness in modern culture, Tylor opted for a progressive evolutionary explanation which considered survivals archaisms

[2] See David Lowenthal, *The Past is a Foreign Country*, Cambridge: Cambridge University Press, 1985.

[3] Laavanyan Ratnapalan, 'E. B. Tylor and the Problem of Primitive Culture', *History and Anthropology* 19/20 (June 2008), 131.

[4] E. B. Tylor, *Primitive Culture: Researches into the Development of Mythology, Philosophy, Religion, Language, Art and Custom*, 2 vols., New York: Henry Holt and Co., 1889, vol. 1, 16.

[5] Tylor, *Primitive Culture*, vol. 1, 16–17.

[6] Tylor, *Primitive Culture*, vol. 1, 71.

in an otherwise teleological advance. It was that explanation, broadly speaking, which would impact upon the study of folk music in Victorian musical culture.

Pitting himself against contemporary degenerationists (who believed that modern savages degenerated from civilized societies and could improve only with external intervention) Tylor placed these archaisms in an anti-degenerational continuum, beginning with savage culture and rising to the incontrovertibly superior heights of civilization. This justified his belief – echoed throughout Victorian musical culture – that 'Civilization, being a process of long and complex growth, can only be thoroughly understood when studied through its entire range; that the past is continually needed to explain the present, and the whole to explain the part'.[7] Here and elsewhere Tylor echoes Comte, concluding that the 'civilization of any age is not a new creation to meet the wants of that age, but is a result of past times, modified to meet new conditions of life and knowledge'.[8] Tylor's problem was more than one of language; it was one of belief. Juggling the teleology of progression and the natural selection of Darwin Tylor – unpopularly for some – dismissed civilized religion as primitive superstition, and religious superstition as a persistent competitor for the modern, scientific minds of man. Religion, like other forms of culture such as folk music, was a survival:

> In the workings of the minds of these early tribes, we trace a childlike condition of thought in which there is a wonderful absence of definition between past and future, between fact and imagination, between last night's dream and to-day's waking. Out of this state of mind we find arising all over the world a consistent, intense, and all-pervading spiritualism to form a basis upon which higher intellectual stages have been reared. In this low and early mental state there reigns supreme the faculty of association of thoughts. Out of this, when unchecked by experience, arise those delusions of sorcery which pervade and embitter the whole life of the savage, and carry a stream of folly far on into the culture of the higher races. But through age after age there has gone on a slow progress of natural selection, ever tending to thrust aside what is worthless, and to favour what is strong and sound.[9]

[7] E. B. Tylor, *Researches into the Early History of Mankind and the Development of Civilization*, London: John Murray, 1870, 2.

[8] E. B. Tylor, 'On the Survival of Savage Thought in Modern Civilization', *Notices of the Proceedings at Meetings of the Members of the Royal Institution of Great Britain* 5 (1869), 533.

[9] E. B. Tylor, 'On Traces of the Early Mental Condition of Man', *Notices of the Proceedings at Meetings of the Members of the Royal Institution* 5 (1867), 92.

Tylor may have believed that 'there is not much difference between the old heathenism and the new Christianity'[10] but equally his Darwinian credo lacked conviction, or at the very least systematic expostulation. Survivals like traditional ceremonies or musical onomatopoeia might have been recognized as 'obsolete stock that had failed to be thrown out'[11] but they were also doggedly persistent – by definition they were fittest survivors, not just survivals. In Darwinian terms – in the terms of natural selection – they must be fit, competitive, adaptive, yet paradoxically in developmental terms they evoked the superstitious forms and belief structures of primitive man. Tylor never resolves that tension; indeed, some have argued that he studiously deflects attention from it, preferring instead to elucidate patterns of development rather than mechanisms of change. In her astonishingly still-current article on the history of survivals Margaret Hodgen attributes this to differences in biological and anthropological method. Darwin focuses on living things – biological organisms, creatures adapting to survive; Tylor, however, studies human culture – customs, beliefs and performative processes like music, dance, drama and poetry surviving effectively un-adapted: 'while the concept of survival solved a problem for Darwin, and referred to the living, the new, organically fit, appropriate, consistent, and harmonious; survival in culture remained for Tylor a problem to be solved, and referred to the non-living, the old, the culturally unfit, inappropriate, inconsistent and illogical'.[12]

Tylor's determination to counter degeneration accounts partially for this inconsistency. Resolving to debunk stalwart degenerationists who would vilify primitive peoples, he hoped to discover even lower forms of culture, conveniently raising the positions of savages and blacks from semi-human ranking. Evidence remained admittedly and consistently wanting, however, and to survive the doctrine of survivals acquired a moralistic agenda. As Hodgen notes, *Primitive Culture* ends optimistically asserting that it is the 'office of ethnography to expose the remains of crude old culture which have passed into harmful superstition and to mark these out for destruction'.[13] Victorian self-improvement saved the day but just who determined

[10] E. B. Tylor, *Anahuac, or Mexico and the Mexicans, Ancient and Modern*, London: Longman, Green, Longman, and Roberts, 1861, 289.

[11] Timothy Larsen, 'E. B. Tylor, Religion and Anthropology', *The British Journal for the History of Science* 3 (September 2013), 474.

[12] Margaret T. Hodgen, 'The Doctrine of Survivals: The History of An Idea', *American Anthropologist* 33 (1931), 307–24, 1, http://aaanet.org/sections/gad/history/042hodgen.pdf, accessed 2/9/13.

[13] Tylor, *Primitive Culture*, vol. 2, 453.

what remained and what was destroyed was another matter. Certainly it was not left to the objective ethnographic verdict of Tylor alone, but to the increasingly professionalized and academically empowered environment of the Folk-Lore Society, and sometime later, its musical counterpart the Folk-Song Society.

Defining Folklore

Tylor's evolutionary dialectic translated easily into the receptive study of folklore and folk song. Though confirmed in his belief that 'the main tendency of culture from primaeval *[sic]* up to modern times has been from savagery towards civilization'[14] he also contended antithetically that 'The history of survival in cases like those of the folk-lore and occult arts . . . has been for the most part been a history of dwindling and decay'.[15] According to Alan Dundes this devolutionary premise subsisted well into the 1930s, if not later, when Ruth Benedict mused that 'in a strict sense folk-lore is a dead trait in the modern world'.[16] What Dundes deems devolutionary, however, others call evolutionary. Tylor was himself uncertain:

> As men's minds change in progressing culture, old customs and opinions fade gradually in a new and uncongenial atmosphere, or pass into states more congruous with the new life around them. But this is so far from being a law without exception, that a narrow view of history may often make it seem to be no law at all. For the stream of civilization winds and turns upon itself, and what seems the bright onward current of one age may in the next spin round in a whirling eddy, or spread into a dull and pestilential swamp. Studying with a wide view the course of human opinion, we may now and then trace on from the very turning-point the change from passive survival into active revival.[17]

When the Folk-Lore Society was founded in 1878 'revival' was as much on the minds of folklore and folk song collectors as 'survival'. The term 'folklore' was coined in 1846 by William John Thoms (folk song would be coined, anonymously, in 1847) in an article in the *Athenaeum* intending to replace 'popular antiquities or popular literature with an English name

[14] Tylor, *Primitive Culture*, vol., 1, 21. [15] Tylor, *Primitive Culture*, vol. 1, 136.

[16] Ruth Benedict, 'Folk-Lore', in *Encyclopaedia of Social Sciences*, Edwin R. A. Seligman and Alvin Johnson (eds.), 15 vols., London: Macmillan Publishers, 1931–67, vol. 5 (1931), cited in Alan Dundes, 'The Devolutionary Premise in Folklore Theory', *Journal of the Folklore Institute* 6/1 (June 1969), 13.

[17] Tylor, *Primitive Culture*, vol. 1, 136–37.

equivalent to the German *Volksmärchen*'.[18] Thoms's seminal article sets the tone for the century: 'No one who has made the manners, customs, observances, superstitions, ballads, proverbs, &c., of the olden time his study, but must have arrived at two conclusions: – the first, how much that is curious and interesting in these matters is now entirely lost – the second, how much may yet be rescued by timely exertion'.[19] The leap Thoms took catapults us from the increasingly superannuated history of literature and myth to the modern, progressive science of culture itself divided across different evolutionary allegiances. Immersed in conflicted Tylorian principles, the folklore grandee Edwin Hartland reflects this tension when after exhausting semantic squabbling he arrives at the following compromise:

> Dealing with thought in its primitive forms, it [folklore] traces it downwards from the higher civilizations where it is exhibited in the conscious logic and historical religions, institutions, arts, science, and literature of the progressive races, to its earliest and lowest manifestations in the forefathers, not only of the Indo-European, but also of the Semitic and Turanian tribes and in the barbarous and savage races of to-day ... Its business is with mankind in its infancy and childhood, when the untrained imagination was dominant and knowledge was purely empirical –, when men could only make futile guesses at the facts of their own natures and of the world about them, and when the organisation of society was as yet more or less rudimentary.[20]

With Thoms Director and Tylor an ordinary member (amongst many others) the Folk-Lore Society was originally founded as a publishing society,[21] an important impetus which would also influence the founding principles of the Folk-Song Society some twenty years later. Initially the Folk-Lore Society aimed to (1) reprint 'scarce books or articles on English Folk-Lore, and the collection and printing of scattered materials now existing in English olden-time literature'; (2) publish 'original communications on Folk-Lore'; (3) print 'accounts of Folk-Lore of the Colonies and of foreign countries'; and (4) collect and print 'the Folk-Lore of savage tribes'.[22] The first annual report was co-signed by the Society's President,

18 Elinor Spalding Lewis, 'Aboriginal Britain: Ethnological Folkoristics and the Origin of the Nation in the Work of Sir George Laurence Gomme, Sir John Rhŷs and Alfred Nutt', PhD diss., University of York, 2010, 3.

19 William John Thoms, 'Folk-Lore', *Athenaeum* 982 (22 August 1846), 862.

20 Edwin Sidney Hartland, 'The Science of Folk-Lore', *Folk-Lore Journal* 3 (1885), 118.

21 Allan Gomme, 'The Folk-Lore Society: Whence and Whither', *Folklore* 63/1 (March 1952), 6.

22 The Folk-Lore Society, First Annual Report of the Council, *The Folk-Lore Record* 2 (1879), 5.

the Earl of Verulam, and G. Laurence Gomme, the Society's Hon. Secretary.

It was Gomme who would become one of the central figures in Victorian folklore studies, becoming President of the Society from 1891 to 1894 and adapting Tylorian thought in essential reading like *Primitive Folk Moots* (1880), *Folklore Relics of Early Village Life* (1883), *The Village Community* (1890), *Ethnology in Folklore* (1892) and *Folklore as an Historical Science* (1908), as well as the Folk-Lore Society's own *Handbook of Folklore* (1890). Gomme's work on folklore, like Engel's work on music, has been the subject of much debate often because it, like Tylor's, was rooted in an increasingly contested and obsolescent yet paradigmatically developmental world view. Rather than studying folklore as contemporaries like Hartland did – as the 'universal psychological phenomena of man'[23] – Gomme claims that 'Folk-lore is a science by itself, with distinct work of its own to accomplish';[24] moreover, it is not

> only another name for anthropology. The sanction at the back of folk-lore is tradition. Thus traditional custom, traditional belief, traditional stories – and no custom or belief originating now, whether in civilized or savage races – can be defined as folk-lore. There can be no modern folk-lore, whereas the psychological phenomena with which anthropology deals exist now, and new facts will present themselves as society progresses.[25]

Cecil Sharp would reach a similar conclusion about the science of folk song: writing in 1907 he considers that 'the work is now being done in the right spirit, scientifically, accurately, and above all with a scrupulous honesty and conscientiousness; and this is scarcely the way in which it would have been approached a century or more ago'.[26] Despite reservations about Gomme's reliance on tradition, the historian of folklorists Richard Dorson admires Gomme's determination to create a new discipline distinct from anthropology.[27] Whether or not he succeeded in creating a new discipline, Gomme succeeded in shaping the study of folklore in the same way Engel shaped Victorian ethnomusicology – with the conflicting evolutionary principles of Tylorian thought. *Ethonology in Folklore*

[23] Lewis, 'Aboriginal Britain', 13.

[24] G. L. Gomme, in E. Sidney Hartland, C. Staniland Wake, Henry B. Wheatley and G. L. Gomme, 'Folk-Lore Terminology', *The Folk-Lore Journal* 2/11 (November 1884), 348.

[25] Gomme, in Hartland, Wake, Wheatley and Gomme, 'Folk-Lore Terminology', 349.

[26] Cecil J. Sharp, *English Folk-Songs: Some Conclusions*, London: Simpkin & Co., Ltd; and Taunton: Barnicott & Pearce, Athenaeum Press, 1907, vii.

[27] Richard Dorson, 'The Great Team of English Folklorists', *The Journal of American Folklore* 64/251 (January–March 1951), 7.

follows Tylor's example in its concern for systematic analysis and organization of material, not merely the subjective accumulation of documents 'arranged for no purpose but that of putting the facts pleasantly before readers'.[28] But for Gomme, as for Tylor, that cultural material is meaningless unless structured biologically. Thus anthropological culture is to Gomme and Tylor what ethnomusicological culture is to Spencer and Engel. Culture is not merely traceable evidence but a living organism:

> Each human group, having its specialised and dormant elements, must be treated as an organism and not as a bundle of separable items, each one of which the student may use or let alone as he desires. That which is anthropological evidence is the indivisible organism, and whenever, for convenience of treatment and considerations of space, particular elements only are used in evidence, they must be qualified, and the use to which they are provisionally put for scientific purposes must be checked, by the associated elements with which the particular elements are connected.[29]

For Gomme and Tylor, the organism of culture cast the material of folklore into disarray unless positioned within an overarching evolutionary context. And that evolutionary context was premised on the metaphysical conflation of space and time which marked the comparative method in which the 'folk' could be transmogrified into 'primitives'.[30] Because of this Gillian Bennett distinguishes Gomme from contemporaries like Hartland, who despite espousing a fundamentally Tylorian orientation, couched folklore in less historical, and more psychological, terms.

In the seventh Annual Report of the Folk-Lore Society Gomme continued to define folklore as 'the science which treats of the survival of archaic beliefs and customs in modern ages',[31] while Hartland, for example, maintained that 'folk-lore is anthropology dealing with the psychological phenomena of uncivilized man'.[32] The intellectual rift engendered by terminological divisions within the Society (resurfacing and unresolved in the Folk-Song Society) owes as much to inconsistencies within Tylorian evolutionary conceptions as it does to disciplinary distinctions materializing through the gradual professionalization of science.

[28] G. Laurence Gomme, preface, *Ethnology in Folkore*, New York: D. Appleton and Co., 1892, v.

[29] G. Laurence Gomme, *Folklore as an Historical Science*, London: Methuen & Co., 1908, 234.

[30] Gillian Bennett, 'Geologists and Folklorists: Cultural Evolution and 'The Science of Folklore', *Folklore* 105 (1994), 32.

[31] G. Laurence Gomme, 'The Folk-Lore Society, Seventh Annual Report of the Council', *The Folk-Lore Journal* 3/4 (1885), 388.

[32] Edwin Sidney Hartland, cited in Gomme, Folk-Lore Society, Seventh Annual Report, 386.

Implicitly (and daringly) because of this, Francis de Caro deprives Gomme of his evolutionism and makes a good case for describing him as an 'ethnohistorian': 'Gomme is most correctly viewed, not as a folklorist with an interest in history, but as an early ethnohistorian'.[33] As such, and unconvincingly for some, *Primitive Folk Moots*, for example, takes descriptions of outdoor meeting grounds amongst savage groups like the Iroquois and Hottentots, ancient and later cultures like the Hebrews and Icelandic, and applies them to written historical documents relating to England as well as verbal folklore such as Celtic folk tales and place names. The order of chapters reflects his Tylorian methodological sequence: (1) introduction, (2) examples from other lands, (3) the evidence of Early English records, (4) the revival of the primitive form, (5) the historical survival in England, (6) open-air courts in Scotland, (7) traditional and philological evidence, (8) evidence of place names and traditions in Scotland, and (9) some further evidence.[34] Later books, such as *Folk-Lore Relics* and *The Village Community*, would adopt much this same approach, substituting the village community for primitive moot – each seeking to locate the primitive or prehistoric in the context of early English cultural life.

Inevitably Gomme was not alone in casting folklore in a Tylorian mould. In *Folk-Lore Relics* Tylor's influence is seconded by another person, Andrew Lang, President of the Folk-Lore Society in 1891. When assembling his 'great team of English folklorists' Dorson makes Lang captain, followed in order of priority by Gomme, Hartland and other volubly Tylorian figures such as David Nutt, Edward Clodd and W. A. Couston. Dorson describes Lang as 'classical scholar, historian, biographer, poet, essayist, journalist, and anthropologist, whose most substantial work nevertheless belongs to folklore'.[35] For him Lang represented an incarnation of Tylorian thought even exceeding that of Gomme: 'the gospel of Tylor according to the exegesis of Lang'.[36] (In the next section the same will be said of Sharp in relation to Engel: 'the gospel of Tylor according to the exegesis of Sharp'.) But as both Dorson and Stocking concede Lang detected and exploited weaknesses in Tylorian thought and pressed beyond them, in Stocking's estimation propelling him from Tylorian folklore to

[33] Francis A. De Caro, 'G. L. Gomme: The Victorian Folklorist as Ethnohistorian', *Journal of the Folklore Institute* 19:2/3 (May–December 1982), 107.

[34] G. Laurence Gomme, *Primitive Folk-Moots; Or, Open-Air Assemblies in Britain*, London: Sampson Low, Marston, Searle & Rivington, 1880, ix–xi.

[35] Dorson, 'The Great Team of English Folklorists', 1–2.

[36] Richard Dorson, *The British Folklorists: A History*, Chicago: University of Chicago Press, and London: Routledge & Kegan Paul Ltd, 1968, 196.

primitive monotheism.[37] Lang's journey begins conventionally enough in Tylorian terms of survival:

Our method throughout will be to place the usage, or myth, which is unintelligible when found among a civilised race, beside the similar myth which is intelligible enough when it is found among savages. A mean term will be found in the folklore preserved by the non-progressive classes in a progressive people. This folklore represents, in the midst of a civilised race, the savage ideas out of which civilisation has been evolved. The conclusion will usually be that the fact which puzzles us by its presence in civilisation is a relic surviving from the time when the ancestors of a civilised race were in the state of savagery.[38]

To test this folklorists ventured 'to compare (with due caution and due examination of evidence) the myths of the most widely [geographically] severed races ... He will not be surprised if Greeks and Australian blacks are in the same tale'.[39] To test this on music Victorian ethnomusicologists ventured to compare the musics of the most widely severed, but their counterparts amongst folk song collectors would look more locally to the geographically proximate races already living in Britain.

Folk Song as Survival

Despite their differences Lang, Gomme and other survivalists presented conceptually unified, if theoretically divergent, methodologies reminiscent of the non-Darwinian revolution. Divergence would provide its own evolutionary advantage, however, as survivalist folklore studies eased into the often contradictory scientific truisms of the Victorian cultural mindset. Itself the product of often frictional ideological contradictions Tylorian survivals blossomed, meeting little resistance until Darwinian modes of evolution began to have greater purchase on the scientific imagination after the First World War. By the later part of the Victorian period survivals became *de rigueur*, not least amongst musical followers of folklore studies anxious to test the survivalist waters themselves. The Folk-Song Society was born a child of this same conflicted evolutionary predilection.

[37] George W. Stocking, Jr, *After Tylor: British Social Anthropology 1888–1951*, Madison: University of Wisconsin Press, 1995, 50–63.

[38] Andrew Lang, 'The Method of Folklore', *Custom and Myth*, London: Longmans, Green and Co., 1884/1910, 25.

[39] Lang, 'Method of Folklore', 23.

Like their counterparts in folklore early Victorian 'folk song' publications were actually often diverse collections lacking any substantive theoretical content. Admittedly debatable, yet functionally defined by A. David Gregory as 'vernacular', collections included popular music (especially street music), narrative and non-narrative music (broadly interpreted as folk song) and national song.[40] Popular street music comprises vocal and/or instrumental, and either an ephemeral yet popular song or a song which has endured generations. To be street music a song must enter oral tradition, urban or rural. Classed as folk song are narrative music – essentially ballads – tell stories; non-narrative music – primarily the folk lyric – including love songs, laments, songs of travel, drinking, sport and comic songs; and occupational songs. National songs are either patriotic in character, such as 'The British Grenadiers' and 'The Roast Beef of Old England', or lyrical in nature like 'Greensleeves' or 'The Miller of Dee'. Inevitably, early Victorian collections further blur already smudged taxonomical lines, as for example William Chappell's transformative volume *A Collection of National English Airs* (1838–40) later reworked as *Popular Music of the Olden Time* (1855–59). Founder of the Musical Antiquarian Society and Percy Club (a society interested in the collection of rare vernacular poems and songs) in 1840, Chappell was the first Victorian to conduct a systematic study of the history of traditional English tunes. In what is effectively a study of English vernacular music history from its origins to the eighteenth century, going 'beyond folk song to embrace all national songs'[41] Chappell organized material in roughly chronological structure, focussing separately on text and tune, and disrupting the chronology only to elucidate complications arising from alternative settings of either or both.

Not dissimilar to Chappell's volumes bar one crucial difference is the Rev. John Broadwood's *Old English Songs* (1847). Unlike Chappell's scholarly yet essentially historiographical compilation, *Old English Songs* is amongst the first to involve fieldwork. Precisely because it was compiled entirely through fieldwork rather than published sources it has been historically described as the first collection of folk song.[42] A more accurate estimation would consider it the first collection based entirely on fieldwork,

[40] E. David Gregory, *Victorian Songhunters: The Recovery and Editing of English Vernacular Ballads and Folk Lyrics, 1820–1883*, Lanham, MD, Toronto and Oxford: The Scarecrow Press, Inc., 2006, 3–4.

[41] Gregory, *Victorian Songhunters*, 80.

[42] Stanley Godman, 'John Broadwood, the Earliest English Folksong Collector', *West Sussex Gazette* (30 January 1964), 106–8.

as a number of Broadwood's contemporaries also collected songs through fieldwork. Influential not only amongst folk song collectors Broadwood's fieldwork also attracted the interest of his family, in particular his niece Lucy who would go on to become one of the founding members of the Folk-Song Society. An eager folk song enthusiast Lucy began attending meetings of Folk-Lore Society in 1892, and in May of that year visited the Eyre Assembly Rooms to see the Burne family dance Oxfordshire Morris dances. It was there that she met Folk-Lore Society figures Charlotte Burne, Laurence and his wife Alice Gomme and folk-drama specialist T. Fairman Ordish.[43] Laurence Gomme was of course interested in English villages; his wife, children's singing games and their music. Attending meetings regularly from that time, by 1898 Lucy had begun to ruminate over the place of folk song in the Folk-Lore Society, and in January of that year met Alfred Nutt, then President of the Folk-Lore Society, to discuss her concerns. Her loyalties divided, Lucy arranged first to meet Alice, Nutt and Marion Ralfe Cox, and later that day attended a meeting of the Irish Literary Society hosted by Alfred Graves, an authority on Irish folk song. Graves convened the meeting to discuss the possibility of restoring folk song. Lucy was joined by fellow Folk-Song Society founders Kate Lee, Fuller Maitland, Barclay Squire, Francis Korbay, Edgar F. Jacques and W. H. Gill, and according to her diaries she, Fuller Maitland and Lee expressed interest in establishing a continuous tradition through fieldwork and collecting.[44] The Folk-Song Society was born, amidst heated debate over its purpose and meaning.

The Folk-Song Society may have been brand new but by then the term 'folk song' had been around for some time. 'Folk song' did not evidently come into common parlance until the 1860s, with John Williams Palmer's *Folk Song* (1865). It first appears in 1847, a year after Thoms coined the term 'folklore', in *Howitt's Journal of Literature and Popular Progress*, describing Mary Howitt's translation of 'The Three Little Roses. A German Folk's song'.[45] In England 'folk song' is not associated with any collection of music until William E. A. Axon's *Folk Song and Folk-Speech in Lancashire*, usually dated 1870 or 1871.[46] From that time the

43 Dorothy De Val, *In Search of Song: The Life and Times of Lucy Broadwood*, Farnham: Ashgate, 2011, 68–69.

44 De Val, *In Search of Song*, 71.

45 'Folk-Song', *Oxford English Dictionary*, Oxford University Press, 2016, www.oed.com.ezphost.dur.ac.uk/view/Entry/72553?redirectedFrom=folksong#eid, accessed 5/9/13.

46 E. David Gregory, *The Late Victorian Folksong Revival: The Persistence of English Melody, 1878–1903*, Lanham, MD, Toronto and Plymouth: The Scarecrow Press, Inc., 2010, 4.

term begins to gather pace, and alongside it theories of folk song (like theories of folklore) began to coalesce. By the time the Folk-Song Society was founded in 1898, after some considerable prevarication, the term had begun to divide across ideological lines, reflecting theoretically unresolved tensions inherent within Tylorian concepts of survival. Methodologically, the leap from *Old English Songs* (1847) to Folk-Song Society is not unlike the same leap from Thom's to Folk-Lore Society. Although much of Thom's work derives from printed sources he clearly advocates collection *in situ*: 'Can no Devonshire correspondent furnish new and untold stories of his native Pixies? Are there no records of a fairy pipe-manufactory to be gathered at Swinborne, in Worcestershire?–In the mining and mountainous districts of Derbyshire are all "such antique fables and fairy toys" entirely extinct? . . . These are all subjects not undeserving attention: and it should be remembered that legends and traditions which are considered trifling, in the localities to which they more immediately relate, assume an interest in the eyes of strangers to whom they are not familiar–and an importance when placed in apposition with cognate materials, by the light which they receive and furnish from such juxtaposition'.[47] Broadwood trumpets this same approach on the collection's cover and typographically identical title page:

> Old English Songs as now sung by the peasantry of the Weald of Surrey and Sussex and collected by one who has learnt from them by hearing them Sung every Christmas from early childhood by the country people who go about to the neighbouring houses, singing 'Wassailing' as it is called, at that Season. The Airs are set to music exactly as they are now sung, to rescue them from oblivion and to afford a specimen of genuine Old English Melody: and the words are given in their original rough state with an occasional slight alteration to render the sense intelligible.[48]

Certain features of Broadwood's original publication (Figure 3.1) emphasize key words typographically in a way that later editions disappointingly do not: highlighted by size are 'Old English Songs'; 'Weald or Surrey and Sussex'; 'The Country People'; '"Wassailing" as it is called at that season'; 'The Airs are set to Music exactly as they are now Sung'; and 'The words

[47] William Thoms, 'Folk-Lore', *The Athenaeum* 983 (29 August 1846), 886.

[48] Rev. John Broadwood, cover and title page, *Old English Songs as now sung by the peasantry of the Weald of Surrey and Sussex and collected by one who has learnt from them by hearing them Sung every Christmas from early childhood by the country people who go about to the neighbouring houses, singing "Wassailing" as it is called, at that Season. The Airs are set to music exactly as they are now sung, to rescue them from oblivious and to afford a specimen of genuine Old English Melody. The words are given in their original rough state with an occasional slight alteration to render the sense intelligible. Harmonized by G. A. Dusert*, London: Balls, 1847. The cover and title page are typographically identical; the cover on blue board, the title page on cream paper.

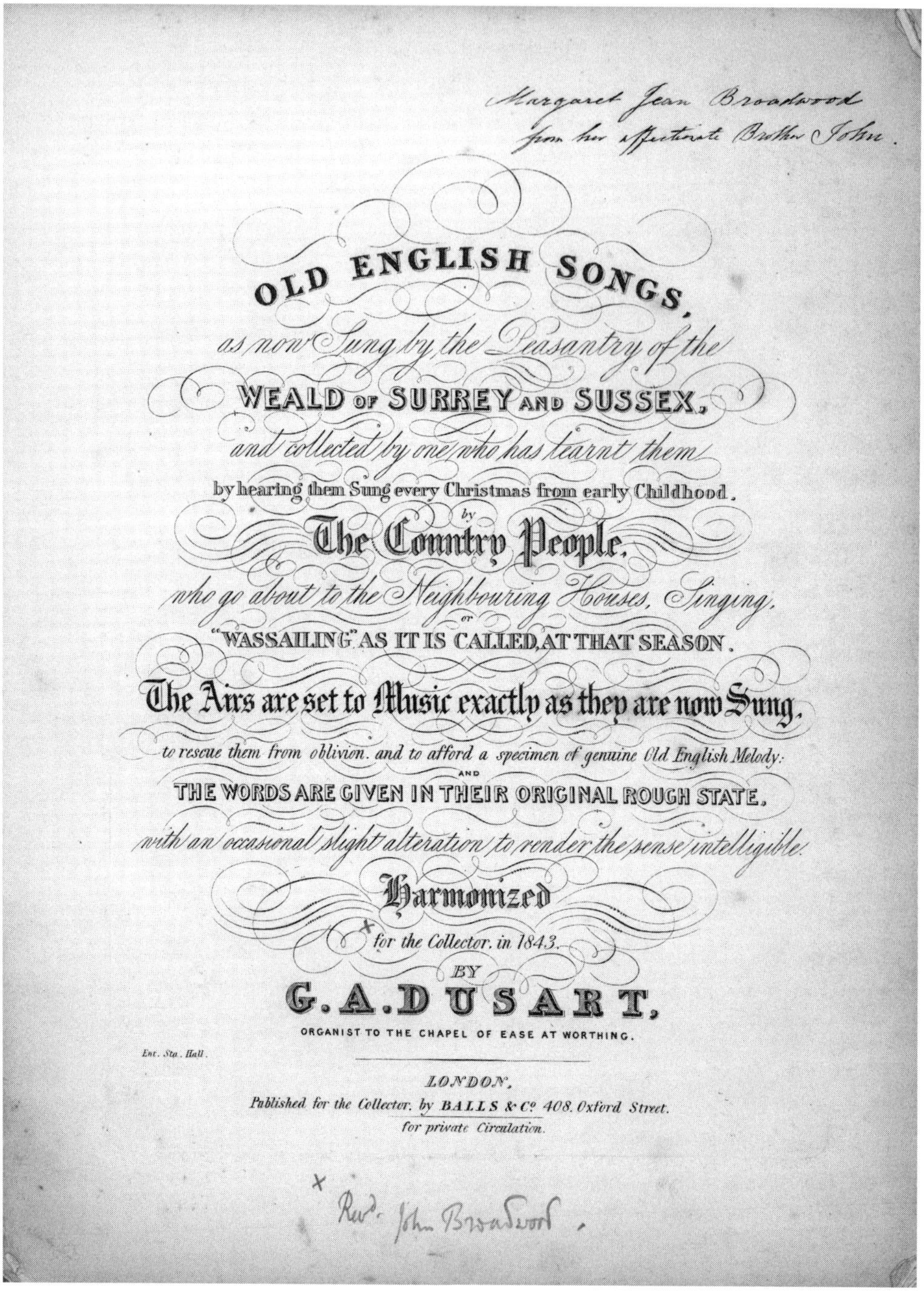

Margaret Jean Broadwood
from her affectionate Brother John.

OLD ENGLISH SONGS,
as now Sung by the Peasantry of the
WEALD OF SURREY AND SUSSEX,
and collected by one who has learnt them
by hearing them Sung every Christmas from early Childhood,
by
The Country People,
who go about to the Neighbouring Houses, Singing,
or
"WASSAILING" AS IT IS CALLED, AT THAT SEASON.
The Airs are set to Music exactly as they are now Sung,
to rescue them from oblivion, and to afford a specimen of genuine Old English Melody:
AND
THE WORDS ARE GIVEN IN THEIR ORIGINAL ROUGH STATE,
with an occasional slight alteration to render the sense intelligible.
Harmonized
for the Collector, in 1843.
BY
G. A. DUSART,
ORGANIST TO THE CHAPEL OF EASE AT WORTHING.

Ent. Sta. Hall.

LONDON,
Published for the Collector, by BALLS & C° 408, Oxford Street.
for private Circulation.

Rev. John Broadwood.

Figure 3.1 Title page, John Broadwood, *Old English Songs* (1847), reproduced by permission of Surrey History Centre, Copyright of Surrey History Centre

are given in their original rough state'. In italics are 'as now sung by the peasantry of the'; 'and collected by one who has learnt from them'; 'who go about to the neighbouring houses, singing'; 'to rescue them from oblivion and to afford a specimen of genuine Old English Melody'; 'with an occasional slight alteration to render the sense intelligible'. The larger lettering indicates source and badge of authenticity – i.e., survival; smaller italicized lettering, authenticity and editorial method – i.e., revival.

Progression from survival to revival is something inherent within Tylorian evolutionary survival; as he says, 'Sometimes old thoughts and practices will burst out afresh, to the amazement of a world that thought them long since dead or dying; here survival passes into revival'.[49] Late nineteenth-century collections reflect this characteristic prominently, building upon Broadwood's methodological approach. *Songs of the West: A Collection Made from the Mouth of the People* (1891) exhibits it not only in its title but also in the Rev. Sabine Baring Gould's definition of folk song: 'an heirloom of the past, from a class of musicians far higher in station and culture than those who now possess the treasure . . . It must be born in mind that folk-music is nowhere spontaneous and authochthonous. It is always a reminiscence, a heritage from a culture past. The yokel is as incapable of creating a beautiful melody as he is of producing a beautiful sculpture, or of composing a genuine poem'.[50] Baring Gould's definition is significant for two perhaps contradictory reasons. As an heirloom reminiscence and heritage from a past culture it clearly situates folk song within doctrinal Tylorian survival, but at the same time folk song subsists as a remnant of a previously higher culture – in other words, as a musically degenerated relic of a former, more evolved age. Baring Gould's degenerationism would also seem to be inimical to Tylorian theory, which is predicated on progressive instantiations of cultural evolution. Indeed Baring Gould is unrelenting: folk song and its custodians are degenerated relics. This reveals not only a terminological departure from Tylor but also a cultural survival itself. A sub-strata of degenerationism continues to play a significant part in the discourse of folk song despite Tylor's best efforts to supersede it, although its persistence is not altogether evident in late-century folk song publications. In the generality of collection prefaces editorial emphasis is not theoretical but practical, not survival but revival. But in the same way degenerationism haunted Tylorian survivals it also placed a recalcitrant

[49] Tylor, *Primitive Culture*, vol. 1, 16–17.

[50] Rev. Sabine Baring Gould and H. Fleetwood Sheppard, *Songs of the West: A Collection Made from the Mouths of the People*, London: Methuen & Co., 1891, ix–x.

degenerational elephant in the otherwise progressive Tylorian room of folk song collection. Cleansing corruptions posed an inevitably irksome task, for example, not simply because it created the illusion of a mythic urtext but because it deprived songs of their own narrative evolutionary journey. It also separated antiquarian from evolutionary methodologies. In *Traditional Tunes: A Collection of Ballad Airs* (1891) Frank Kidson opts to admit corruptions lest the authenticity of transcription be compromised:

> He has endeavoured to set down the airs as far as musical notes will permit with the utmost fidelity, scrupulously avoiding any attempt at arrangement or emendation, for, however desirable it might be in some instances to amend the airs from corruptions which have crept in, the Editor has considered that he would be scarcely justified in tampering with them, as such an attempt at revisal would greatly detract from the antiquarian value of the whole.[51]

Old English Country Dances (1890), however, preserves 'the evolution of numbers of our popular airs . . . [I]n this volume the editor has attempted to show the growth of the tune'.[52]

The seeping degenerational subtext of late Victorian editorial practice is reiterated in the first published writings of the Folk-Song Society in the words of C. Hubert H. Parry's very own inaugural address. Unexpectedly, however, Parry's degenerationism is directed not against the systemically incorruptible rural peasantry but its aesthetically fraudulent urban counterpart. Caught between the country rock of evolution and the urban hard place of degeneration Parry describes the demographically ambiguous folk song audience as 'unregenerate',[53] possibly akin to what today might be called 'unreconstructed'. For Parry popular music is a plague upon folk song not because it is inherently dishonest but because degeneration breeds virally to ensure its own survival. Converting his anxiety into a musicological analogy Parry defends countrymen by defining them as primitive, and re-casting the primitive in Tylorian cultural terms. Parry deploys a subtle rhetorical manoeuvre to anticipate and deflect developmental criticisms by ascribing degenerational qualities to geographical places rather than evolutionary positions – to city, rather than country, and primitive, rather than savage. Parry's redefinition of primitive comes directly from earlier Tylorian readings of culture in which primitive cultural survivals not only denote a progression from savagery but crucially originate

[51] Frank Kidson, *Traditional Tunes: A Collection of Ballad Airs*, Oxford: Charles Taphouse & Son, 1891, v.

[52] Frank Kidson, *Old English Country Dances*, London: William Reeves, 1890, iii.

[53] C. Hubert H. Parry, 'Inaugural Address', *Journal of the Folk-Song Society* 1/1 (1899), 1.

national musical identity. In *The Art of the Evolution of Music* (1893/1896) folk music is sandwiched liminally between chapters on scales and incipient harmony in what is a superficially chronological yet ideologically teleological narrative privileging the primitive over the savage, and the nation over the individual.

Parry's Great Chain of Musical Being paraphrases Tylorian survival in two ways: first, it sets in motion an optimistically unilinear projection from beginnings in abject savagery; and second it understands modern savagery to correspond to the condition of early man (i.e., the comparative method):

> The basis of all music and the very first steps in the long story of musical development are to be found in the musical utterances of the most undeveloped and unconscious types of humanity; such as unadulterated savages and inhabitants of lonely isolated districts well removed from any of the influences of education and culture. Such savages are in the same position in relation to music as the remote ancestors of the race before the story of the artistic development of music began; and through study of the ways in which they contrive their primitive fragments of tune and rhythm, and of the principles upon which they string these together, the first steps of musical development may be traced. True folk-music begins a step higher, when these fragments of tune, as nuclei, are strung together upon principles which give an appearance of orderliness and completeness; but the power to organise materials in such a manner does not come to human creatures till a long way above the savage stage.[54]

As this suggests the atomistic nature of savage musical expression corresponds to preliminary stages in intellectual development, and as the intellect evolves so too does the coherence and organization of music. Folk music appears when suddenly thematic reiterations are intentionalized: 'As the standard of human organisation improves, the capacity to balance things more regularly becomes evident; and the power to alternate simple figures more systematically immediately produces a primitive type of more definite character'.[55] The decisive transition from repetition to reiteration also signals progression from incompleteness to completeness, incoherence to coherence, child to adult, expression to perception, and – following Spencer more directly – ultimately language to music. By gradually controlling the forces of music man ensures that music itself becomes potentially more permanent, and as it sheds the flaws of earlier stages it acquires the longevity more commonly associated with more

[54] C. Hubert H. Parry, *The Evolution of the Art of Music*, ed. H. C. Colles, New York: Greenwood Press, 1893/1968, 61.

[55] Parry, *Evolution of the Art of Music*, 68.

advanced art music. Indeed the musical step change from savage to folk engenders hereditable traits – or survivals. The best art music must therefore stem from the folk music with the indisputably best genes. Unsurprisingly given his national predilections Parry's best are German and English; they alone achieve the most considerable 'concentration of faculty and self-restraint'.[56] By definition, however, folk music is self-surviving whether produced by highly sensitive races, placid races, races of moderate intelligence, or serious and strong races: 'the same principles which appear in folk music were instinctively adopted in all the forms of mature art'.[57]

As instinct – as deeply embedded heritable character – folk music represents the most naturally integral form of cultural survival and the earliest writings of the Folk-Song Society reflect this in their appropriation of Tylorian language. Poignantly the first article to follow Parry's inaugural address is 'Modal Survivals in Folk-Song' by Edgar Jacques, and that is followed in the next issue by Francis Cohen's 'Folk-Song Survivals in Jewish Worship-Music' (1900). On the surface Jacques would seem to come with excellent Tylorian credentials, revealed in an earlier article for the Musical Association called 'The Laws of Progress in Music' (1888–89). But half- or even semi-Darwinian 'The Laws of Progress in Music' encapsulates the confused theoretical allegiances of the non-Darwinian revolution. Part Darwinian, part Spencerian, Jacques's law of progress represents a theoretical conflation of the immutability of nature (law) and the mutability of man under the influence of that law (progress). Music, like all life, exists on a sliding scale from 'natural inflections' to idealisation, and any position in that scale should expose living paleontological fossils jutting up from the developmental rock face as evidence of increased differentiation and increased integration – as survivals amassing in later, more progressed generations. Rather than test the mettle of his Darwinian convictions, like so many of his contemporaries Jacques unenergetically defaults to the 'devil-you-know', importing from Spencer a belief that music is strictly the product of civilization. And because Spencerian folk song fails to occupy even a liminal developmental position, paradoxically it also fails to qualify as music. In his later article that tension remains unresolved because confusion persists over basic differences between Spencerian survival (which Darwin adopts) and Tylorian survivals.

[56] Parry, *Evolution of the Art of Music*, 91. [57] Parry, *Evolution of the Art of Music*, 101.

Tylorian survivals – living fossils – are not the same as Spencerian fittest survivors because fittest survivors have by definition changed to adapt to the exigencies of their environment. Survivals, in other words, do not necessarily evolve. This disparity poses a number of semantically imponderable questions. If, for example, as Folk-Song Society members have us believe folk song changes, why does it not also evolve? And if it does not evolve, how do we account for different versions? And if there are a variety of legitimate versions which one represents the earliest and how can we tell?

The problem in answering these questions lies with what might be called the 'non-Tylorian revolution' and the ideational confusion over survival. In fact Tylor developed his ideas independent of Darwinian or Spencerian influences; indeed the absence of particular reference to Darwin and Spencer 'is accounted for by the present work, arranged on its own lines, coming scarcely into contact of detail with the previous work of these eminent philosophers'.[58] As explained previously, whereas Darwinian natural selection refers to survival of the fittest, Tylorian survival connotes survival of the unfittest.[59] For Jacques folk song occupies a uniquely neither Spencerian nor Darwinian Tylorian ontological space, their modes 'so alien to modern ideas of tonality that, to the uninformed, they "sound wrong"'.[60] But in reality Jacques's folk song is typical of a non-Tylorian revolution which esteems unfitness in a way which Spencerianism conversely regards its opposite. Of course the rise of selective unfitness – exemplified in the work of the Folk-Lore Society and the Folk-Song Society – coincides with late Victorian re-estimations of tradition as culture, fuelled in no small part by Tylor himself. If folk song is tradition and tradition is culture then *ipso facto* folk song is culture. Jacques suggests this when he roots folk song in the same communal pool of ancient modes as Cohen. Christian music, Jewish music – they are all brought together in the cultural fraternity of folk song. Survival is not merely survival in the present but shared ancestry in the past, and as the non-Darwinian, non-Tylorian revolutions teach, the past is not as foreign a country as Victorians would have themselves believe, for national identities are forged there and the refining fires of folk song are just one of many indicators of its birth.

[58] Tylor, *Primitive Culture*, vol. 1, vii–viii.

[59] Georges Did-Huberman, 'The Surviving Image: Aby Warburg and Tylorian Anthropology', *Oxford Art Journal* 25/1 (2002), 67.

[60] E. F. Jacques, 'Modal Survival in Folk-Song', *Journal of the Folk-Song Society* 1/1 (1899), 4.

Cecil Sharp

Critical Background

The non-Tylorian revolution comes to a musicological head in the writings of Cecil Sharp (1859–1924). Like Darwin or Tylor, Sharp was a theorist in command of his material and like them both he often divided opinion. Renowned as a folk song 'missionary'[61] Sharp was a collector, editor and popularizer. Inspired by displays of Morris dancing in Oxford he entered the fray in 1899, just a year after the founding of the Folk-Song Society and was not long afterward seen collecting folk songs in Somerset in 1903. His early collections were published as *Folk Songs from Somerset* (1904–9), and followed soon thereafter by his revolutionary volume *English Folk Song: Some Conclusions* (1907). Important contributions continued into the 1920s, with *English Folk Songs for Schools* (1905), *Folk-Songs from Dorset* (1908), *English Folk-Songs of English Origin Collected from the Southern Appalachians* (1912), *Folk-Singing in Schools* (1912), *Folk Songs from Various Counties* (1912), *Folk-Dancing in Elementary and Secondary Schools* (1912) and *English Folk Songs* (1920).

Although Sharp's reputation was laudatory, by and large many people disdained his sudden and perhaps uncontested rise to prominence. While Lucy Broadwood gently mocks Sharp's early 'pricks and pokes' at the seeming inactivity of the Folk-Song Society[62] Sydney Nicholson complains that 'He puffed and boomed and shoved and ousted, and used the Press to advertise himself'.[63] Unsurprisingly for such a purist Sharp simply dug his heels in, dividing opinion and eliciting severe criticism at times. A. H. Fox Strangways describes Sharp: 'A man of such distinct flavour could not fail to make some enemies . . . [some] thought him arrogant . . . and even his best friend called him "confoundedly dogmatic"'.[64] Even after his death Sharp continued to excite controversy, not least from the 1960s when a second folk song revival began reinterpreting his work in light of Marxist philosophical critique. That often toxic yet compelling critique forms the basis of several key studies, amongst them A. L. Lloyd's *Folk Song in*

[61] Frank Howes: 'Sharp, Cecil (James)', Grove Music Online, *Oxford Music Online*, www.oxfordmusiconline.com/subscriber/article/grove/music/25594?q=cecil+sharp&search=quick&pos=1&_start=1#firsthit, accessed 15/4/16.

[62] Maud Karpeles, *Cecil Sharp: His Life and Work*, London: Routledge and Kegan Paul, 1967, 48.

[63] Sydney Nicholson, cited in Ian Russell, *Popular Music in England 1840–1914*, Manchester: Manchester University Press, 1987, 46.

[64] A. H. Fox Strangways, *Cecil Sharp*, in collaboration with Maud Karpeles, Oxford: Oxford University Press, and London: Humphrey Milford, 1933, 210.

England (1967), David Harker's *Fakesong: The Manufacture of British 'Folk Song', 1700 to the Present Day* (1985) and Georgina Boyes's *The Imagined Village: Culture, Ideology, and the English Folk Revival* (1993). These portray Sharp as an upper-middle-class cultural interloper, with unreconstructed class values at odds with his project's principal musical informants, the folk song singers themselves.

Sharp saw folk song as denoting 'the song which has been created by the common people ... in those country districts, which, by reason of their remoteness, have escaped the infection of modern ideas', and folksingers as 'the remnants of the peasantry, which originally consisted of those of the "common people", who resided in the country and subsisted on the land'.[65] Lloyd describes this attitude as 'an ideology of primitive romanticism with a vengeance',[66] and Harker as 'de-humanizing and de-individualizing'.[67] Boyes draws these criticisms together, claiming that Sharp and his upper-middle-class minions took the 'cultural products of the rural working class' and for financial gain 'daintily and selectively re-worked [them] for school and drawing room performance'.[68] As C. J. Bearman has shown, however, while 'Sharp defined folk music in terms of his sources', his 'Marxist critics define it in terms of their *class*', identifying Sharp 'as part of a classic pattern of expropriation'.[69] That confusing political subterfuge which has dominated criticism of the folk song revival since the 1960s has begun to unravel as studies in social demography clarify the true nature of Sharp's folksinging contingent. As Bearman says: 'For the sake of the material as well as of historical understanding, it is necessary to reject these unfounded and limiting assumptions and to return to the view of folk music as an art form and national treasury, the property not of a class but of the English people, and beyond them, of the artistic world'.[70] The next section of this chapter does just that, reaching towards Britain's cultural and ideological landscape, to the intellectual ecology of Sharp's folk song revival. Suspending Marxist critique over culture and class it looks more holistically at the variegated nature of anthropological knowledge of the time to explain (but

65 Sharp, *English Folk-Songs*, 3–4.

66 A. L. Lloyd, *Folk Song in England*, London: Panther Arts, 1967, 17.

67 David Harker, *Fakesong: The Manufacture of British 'Folksong', 1700 to the Present Day*, Milton Keynes and Philadelphia: Open University Press, 1985, 189.

68 Georgina Boyes, *The Imagined Village: Culture, Ideology and the English Folk Revival*, Manchester and New York: Manchester University Press, 1993, 47.

69 C. J. Bearman, 'Who Were the Folk?: The Demography of Cecil Sharp's Somerset Folk Singers', *The Historical Journal* 43/3 (September 2000), 755.

70 Bearman, 'Who Were the Folk?', 755.

perhaps not exonerate) Sharp's seemingly de-humanizing attitude and approach. It is not by imposing upon Sharp a Marxist dialectic that his true nature emerges but by situating him against the multivalenced, sometimes mutually opposed anthropologies of contemporary Tylorian and non-Tylorian paradigms.

Some Conclusions

Linguistically Darwinian yet spiritually Spencerian, Sharp explains and elaborates these paradigms in his seminal treatise *English Folk-Song: Some Conclusions* (1907). Chapters divide into roughly four sections, representing theoretical formulations (definition, origin, evolution and what he calls 'conscious and unconscious music'); technical analyses of folk song (modes, folk scales, rhythmical forms and melodic figures, and folk poetry); social demography (folk singers and their songs); and historical and broadly current issues (the decline, antiquity and future of the folk song). Like Tylor, Sharp tacitly premises his work on three complementary developmental ideas: first, that all forms of mental activity are due to the development and specialization of natural inborn qualities; second, that man's natural qualities become proportionally *less* conspicuous as he progresses in development from primitiveness to civilization; and last, that people least influenced by contact with civilization are most capable of expressing a people's (or nation's) rudimentary characteristics. Thus, 'in every land we do find music of a very beautiful quality, prevalent among the unlettered classes; bound, it is true, by certain limitations, but of a beauty and character of its own which colour although they may not be entirely shared by the educated or art-music of the same nation. This spontaneous utterance is called *folk song*'.[71] Sharp describes folk song as emanating from 'the common people' – non-educated (as opposed to uneducated), intuitive and spontaneous rural peasants, largely unaffected by contact with civilization – not the educated, conscious and intentionally trained urbanites, and so folk song is generically distinguishable from art music. As 'the unaided composition of the unskilled',[72] it is authored by a community rather than by an individual. The process of communal composition, however, 'is not fundamentally different from the method of every individual creative artist. One is only a magnified reproduction of the other, on an immense scale. In principle they are identical ... in both

[71] Sharp, *English Folk-Songs: Some Conclusions*, 1.
[72] Sharp, *English Folk-Songs: Some Conclusions*, 7.

cases they record the stages in a process of evolution'.[73] Passing through 'ceaseless and continuous change', folk song, in its successive forms, ultimately assumes 'a shape that bears little or no similarity to the form which bore its inception'.[74] As this suggests, Sharp's people appear static and undeveloping, yet their music is constantly evolving and varying. In addition, their music arises communally, but resembles the work of an individual composer.

Sharp's view of folksingers, like Tylor's observations of savages, is drawn from a fountainhead of anthropological tradition dating back to the late eighteenth century according to which human progress is universal: 'Savage races may become civilized', William Godwin opines, 'for this has already occurred – the most cultivated nations of modern times are the descendents of savages'.[75] Essentially identical some hundred years later, developmentalism came to characterize the modern savage as ossified, settling comfortably into armchair developmental beliefs of cultural stasis and fossilization. This is exemplified across comparative scientific culture in works like John Lubbock's tellingly entitled *Prehistoric Times, as Illustrated by Ancient Remains and the Manners and Customs of Modern Savages* (1895). For Sharp and other high Victorian musicologists the comparative method was learnt (as previously discussed) through classic Tylorian Carl Engel. Tylor delineates 'lower tribes as related to the civilization of the higher nations';[76] for Engel the music of uncivilized nations 'is capable of yielding important suggestions for the science and history of music, just as the languages of savage nations are useful in philological and ethnological inquires'.[77] Yet both condemn denigrating moral judgments. Tylor, for example, insists that a belief in 'any known savage tribe' being 'improved by judicious civilizations, is a proposition which no moralist would dare to make';[78] and Engel writing of Hawkins's *A General History of the Science and Practice of Music* (1776) is scathing:

[73] Sharp, *English Folk-Songs: Some Conclusions*, 13.
[74] Sharp, *English Folk-Songs: Some Conclusions*, 14.
[75] William Godwin, *Enquiry Concerning Political Justice and Its Influence on Morals and Happiness: 1793*, Oxford: Woodstock Books, [1793]/1992, cited in Robert Nisbet, *History of the Idea of Progress*, London: Heinemann, 1980, 275.
[76] Tylor, *Primitive* Culture, vol. 1.
[77] Carl Engel, *The Music of the Most Ancient Nations, particularly of the Assyrians, Egyptians, and Hebrews; with Special Reference to Recent Discoveries in Western Asia and in Egypt*, London: John Murray, 1864, v.
[78] Tylor, *Primitive Culture*, vol. 1, 31.

Hitherto it [the study of national music] has been almost entirely disregarded by musical savants. Sir John Hawkins, in the preface to his 'History of Music', says: 'The best music of barbarians is said to be hideous and astonishing sounds. Of what importance then can it be to inquire into a practice that has not its foundation in science or system, or to know what are the sounds that most delight a Hottentot, a wild American, or even a more refined Chinese?' I have transcribed Hawkins's own words, because he precisely expresses the prevailing opinion, not only of his own day, but also of the present time. I think, however, a few moments' reflection will convince the reader of its fallacy.[79]

Buoyed by this more ideologically progressive stance Engel and Tylor created the very first British methodological template for the study of non-Western and indigenous music – the first British, professionally ratified ethnomusicological framework for studying music in the field. Engel's extensive contribution to *Notes and Queries on Anthropology, For the Use of Travellers and Residents in Uncivilized Lands* (1874) produced under the aegis of Tylor and reprinted in different editions until 1899 encapsulates Tylorian moral disinterest, focussing on the individual difference of national musical expression and the importance of absolutely faithful musical transcription. As quoted previously in Chapter 2:

The music of every nation has certain characteristics of its own. The progression of intervals, the modulations, embellishments, rhythmical effects, &c. occurring in the music of extra-European nations are not unfrequently too peculiar to be accurately indicated by means of our musical notation. Some additional explanation is therefore required with the notation. In writing down the popular tunes of foreign countries on hearing them sung or played by the natives, no attempt should be made to rectify any thing which may appear incorrect to the European ear. The more faithfully the apparent defects are preserved, the more valuable is the notation. Collections of popular tunes (with the words of the airs) are very desirable. Likewise drawings of musical instruments, with explanations respecting the constructions, dimensions, capabilities, and employment of the instruments represented.

As this suggests although Engel can be characterized as understanding music through the 'generally simple and unpretending' culture of the 'uneducated or simple-minded [rural] man'[80] he also eschews the conservative attitude of more developmentally entrenched predecessors – even possibly Tylor himself, who recognizes that the progress is hypothetical: 'The educated world of Europe and America', he claims, 'practically settles

[79] Engel, *Music of the Most Ancient Nations*, 3. [80] Harker, *Fakesong*, 142.

a standard by simply placing its own nations at one end of the social series and savage tribes at the other, arranging the rest of mankind between these limits according as they correspond more closely to savage or to cultured life'.[81] While advocating developmentalism Engel nonetheless resists its invidious teleological orientation which locates isolated rural peoples, amongst others, at a point of stymied cultural nascence. Indeed while he seems to maintain a classically 'de-humanizing' developmental brief Engel also maintains recognition of the universality of individual difference: 'The similarities [between the music of separate nations] are often of such a nature that they cannot possibly be explained as accidental coincidences, but must either have originated in a former connexion between the nations, or must have been derived from a common source which perhaps no longer exists'.[82] It is this universality, tensely framed within concepts of national individual difference, which Boyes decries and reinterprets as imperial when describing C. Hubert H. Parry's opening presidential address to the Folk-Song Society in 1898, complaining that 'Parry's suggestion that folk culture offered the sophisticated and vulgar a means of salvation was an appeal to *communitas*'.[83] Parry certainly reflects a broadly imperialist developmental attitude, writing unflinchingly that English folk songs 'are the characteristic of the race, of the quiet reticence of our country folk, courageous and content',[84] and in this way he is even more dogmatically unreconstructed than contemporaries Engel and Tylor.

Yet it is in criticism of Parry and views like his that neo-socialist authors of the second folk song revival conflate concepts of culture and class. For Parry, Engel and Tylor, culture is not class but civilization, and civilization is not exclusive but universal or individually different, at whatever point of historical progress; to reiterate Tylor: 'Culture or Civilization, taken in its wide ethnographic sense, is that complex whole which includes knowledge, belief, art, morals, law, custom, and any other capabilities and habits acquired by man as a member of society'.[85] For Sharp this distinction did not prompt a change in attitude towards the rural, communal, origin of folk song, as the second folk song revivalists might have liked, but it did affect the way he would have seen his own rural folksingers. So rather than perpetuating 'an ideology of primitive romanticism' he held a progressive developmental idea of human growth; rather than having a 'dehumanized and de-individualized' perspective on his informants he would have seen

[81] Tylor, *Primitive Culture*, vol. 1, 26.
[82] Engel, *Introduction to the Study of National Music*, 317. [83] Boyes, *Imagined Village*, 36.
[84] Parry, 'Inaugural Address', 3. [85] Tylor, *Primitive Culture*, vol. 1, 1.

them in some way or other as 'cultured' and 'civilized'; and rather than capitalizing on the 'cultural products of the rural working class' he would have seen himself as salvaging and disseminating an increasingly vulnerable repertoire. In other words, while Sharp's detractors are not wrong in their criticism of his rural favouritism they are wrong to portray him, because of this, as an unreconstructed developmental recidivist. In fact Sharp held modern ideas pointing to an even more progressive, ethicized, musicological knowledge but like most of his contemporaries his brand of developmentalism was conflicted – not between culture and class but between inherited developmental certainties and increasingly learnt anthropological and evolutionary scepticism. This is intimated when he speculates on the expressive nature of man, revealing a typical (comparative-methodological) ambiguity over the interrelationship of man's diachronic and synchronic histories: 'The dominant feeling of mankind, civilized or uncivilized, is the desire for self-expression ... The mental life of the brute is ... a continuous series of disconnected sensations. But the mental mind of man, even of primitive man, is stored with pictures of his past experience, which he can produce at will'.[86] As this shows while Sharp – like Engel and Tylor – technically fell short of condemning native rural inhabitants to exclusion from the ineluctable force of human progress towards civilization he also maintained that man, no matter how primitive, was through his mind capable of recapitulating earlier stages in his development.

As such the real developmental issue in folk song is not between the civilized and uncivilized, or town and country, but the putatively antithetical relationship between 'that which is the product of the spontaneous and intuitive exercise of untrained faculties, and that which is due to the conscious and intentional use of faculties which have been especially cultivated and developed for the purpose'.[87] In other words, for Sharp the real question is not about class (as he says, if rural people 'can be called a class at all')[88] but whether as ideas consciousness and unconsciousness in music are mutually opposed or in fact complementary. This question finds its greatest exploration in the work of Sharp's slightly younger contemporary Charles Samuel Myers (discussed in Chapter 2). In 'Instinct and Intelligence' (1910) Myers creates a sliding scale of evolution from consciousness to instinct and instinct to intelligence and he begins by rejecting

[86] Sharp, *English Folk-Songs: Some Conclusions*, 5.
[87] Sharp, *English Folk-Songs: Some Conclusions*, 4.
[88] Sharp, *English Folk-Songs: Some Conclusions*, 4.

views which equate instinct with unconsciousness and fixity and intelligence with consciousness and plasticity:

It is universally admitted that intelligence and instinct are distinguished from each other by two principal characters. One of these consists in consciousness or unconsciousness of end, the other in plasticity or fixity of reaction. The common assumptions are (1) that in typically instinctive behaviour the organism is wholly unaware of the end thereby to be attained, and (2) that such behaviour is unalterable and from the very outset perfect.[89]

Here, and elsewhere, Myers argues that instinct always involves some element of consciousness, and that this consciousness derives from past rudimentary experience.[90] When a chick pecks for the first time, for example, it has already acquired 'a vague awareness of the result of its first peck, before it has actually performed the action';[91] it has a consciousness which is 'the embryonic representative of meaning'.[92] For Myers, with whom Sharp was known to have been in contact,[93] that belief in an embryonic representative of meaning led to the evolutionary model of human growth called accrescence. This claims that man gradually acquired developmentally positive characteristics through a teleological progression expressed in universal human bifurcations, such as body to mind, instinct to intelligence, and unconsciousness to consciousness. Propelling accrescent teleology were two key concepts: first, belief in the necessity and evolutionary function of individual human differences – known as differential psychology;[94] and second, the concept (admittedly waning by Myers's time) of a biological relationship between phylogeny (the growth of the race) and ontogeny (the growth of an individual) – recapitulationism.

These synthesize with unwavering gestalt when Myers writes that life comprises the 'lives of its several parts and of the "life" of the unitary "individual", which is more than the sum of the life of its several

[89] Charles S. Myers, 'Instinct and Intelligence', *British Journal of Psychology* 3/3 (October 1910), 210.

[90] Bennett Zon, *Representing Non-Western Music in Nineteenth-Century Britain*, Rochester, NY: University of Rochester Press, 2007, 198.

[91] Zon, *Representing Non-Western Music*, 211. [92] Zon, *Representing Non-Western Music*, 211.

[93] Letter from Myers to Sharp, Cecil Sharp House, no shelf mark.

[94] Differential psychology is 'the branch of psychology that considers how (and equally importantly *why*) people are psychologically very different from one another. It also considers how such individual differences may be measured by use of psychological tests and other techniques' (Colin Cooper, *Individual Differences*, 2nd ed., London: Arnold, and New York: Oxford University Press, 2002, [ix]).

parts'.[95] Sharp echoes this when he says that 'although folk-tunes of different nations differ from one another, and they all differ in certain respects from art-melodies, yet they are one and all constructed upon the same fundamental and scientific principles'.[96] Concepts balancing individual differences and universality, borne of embryological science, had become abundant at this time and Sharp was therefore not unusual in exploiting them as he did. Where this balance is most evident is in the communal origin of folk song:

> the most typical qualities of the folksong have been laboriously acquired during its journey down the ages, in the course of which its individual angles and irregularities have been rubbed and smoothed away, just as the pebble on the seashore has been rounded by the action of the waves; that the suggestions, unconsciously made by individual singers, have at every stage of the evolution of the folksong been weighed and tested by the community, and accepted or rejected by their verdict; and that the life history of the folksong has been one of continuous growth and development, always tending to approximate to a form which should be at once congenial to the taste of the community and expressive of its feelings, aspirations, and ideals.[97]

This long-held view emerged in the 1860s through Ernst Haeckel's recapitulationary axiom discussed in the Introduction to this book: 'Ontogenesis is a brief and rapid recapitulation of phylogenesis, determined by the physiological functions of heredity (generation) and adaptation (maintenance)'.[98] Thus as we have seen previously, for Sharp the process of communal composition 'is not fundamentally different from the method of every individual creative artist. One is only a magnified reproduction of the other, on an immense scale. In principle they are identical . . . in both cases they record the stages in a process of evolution'.[99] Again, passing through 'ceaseless and continuous change', folk song in its successive forms ultimately assumes 'a shape that bears little or no similarity to the form which bore its inception'.[100] In metaphorically Haeckelian terms, therefore, the composition of a folk song, as an ontological process, effectively

[95] Charles Samuel Myers, 'The Absurdity of any Mind–Body Relation' (1932), in Charles Samuel Myers (ed.), *In the Realm of Mind: Nine Chapters on the Application and Implications of Psychology*, Cambridge: Cambridge University Press, 1937, 202.
[96] Sharp, *English Folk-Songs: Some Conclusions*, 35.
[97] Cecil Sharp, *One Hundred English Folk Songs*, Boston: Oliver Ditson Co., 1916, xiv.
[98] Ernst Haeckel, *Generelle Morphologie des Organismen*, 2 vols., Berlin: Georg Reimer, 1866, cited in Ernst Haeckel, *Riddle of the Universe at the Close of the Nineteenth Century*, trans. Joseph McCabe, New York and London: Harper and Brothers Publishers, 1900, 81.
[99] Sharp, *English Folk-Songs: Some Conclusions*, 13.
[100] Sharp, *English Folk-Songs: Some Conclusions*, 14.

recapitulates the composition of an art song as a phylogenic process. Thus folk song recapitulates art song, as ontogeny recapitulates phylogeny: '*Art music*, then is the work of the individual, and expresses his own personal ideals and aspirations only; it is composed in, comparatively speaking, a short period of time, and, by being committed to paper, it is for ever fixed in one unalterable form. *Folk-music*, on the other hand, is the product of the race, and reflects feelings and tastes that are communal rather than personal; it is always in solution; its creation is never completed; while, at every moment of its history, it exists not in one form but in many'.[101]

For Sharp recapitulationism offered to elevate man (and folk song) to an anthropological pinnacle, at least conceptually, and provided a remission from more linearly conceived understandings of human development. As the nineteenth century wore on, however, criticisms gradually eroded the popularity of recapitulation in the face of more convincing evolutionary thinking. Even rank and file recapitulationists found themselves being drawn ineluctably towards Darwinian explanations of anthropological development. Sharp's progression from recapitulation to evolution, within the confines of *Some Conclusions*, encapsulates that more general movement towards evolutionism. Examining the origin of folk song, for example, he begins with a linear model akin to a story being told to individuals in succession. A invents a story, and tells B; then B tells C, and so on, until Z, at which point Z's story barely resembles A's. 'Thus, the authorship', he suggests, 'originally individual, has become communal'.[102] Slightly later the story changes, and rather than each person telling another in turn each may tell more than one person. A invents a story and tells B, but B then tells C1 and C2, and C1 tells D1, D2, D3 and D4, for example, until by the time any particular Z reaches the story there may be innumerable versions yet all stemming from a single source. Thus 'These several forms are not corruptions in varying degree of one original ... [but] changes which, in the mass, engender growth and development'.[103] So Sharp has moved seamlessly from a linear developmental model to a model more akin to Darwin's own conception of evolutionary movement.

As such Sharp's approach approximates very nearly Darwin's own visualisation of evolution in his phylogenic tree in Beagle Notebook B (see Introduction, Figure I.8). The mechanism for these changes is Darwinian natural selection which radically distinguishes it from earlier

101 Sharp, *English Folk-Songs: Some Conclusions*, 15.
102 Sharp, *English Folk-Songs: Some Conclusions*, 11.
103 Sharp, *English Folk-Songs: Some Conclusions*, 11.

developmental models of the kind Tylor and Engel espoused. Gone are the certainties of teleology, substituted instead with the randomness and uncertainty of chance, natural selection and variation. As Stephen Jay Gould writes, Darwin postulated 'a causal theory stripped of such conventional comforts as a guarantee of progress, a principle of natural harmony, or any notion of an inherent goal or purpose. Darwin's mechanism can only generate local adaptation to environments that change in a directionless way through time, thus imparting no goal or progressive vector to life's history'.[104] For Sharp's Marxist critics this was a radical departure, a step too far perhaps, exonerating him from what remnant of ethics he retained in his developmental treatment of his musical informants, 'so as to minimize human agency as much as possible in his "scientific" definition'[105] as Harker says. So while Sharp is criticized for maintaining a largely superannuated developmental model of his informants, he is also castigated for yielding to an apparently unmoral scientific template for their music.

Nevertheless, Sharp's evolutionary model is based unrepentantly on clearly formulated Darwinian principles, dividing folk song evolution into continuity, variation and selection. Continuity refers to the existence and perpetuation of individual folk songs in (at times) remarkably unchanged forms. Variation occurs when melodies are subject to change, due to exigencies of time and place, differences of text, and improvisation and ornamentation. Selection preserves those variations which are advantageous to the taste of the community over the individual:

> Without the first, *Continuity*, no evolution can take place. Its function is to prepare the way. It is a passive rather than active agent; a condition, not a cause. The second principle, *Variation*, creates the material which renders development possible. Variation, or itself, does not necessarily lead to development. Change may produce growth, or it may be sterile; or, again, it may lead to corruption. The function of the third principle, *Selection*, is to ensure that variation shall, in certain cases, result in organic grown and development. Of itself, variation merely provides the building material, the bricks and mortar. The moulding of that material, the business of construction, the determination of the form that the building shall take, these are the work of *Selection*.[106]

As Peter Bowler points out what makes a Darwinian formulation different from its antecedents is the absence of an 'ideal type for the species on

[104] Stephen Jay Gould, 'Introduction', in Carl Zimmer (ed.), *Evolution: The Triumph of an Idea*, London: William Heinemann, 2002, xi–xiii.

[105] Harker, *Fakesong*, 187.

[106] Sharp, *English Folk-Songs: Some Conclusions*, 29.

which all the individuals are modeled'.[107] For Darwin 'Natural selection acts solely through the preservation of variations in some way advantageous, which consequently endure'.[108] For Sharp 'Manifestly, those alterations will alone survive which commend themselves to other singers and narrators and are imitated by them . . . the folk-tale or song, through its life-history, will always be approaching a form which will accurately express the taste and feeling of the community; what is purely personal will be gradually but surely eliminated'.[109] There were of course competing visions of selection, most notably that of Herbert Spencer. It should be remembered that it was Spencer who coined the phrase 'survival of the fittest',[110] not Darwin who later embraced it, and with it adapted one of von Baer's principle axioms, namely that 'Less general characters are developed from the most general, and so forth, until finally the most specialized appear'.[111] For musicologists, especially those influenced by C. Hubert H. Parry, this combination sustained conceptual thought until well into the 1920s – well into Sharp's formative period of activity. As mentioned previously for Parry it is impossible for us

> to understand its qualities and characteristics, or to realise justly the light it throws upon the state of music in our own time, without tracing the conditions which led to it, and following the steps from the small and insignificant beginnings to the masterpieces which we regard as triumphs of our art . . . The study of the steps from elementary simplicity up to our complex condition of art shows how progression after progression became admissible by being made intelligible . . .[112]

107 Peter J. Bowler, *Evolution: The History of an Idea*, 3rd ed., Berkeley, Los Angeles and London: University of California Press, 2003, 156.

108 Charles Darwin, *On the Origin of Species by Means of Natural Selection, or the Preservation of Favoured Races in the Struggle for Life*, London: John Murray, 1859, 109.

109 Sharp, *English Folk-Songs: Some Conclusions*, 11–12.

110 The origin of Spencer's phrase 'survival of the fittest' is widely disputed. As David Crook, *Darwinism, War and History*, Cambridge: Cambridge University Press, 1994, 216, points out, some consider the phrase as having originated in Spencer's 'A Theory of Population, Deduced from the General Law of Animal Fertility', *Westminster Review* 57/112 (April 1852), 468–501. James Allan Rogers, 'Darwinism and Social Darwinism', *Journal of the History of Ideas*, 33/2 (April 1972), 266, reiterates this view. According to Diane B. Paul and John Offer it is now accepted that it was first used in 1864 in Part 3, chapter 12, volume 1 of Spencer's *Principles of Biology*; see Diane B. Paul. 'The Selection of "Survival of the Fittest"', *Journal of the History of Biology* 21/3 (September 1988), 412–14, and John Offer, *Herbert Spencer: Critical Assessments*, London and New York: Routledge, 2000, 3.

111 von Baer, cited in Panchen, *Classification, Evolution, and the Nature of Biology*, 20.

112 C. Hubert and H. Parry, 'On Some Bearings of the Historical Method upon Music', *Proceedings of the Musical Association* 11 (1884–85), 3–4.

For Sharp

> the first musical efforts were of necessity very simple and rudimentary ... The second stage was reached when a few sounds, connected by simply related intervals were strung together ... the way that savages at the present day affect ... A further advance was made when two of these phrases, having nothing in common, were sung alternately ... The fourth and final stage was attained when two or more phrases, involving the principles of contrast and repetition, were so combined as to form an intelligible pattern; a whole as distinct from its parts.[113]

It is a testimony to Sharp's resistance – or selection process – however, that despite Parry's conceptual hegemony he did not accept a uniformly Spencerian perspective on evolution. For Sharp might well have accepted Spencerianism in regard to musical origins, yet in his interpretation of folk music he is unremittingly Darwinian in denying the concept of an ideal type. Whatever survives does so because it is perfectly adapted to its environment, not because it is in some way perfect: 'only those [changes] that win general approval are perpetuated; the rest, being ignored, pass into oblivion'.[114] Like Darwin, who writes that 'Natural selection will not necessarily produce absolute perfection'[115] he admits that 'Folk-melodies may not always appeal to us with irresistible force, or strike us as transcendently beautiful; but they are always sincere, and free more the pretence and affectation'.[116] As such folk music is also 'the germ of art-music',[117] becoming the locus of national identity, 'always to be found in its purest, as well as in its most stable and permanent form, in the folk-arts of a nation'.[118] It is, in a sense, the origin of musical species.

Conclusion

Amongst writers of the first folk song revival Sharp's evolutionary considerations were often praised for their theoretical imagination yet comparatively marginalized in relation to the greater importance of the folk music itself. Frank Kidson, for example, wonders whether despite *Some*

[113] Sharp, *English Folk-Songs: Some Conclusions*, 32–33.
[114] Sharp, *English Folk-Songs: Some Conclusions*, 30. [115] Darwin, *Origin of Species*, 224.
[116] Darwin, *Origin of Species*, 34.
[117] Cecil Sharp, 'General Preface to the Folk-Song Series', *Folk-Songs from Various Counties*, London: Novello and Co., Ltd, n.d., 1.
[118] Olive Dame Campbell and Cecil Sharp, *English Folk Songs from the Southern Appalachians*, London: G. P. Putnam's Sons; and New York: The Knickerbocker Press, 1917, xx.

Conclusions being a 'well-considered, scientific treatise ... all will agree with his [Sharp's] deductions'.[119] More informed thinkers went further in their criticism yet continued to laud Sharp's effort. Perhaps unsurprisingly the more scientifically accomplished Charles Samuel Myers takes umbrage, complaining that, disappointingly, *Some Conclusions* exposes genuine lacunae in Sharp's knowledge of primitive music and that as a result his occasionally simplistic 'conclusion may at first appear satisfactory, but on closer observation it is open to the objection that not every musical creation of the untrained mind constitutes folk song. Indeed, Mr. Sharp lays stress on this point elsewhere in his book'.[120] Amongst writers of the second folk song revival criticism changes in both form and content. Whereas earlier critics may have complained about Sharp's bullish, self-aggrandizing intellectual swagger or the somewhat unsubstantiated nature of his evolutionary formulations, the tenor of their criticism, albeit personal, was seldom pitched at a sociological level. For Harker, Boyes and Lloyd that changed and with it our previous understanding and appreciation of Sharp's work. Sociologically conflicted as it may well be the seminal value of his work remains uncontested, and it is our task in the wake of the second folk song revival to revisit it without prejudice. Revisiting Sharp without prejudice does not necessarily mean criticizing the critics, be they first or second folk song revivalists, but it does mean returning to Sharp's cultural and intellectual horizon to understand the context of his own time and place. This is not to exonerate him or to rationalize his own prejudices but to conceptualize the reasons for his own ideological beliefs and orientations.

These beliefs and orientations are cemented by often patently self-conflicted, multivalenced, anthropological convictions of his time. Those convictions, formulated in the nascence of high Victorian and Edwardian science, were in themselves a reflection of science in transition from Enlightenment principles of developmentalism to embryological notions of recapitulation, and then to advanced understandings of Darwinian evolution. As such Sharp viewed *Some Conclusions* as pointing specifically towards evolution, maintaining that 'The main thesis of this book is the evolutionary origin of the folk song'.[121] A close look at the book, however, reveals a much more selective process, with Sharp choosing strategically between anthropological theories as he saw fit and necessary. Far from consistently evolutionary

[119] Frank Kidson, 'English Folk-Song', *The Musical Times* 49/779 (1 January 1908), 23.

[120] Charles Samuel Myers, review of *Some Conclusions*, *Folklore* 20/1 (30 March 1909), 97–98.

[121] Sharp, *English Folk-Songs: Some Conclusions*, x.

Some Conclusions is theoretically prismatic – it is the very paradigm of the non-Tylorian revolution, arguably as well as the non-Darwinian and non-Spencerian revolutions. Like a Victorian mezzotint it colours with evolution over the black and white developmental outline. Like a mezzotint Sharp's work is also a form of reproduction – in this case a reproduction of the type of late nineteenth- and early twentieth-century knowledge cast through the telescopic lenses of anthropology, folklore and folk song – from Tylor, Gomme, Lang and Parry to Darwin, Spencer, Myers and Sharp himself. And it is perhaps this which irritates second folk song revivalists so much: seen in his own terms Sharp is an undeniable progressive; seen in the harsh light of Marxist criticism he reproduces the worst ideological doctrine of bourgeois imperialist Britain. Yet Sharp's evolutionary legacy is tarred with the same Marxist brush, dismissing his entire mission as retrograde classist bumph. That it may well be from a Marxist standpoint but it does not help to broaden our understanding of Sharp nor his ideological mindset. It only serves to blur the more progressive aspects of his nature which distinguish him from his less progressive contemporaries.

Sharp's contemporaries may have been slightly flummoxed by his scientific aspirations but his immediate successors continued in that vein. Vaughan Williams encapsulates this, notably throughout in *National Music* (1934) and many other writings. Echoing Sharp in an early article of *The Musical Times* (1911) he writes that 'This, then, is what I mean by folk song – a spontaneous musical utterance by unlettered people which, by the process of evolution, has become the property of that people'.[122] Conversely Sharp's less progressive developmental ideas evanesced as anthropological knowledge fixed increasingly on a rigorously tested Darwinian model. Frank Howes questions these, for example, in 'Recent Work in Folk-Music', claiming that 'On the question whether folk-art represented the last stages of decline of an aristocratic art or whether on the other hand it was of popular origin ... there is a stalemate'.[123] Likewise Samuel Bayard later maintains that 'We can often see what look like links between widely divergent forms of the same tune; what may *seem* to be stages of development between those forms ... are actually not *stages* [both emphases Bayard's] in the formal development of the tunes, so far as we

[122] Ralph Vaughan Williams, 'English Folk-Song', *The Musical Times* (1 February 1911), 102.

[123] Frank Howes, 'Recent Work in Folk-Music', *Proceedings of the Musical Association*, 64th session (1937–38), 59.

really know'.[124] Whether in retrospect these and later criticisms conceptually undermine Sharp's evolutionary stance is anybody's guess but what is certain is that his views, in their own terms and in the context of the time, are not 'an ideology of primitive romanticism with a vengeance', nor 'de-humanizing and de-individualizing'.[125] Neither does Sharp exploit the 'cultural products of the rural working class'. In fact although for critics of the second folk song revival Sharp indulged a scientific veneer concealing deeper uncomfortable truths for others his formulation, like Engel and Tylor's, brought clarity and depth to part of culture on the verge of extinction. Like Darwin he also theorized origins without scientific certitude, and while genetic science has latterly proved Darwin right, Sharp's scientific veracity is considerably less assured. Despite that Sharp's legacy remains and with it the possibility of reformulating not only our understanding of Sharp himself but more importantly the time in which he lived. That time was a period in flux, a time of continuity, variation and selection, all pointing towards an evolutionary ideology of which Sharp's work is merely one expression. As Philip Bolman says, '[I]t was not unusual when Cecil Sharp stated baldly that "the life history of a folk song . . . is clearly a case of evolution"'.[126]

[124] Samuel Bayard, 'Prolegomena to a Study of the Principal Melodic Families of British-American Folk Song', *The Journal of American Folklore* 63/247 (January–March 1950), 5.

[125] Harker, *Fakesong*, 189.

[126] Philip V. Bohlman, *The Study of Folk Music in the Modern World*, Bloomington and Indianapolis: Indiana University Press, 1988, 7.

4 | Music Pedagogy

Introduction

The Victorian child is not a child, but a Child, 'a timeless figure of essential childhood'[1] recapitulating the totality of human experience and understanding. According to Victorian educationist Alexander Chamberlain

> The child, in the helpless infancy of his first years, in his later activity of play, in his *naïveté* and genius, in his repetitions and recapitulations of the race's history, in his wonderful variety and manifoldness, in his atavisms and his prophecies, in his brutish and in his divine characteristics, is the evolutionary being of our species, he in whom the useless past tends to be suppressed and the beneficial future to be foretold. In a sense, he is all.[2]

This made educating Victorian children doubly difficult because it required more than just education; it required Education – a system of teaching and learning which recognized and developed the individual as a living embodiment of the species. 'If the education of the centuries to come be cast in the spirit of wisdom', Chamberlain opines, 'the childlike elements necessary to the race's full development will persist to the greater glory of the individual and the perfection of mankind'.[3] This chapter tests Chamberlain's theory, and aims to prove that the musical child was also a Child, and similarly Educated. An introductory section explores recapitulation in Victorian educational theory and practice, covering basic psychological issues and the pedagogical genealogy of Victorian education through influential educationists Johann Heinrich Pestalozzi, Friedrich Froebel and Herbert Spencer. The main section divides into two separate but not unrelated studies of vocal and piano pedagogy, examining the role of recapitulation in the metaphorical discourse of Tonic Sol-fa and

[1] Judith Plotz, *Romanticism and the Vocation of Childhood*, London: Palgrave: New York and Houndsmill, 2001, 5.

[2] Alexander Francis Chamberlain, *The Child: A Study in the Evolution of Man*, London: The Walter Scott Publishing Co., Ltd, and New York: Charles Scribner's Sons, 1902, 464.

[3] Chamberlain, *The Child*, 464.

two well-known piano methods. A conclusion traces the fortunes of recapitulationary theory through musical pedagogy and speculates on its legacy in post-Victorian thought.

Pedagogy

The word 'pedagogy' derives from the Greek παιδαγωγέω (paidagōgeō, child) and άγω (ágō, lead). Leading the child anywhere in Victorian Britain required a special form of anthropological gift because the Victorian child occupied a uniquely recalcitrant, yet progressive, imaginary place fixed liminally (not unlike birds) at so many thresholds: ignorance and knowledge, slavery and freedom, poverty and wealth, bad and good, negative and positive, ugliness and beauty, darkness and light, reality and fantasy, secular and sacred, nature and divinity, heathen and Christian, emotion and intellect, matter and mind, immanence and transcendence, animal and human, savage and civilized, nation and empire, general and particular, simplicity and complexity, being and becoming; and past and future pressed into a seemingly perpetual present.[4] Like all things Romantic the Victorian child was stalked by its own symbolism and that symbolism projected onto children everything adults already knew or felt about themselves. Childhood embodied the contingent realities of life – the known and the unknown, the determined and undetermined, the growth of the individual and the growth of the species. In all its recapitulationary majesty Victorian childhood represented the philosophically invigorating ontological contraction of certainty and uncertainty, life and living, unity and the multiplicity of being.[5] Early child psychologist James Baldwin reflects this in the recapitulationary title of his important study *Mental Development in the Child and the Race: Methods and Processes* (1895). For Baldwin there is 'the child-mind . . . [and] the revelations of the man-mind', a process of '[individual] involution as well

[4] For salient expositions on the idea of the Victorian child, see Laura C. Berry, *The Child, the State, and the Victorian Novel*, Charlottesville and London: University of Virginia Press, 1999; Susan Honeyman, *Elusive Childhood: Impossible Representations in Modern Fiction*, Columbus: Ohio State University Press, 2005; Judith Plotz, *Romanticism and the Vocation of Childhood*, London: Palgrave: New York and Houndsmill, 2001; and Carolyn Steedman, *Strange Dislocations: Childhood And The Idea Of Human Interiority, 1780–1930*, Cambridge, MA: Harvard University Press, 1995.

[5] Isaiah Berlin, *The Roots of Romanticism: The A. W. Mellon Lectures in the Fine Arts, 1965, The National Gallery of Art, Washington, DC*, ed. Henry Hardy, London: Pimlico, 2000, 16–18.

as of [species] evolution',[6] and it is the role of the educationist to release this developmental process in his subject.

Precisely because of its liminality childhood attracted developmental explanations rooted in an evolutionary science optimizing unity between individual and species, content and form. Like so many classic recapitulationists Baldwin maintains that 'The individual in embryo passes through stages which represent morphologically, to a degree, the stages actually found in the ancestral animal series'.[7] But as recapitulation teaches, analogy never stops at the level of embryo and for Baldwin it reached from the deepest core of physical and mental child development to the highest and most advanced state of human evolution: one can expect, he surmises, 'general analogies to hold between nervous development and mental development, one of which is the deduction of race history epochs from individual history epochs through the repetitions of phylogenesis in ontogenesis, called in biology "Recapitulation"'.[8] There is a self-acknowledged socio-educational imperative at the root of Baldwin's psychological research, contrasting his work with more pronouncedly Darwinian psychologies. Restricting 'man and brutes', one of Darwin's ideological apprentices, George Romanes, may echo the content of Baldwin's speculations on the evolution of mental faculties but in form eschews anything less orthodox than Darwinian mechanisms of change: 'the process of organic and mental evolution has been continuous through the whole region of life and of mind'.[9] Like all good Darwinians Romanes separates the evolution of man from brute and in the process ideologically uncouples the very developmental links which Baldwin ethicizes between lower and upper forms of life. As in so many areas of Victorian culture, shorn of its capacity to reconcile, navigate and interpret the implications of ontogenic and phylogenic relationships Darwinian science spectated while recapitulation played. Indeed the world of Victorian education both generally and musically was dominated by recapitulation, not least because, unapologetically, recapitulation spoke, sang and played a language Victorians could understand. William Forbush speaks consensually when at the end of the Victorian period he claims that

[6] James Mark Baldwin, *Mental Development in the Child and the Race: Methods and Processes*, New York and London: Macmillan and Co., 1895, 2, 3.

[7] Baldwin, *Mental Development in the Child and the Race*, 15.

[8] Baldwin, *Mental Development in the Child and the Race*, 14.

[9] George John Romanes, *Mental Evolution in Man: Origin of Human Faculty*, New York: D. Appleton and Co., 1889, 4.

> The birth of a boy is not his beginning. The pre-natal child passes up through every grade of animal life from the simplest and lowest to the highest and most complex. Over one hundred and forty useless organs appear, grow and are done away, like leaves upon the tree of life, in this miracle of child-evolution. After his birth this 'candidate for humanity!' continues this evolution, in which he has already repeated the history of the animal world, by repeating the history of his own race-life from savagery unto civilization.[10]

The birth of Victorian general and musical education is also not its beginning; indeed their recapitulationism had a head start in British pedagogical life under the British acolytes of Swiss educationist Johann Heinrich Pestalozzi (1746–1827). One of Britain's foremost educational influences Pestalozzi was constantly re-imagined by Victorian educationists, and it is easy to see why, given their receptivity to recapitulationary thought. In *Enquiries concerning the Course of Nature in the Development of the Human Race* (1797) Pestalozzi proffers a three-fold development of man, from animal to social to moral. Not unlike Rousseau Pestalozzi conceives of primitive man in an original state of perfection, only to be paradoxically and irretrievably contaminated by the corrupting experience of society. Man remembers his perfection, however, and in his resolution to overcome himself achieves a state of moral consciousness animating moral behaviour. Freed from the shackles of primitive impurity moral man triumphs over his immoral self, and sound (leading to language and then song) is an important component in that triumph. As Geoffrey Bantock explains, to elucidate this process Pestalozzi identifies 'two elements in man, the empirical and the ability to transcend the empirical through the force of the autonomous will'[11] not dissimilar to Kantian principles of the phenomenal and the noumenal. Accordingly, the empirical recapitulates the transcendent will: 'Sensual enjoyment, social law, and morality, seem to bear the same relationship to one another as do childhood, adolescence and manhood'.[12] *How Gertrude Teaches Her Children* (1801), Pestalozzi's epistolary sequel to the successful novel *Leonard and Gertrude*, applies this principle to sound and music and puts developmental theory into pedagogical practice. In it certain recapitulationary precepts are held sacrosanct. Perhaps most importantly, all things progress from simplicity to complexity:

[10] William Byron Forbush, *The Boy Problem: A Study in Social Pedagogy*, Boston, MA: The Pilgrim Press, 1901, 9.

[11] Geoffrey Herman Bantock, *Studies in the History of Educational Theory*, 2 vols., London and New York: Routledge, 1984, vol. 2, 67.

[12] Pestalozzi, cited in Bantock, *Studies in the History of Educational Theory*, 66.

'Learn therefore to classify observations and complete the simple before proceeding to the complex. Try to make in every art, graduated steps of knowledge, in which every new idea is only a small, almost imperceptible, addition to that which has been known before, deeply impressed and not to be forgotten'.[13] Child development incorporates three sequentially recapitulated stages corresponding to sound, form and number: from the power of making sounds comes the origin of language and eventually music; from 'indefinite, simple sensuous-power of forming images' arises the consciousness of forms; and from the 'definite, no longer merely sensuous-power of imagination' evolves consciousness of unity as well as calculation and arithmetic.[14]

The body and mind share a significant partnership in Pestalozzi's recapitulationary programme because, as this suggests, development begins in the physical and material and culminates in the mental and spiritual: 'The same laws of the physical mechanism that develop in me the sense-foundations of knowledge, are also the sense-means of facilitating my virtue'.[15] That sound originates this process is significant because it helps the child progress 'from vague sense-impressions to clear ideas'.[16] Harnessed into language (from sound to word to language, and eventually song) it symbolizes man's innermost yearnings to overcome the limitations of his own corporeal body, and evolve into a fully formed spiritual human being. If the spirit of morality germinates in sound it is the artful pedagogue who releases it and literally leads it from the shackles of its own body: 'As the first germination of love, gratitude, trust, and obedience was a simple result of the *coincidence of instinctive feelings* between mother and child, so the *further development* of these germinated feelings is a *high human art*'.[17] If according to Pestalozzi sound recapitulates morality, then the evolution of music by its very nature expresses a deep phylogenic truth about the ontogenic development of man. As an art music reclaims man's primitive purity and stimulates morality through its beauty. The nursery song, for example, is national song in embryo:

[13] Johann Heinrich Pestalozzi, *How Gertrude Teaches Her Children: An Attempt to Help Mothers Teach Their Children and An Account of the Method by Johann Heinrich Pestalozzi*, trans. Lucy E. Holland and Francis C. Turner, and ed. with introduction and notes by Ebenezer Cooke, London: Swann Sonnenschein, and Syracuse: C. W. Bardeen, 1894, 78.

[14] Pestalozzi, *How Gertrude Teaches Her Children*, 89.

[15] Pestalozzi, *How Gertrude Teaches Her Children*, 181.

[16] Pestalozzi, *How Gertrude Teaches Her Children*, 111.

[17] Pestalozzi, *How Gertrude Teaches Her Children*, 185.

With song the mother lulls her babe to sleep, but here, as in everything else, we do not follow the law of Nature. Before the child is a year old, his mother's song ceases; by that time she is, as a rule, no longer a mother to the weaned child ... Why has not the Art of ages taught us to join the nursery lullabies to a series of national songs, that should rise in the cottages of the people, from the gentle cradle song to the sublime hymn of praise?[18]

While Pestalozzi contends that song returns us to a state of original purity, teaching music is an art which must be systematically developed from nature to nurture: 'song proper', he claims, 'cannot be regarded as a means of rising from vague sense impressions to clear ideas, but rather as a faculty that must be developed'.[19] How Pestalozzi would develop that faculty is entirely speculative, as it was left to others to build upon his work.[20] And that they did, transforming Pestalozzian pedagogy into a veritably unstoppable educational industry. In music Michael Traugott Pfeiffer and Hans Georg Nägeli were amongst the first off the mark, converting Pestalozzi's triangulation of sound, form and number into rhythm, pitch and dynamics: 'The musical sense, first comprehensively aroused by the rhythmic, melodic, and dynamic courses, then expanded many times over by the combining of the musical elements, should now be raised and refined to genuine artistic sense'.[21] Recognizing its larger significance Pestalozzian advocate Edward Biber includes musical extracts from Pfeiffer and Nägeli in *Henry Pestalozzi and His Plan of Education* (1831), filtering them through an expectedly recapitulationary model:

as an intellectual operation, though of the very lower kind, [the measuring of time] forms an essential ingredient of the internal delight which the movement produces. This internal delight is increased, if, to the harmony of time, that of tone be added; and in their joint effect consists the deep charm of music. The value of music again is raised, if, with its sounds, the feelings of kindred affections, or the higher ones of adoration, be associated in simultaneous expression; and, abstracting from all that is external, or addresses itself to the senses, nothing can be more ravishing that the idea of myriads of spirits whose deepest and unuttered thoughts are united in an

18 Pestalozzi, *How Gertrude Teaches Her Children*, 204.

19 Pestalozzi, *How Gertrude Teaches Her Children*, 95.

20 Bernarr Rainbow, *The Land Without Music: Musical Education in England 1800–1860 and Its Continental Antecedents*, London: Novello and Co., Ltd, 1967, 67.

21 Hans Georg Nägeli and Michael Traugott Pfieffer, *Gesangbildungslehre nach Pestalozzischen Grundsätzen, Pädagogisch Begründet von Michael Traugott Pfeiffer, Methodisch Bearbeitet von Hans Georg Nägeli*, Zurich: H. G. Nägeli, 1810, 120, cited in David Gramit, *Cultivating Music: The Aspirations, Interests, and Limits of German Musical Culture 1770–1848*, Berkeley and Los Angeles: University of California Press, 2002, 101.

everlasting harmony of love and praise to the Father of Spirits. Such is the effect of one and the same feeling at different stages of human development.[22]

When Pestalozzi died in 1827 he was considered to be 'the Continent's pre-eminent educator, his name . . . synonymous with the movement for infant education'[23] yet in Britain his influence was deemed to be relatively modest[24] despite the proselytising efforts of Irishman John Hatch Synge, and later Englishmen James Pierrepoint Greaves and Charles Mayo, all of whom had direct experience of the Pestalozzian system.[25]

Pre-Victorian treatises such as Lant Carpenter's *Principles of Education* (1820) and Samuel Wilderspin's *On the Importance of Educating the Infant Poor* (1823) and *Early Discipline* (1832) feel Pestalozzi's presence without directly acknowledging their source. Pre- and early-Victorian education seems to prove Joseph Payne's implication – made from the vantage point of high Victorian hindsight – that Pestalozzi's contribution was retrospectively honorific.[26] Even direct experience of observing Swiss Pestalozzian schools failed to disentangle pedagogical influences in the mind of Assistant Commissioner to the Central Poor Law Board, James Kay, author of *Training of Pauper Children* (1839). Though reporting widespread recapitulationary influence in Switzerland and some parts of Germany Kay appears to finds direct evidence of Pestalozzian education limited to Holland: 'The methods of Pestalozzi, as reduced to practice by Mr Prinse in the schools of Holland, appear worthy of adoption wherever the simultaneous system of instruction is introduced, both as respects reading, ciphering, and general instruction'.[27] Without acknowledging his source, however, Kay praises pedagogy which 'supposes that the teacher leads the children from the known to the unknown by such gradual steps as to

22 Edward Biber, *Henry Pestalozzi and His Plan of Education, Being an Account of His Life and Writings, With Copious Extracts from His Works, and Extensive Details Illustrative of the Practical Part of His Method*, London: John Souter, 1831, 165.

23 Stephen Tomlinson, 'From Rousseau to Evolutionism: Herbert Spencer on the Science of Education', *History of Education* 25/3, 238.

24 Richard Aldrich, *School and Society in Victorian Britain: Joseph Payne and the New World of Education*, Abingdon: Routledge, 2012, 234.

25 J. E. Latham, *Search for a New Eden: James Pierrepont Greaves (1777–1842)*, Fairleigh Dickinson University Press, 1999, *passim* 45–48.

26 Joseph Payne, *Contributions to the Science of Education*, New York, Cincinnati and Chicago: American Book Co., 1886, *passim* 238–53.

27 James Phillips Kay, *The Training of Pauper Children: A Report Published by the Poor Law Commissioners in Their Fourth Annual Report*, London: William Clowes and Sons, 1839, 30. See also 'Minutes on the Constructive Method of Teaching Reading, Writing, and Vocal Music', *Minutes of the Committee of Council on Education, 1840–41*, London: William Clowes and Sons, 18.

require no effort of analysis on their part, but to render the knowledge of general facts the consequence of an acquaintance with the elements from which they spring'.[28] Ebenezer Cooke compounds this confusing history of attribution in his introduction to *How Gertrude Teaches Her Children* (1894). Tracing Victorian educational genealogy from Pestalozzi to Ruskin to Pre-Raphaelite Rossetti, Cooke portrays Pestalozzi as defining the age: Rossetti 'restored to form its old power of expressing thought, and unconsciously illustrated . . . the Pestalozzian principle; – the development of the individual follows that of the race'.[29] What eventually won Pestalozzi the prize of Victorian meaning was the advent of another protean educational force, also Continental: Friedrich Froebel (1782–1852), Pestalozzi's student, founder of the kindergarten movement and an inveterate advocate of music.

Froebel coined the term 'kindergarten' in 1840 after vast experience teaching across Europe. In 1851 the movement reached England through émigré teachers Johannes and Bertha Ronge made redundant by the mass, politically driven closure of Prussian kindergartens.[30] The Ronges published the first English guide to the system with their lavishly produced *A Practical Guide to the English Kinder Garten* (1855), subtitled 'for the use of mothers, nurses, and infant teachers; being an exposition of Froebel's system of infant training, accompanied by a great variety of instructive and amusing games, and industrial and gymnastic exercises, also numerous songs set to music and arranged to the exercises'.[31] By the late 1890s it had passed through its nineteenth edition, testifying to its enduring popularity and appeal. In 1874 the Froebel Society for the Promotion of the Kindergarten System was formed, and in 1887 the Society founded the National Froebel Union to validate examinations and set standards for the Froebel Teacher's Certificate.[32] In 1892 the Froebel Educational Institute opened – its first Principal was Emilie Michaelis, with whom the musical polymath Henry Keatley Moore translated a number of Froebel's works and

[28] James Kay, *Report to the Secretary of State for the Home Department, From the Poor Law Commissioners, on the Training of Pauper Children; with Appendices*, London: W. Clowes and Sons, 1841, 108.

[29] Ebenezer Cooke, 'Editor's Preface', in Pestalozzi, *How Gertrude Teaches Her Children* (1894), xi.

[30] UCL Bloomsbury Project, Froebelian Kindergarten and Association School, www.ucl.ac.uk/bloomsbury-project/institutions/humanistic_schools.htm, accessed 29/3/13.

[31] Johannes Ronge and Bertha Ronge, *A Practical Guide to the English Kinder Garten (Children's Garden)*, London: Hodson and Son, 1855.

[32] National Froebel Foundation Archive, https://archiveshub.jisc.ac.uk/search/archives/8841a8a6-1d35-352c-aa22-5d77fae8bc82?terms=%22Early%20childhood%22.

(presumably with Keatley Moore) jointly founded the Croydon Kindergarten and Preparatory School Company.[33] By 1891 there were roughly two hundred kindergartens in England, and according to Henry Keatley Moore the number was rapidly increasing.[34]

True to his master Pestalozzi Froebel's ideological apprenticeship begins from a recapitulationary standpoint. The child is an independent yet dependent, separate yet contingent, reflection of the greater whole of nature – a 'life-union' not unlike Pestolozzi's empirical and transcendental will: 'Thus life-union consists of the management of the child and the observation of the human being in and according to all relations in life – to the whole life-power – following the pattern of Nature, who treats the smallest seed and the least plant like the entire world-process and realizes God's image in nature'.[35] This coincides with a three-fold developmental programme from nature to intellect to history mirroring Pestalozzi's animal to social to moral: 'Firstly, and first of all, Nature in its necessary, changeless laws of development and formation, then the intellect in its unchanging, logical laws of thought, and finally history (the history of the inner and invisible, as well as of the outer life) in its actual manifest results'.[36] Froebel elevates the metaphor to an even higher phylogenic level by deliberately aiming development towards a spiritual goal:

> The whole of nature, up to the appearance of man, the whole of the history from the beginning of the human race through all the past up till the present moment, and then still onwards, beyond us, to the next final consummation, when the development of man shall fall from the Tree of Life as a ripened fruit, whose kernel is the All, stands before my soul as a perfectly accurate and so to speak an exhaustive representation of true education.[37]

Froebel's commitment to recapitulation is not simply theoretical but practical as well, and like his master he judges that children learn best through graduated experiences recapitulating progress from sense-impression to

33 For more information on the Froebel movement in England, see Kevin J. Brehony, 'The Fröbel Movement in England, 1850–1911: Texts, Readings and Readers', in Helmut Heiland, Michael Gebel and Karl Neumann (eds.), *Perspektiven der Fröbelforschung*, Würzburg: Königshausen and Neumann Verlag, 2006, 49–64.

34 Keatley Moore, in Fröbel, *Fröbel's Letters on the Kindergarten*, viii.

35 Friedrich Froebel, *Friedrich Froebel's Education by Development: The Second Part of the Pedagogics of the Kindergarten*, trans. Josephine Jarvis, London: Edward Arnold, 1899, 8.

36 Froebel, *Friedrich Froebel's Education by Development*, 29.

37 Friedrich Froebel, *Journal for Friedrich Froebel's Educational Aims* 1, 14, cited in Friedrich Froebel, *Fröbel's Letters on the Kindergarten*, ed., trans. and annotated by Emilie Michaelis and Henry Keatley Moore, London: W. Swan Sonnenschein & Co., 1891, 6.

ideas. To do this the good educator not only commands nature but reveals the omnipotent power of God's intelligent design, ascending the Great Chain of Being from lower forms of fish to higher forms of birds and man:

> Indeed, even the child's own fingers must spread themselves out and become now different children, now little fishes, now little birdies, etc. Thus the child, whose life is a whole in itself, at first always represents life as a whole in the objects around him, since each and everything which has entered existence from the being of life, and the action of God the Creator is a whole, and is also part of the great All-life.[38]

Froebel embodies these principles in what are called his 'gifts', solid objects such as a soft-ball, hard-ball, cylinder, cube and bricks; planes such as tablet-shaped geometrical patterns; lines such as sticks, rings and threads; and points such as stones, shells and sand. Through physical manipulation these increasingly complex 'gifts' have the capacity to release limitlessly creative play. Like educators themselves, children learn to control, manipulate and develop the potential of objects bound by the limitations of their physical being, transforming common, simple shapes into particular, complex objects, from blank canvases to transcendental masterpieces of art. This is the nature of the 'gift': as recapitulation the child is its own gift, capable of infinite self-development: 'For the vivid thought, the eternal itself, as such, requires and conditions free self-activity and self-determination of man, of the being created for freedom, and resemblance to God'.[39] And as so often happens the acquisition of key skills culminates in music because sound recapitulates music in the same way that children recapitulate their gifts. After extensive manipulation of the divisible cube, for example, the children furnish a set of chairs for weary parents and grandparents and a jolly song, 'The Happy Home', marks the occasion of educational achievement.[40] Like Pestalozzi Froebel considers music 'representation through sound', and song 'life-giving word', situating them in a recapitulationary sequence ordered 'playthings, talk, words, song'.[41] Songs and singing occupy a particularly favoured position in Froebel's educational organism, constantly recycling man to his spiritually purer prelinguistic origins: 'The child', Froebel argues, 'at first speechless, begins

[38] Froebel, *Fröbel's Letters on the Kindergarten*, 60.
[39] Friedrich Froebel, *The Education of Man*, trans. Josephine Jarvis, New York: A. Lovell & Co., 1826/1885, 7.
[40] Ronge and Ronge, *A Practical Guide to the English Kinder Garten*, 9, 57.
[41] Friedrich Froebel, *Froebel's Chief Writings on Education Rendered into English*, trans. S. S. F. Fletcher and J. Welton, London: Edward Arnold, 1912, 153, 176, 182.

by singing, and not till later does he form articulate sounds'.[42] *Mother and Nursery Songs* (*Mutter- und Kose-Lieder*, 1844) exemplifies his observational approach, recapitulating songs from the simplest physical gestures to the most complex, game-type movement songs. Thus he begins with 'The Kicking Song' and ends with 'The Little Artist', progressing from nature (e.g. 'The Kicking Song, 'The Weathercock', 'Mowing', 'The Chickens', The Pigeons', 'Fish in the Brook') to child (e.g. 'Children on the Tower', 'Child and the Moon', 'Little Girl and the Stars') to man (e.g. 'The Carpenter', 'The Wheelwright', 'The Joiner') and to God and the transcendental (e.g. The 'Church Door and Window' and 'The Little Artist').

If for Victorian Britain Pestalozzi and Froebel incubated the recapitulationary egg it was Herbert Spencer who hatched it. By 1900 Spencer's epoch-changing treatise *Education: Intellectual, Moral, and Physical* (1861) had sold fifty thousand copies, was translated into fifteen languages, published in thirteen editions in France and in America became 'part of an intellectual watershed, from which the psychological and sociological theories of America's progressive education movement flowed'.[43] Spencer built upon Pestalozzi and Froebel enhancing, their legacy through the authority of his own evolutionary thinking: (1) 'education should proceed from the simple to the complex'; (2) 'The development of the mind, as all other development, is an advance from the indefinite to the definite'; (3) 'our lessons ought to start from the concrete and end in the abstract'; (4) 'The education of the child must accord both in mode and arrangement with the education of mankind, considered historically . . . the genesis of knowledge in the individual, must follow the same course as the genesis of knowledge in the race'; (5) 'we should proceed from the empirical to the rational'; (6) 'the process of self-development should be encouraged to the uttermost'; and (7) 'As a final test by which to judge any plan of culture, should come the question, – Does it create a pleasurable excitement in the pupils?'[44] While critical of the quality which Pestalozzian (and by implication Froebelian) education had attained Spencer credits its pedagogical methods with intuitive psychological accuracy. As with Pestalozzi's empirical sense-impressions the earliest impressions of childhood 'are the undecomposable sensations produced by resistance, light, sound, &c.',[45]

[42] Michaelis and Moore, 'Introduction', in Froebel, *Fröbel's Letters on the Kindergarten*, 9.
[43] Tomlinson, 'From Rousseau to Evolutionism', 236.
[44] Herbert Spencer, *Education: Intellectual, Moral, and Physical*, London: Williams and Norgate, 1861/1893, 65–71.
[45] Spencer, *Education: Intellectual, Moral, and Physical*, 72.

and sound and music provide crucial educational metaphors in his evolutionary armoury. 'That music is a product of civilization', we have heard Spencer claim,

> is manifest; for though some of the lowest savages [children] have their *dance-chants*, these are of a kind scarcely to be dignified by the title musical: at most, they supply but the vaguest rudiment of music, properly so called. And if music has been by slow steps developed in the course of civilization, it must have been developed out of something.[46]

Like Pestolozzi and Froebel that something is advanced only by application of a type of self-instruction akin to the transcendental will:

> We believe that on examination they [perceptions in early childhood] will be found not only to progress from the simple to the complex, from the concrete to the abstract, from the empirical to the rational; but to satisfy the further requirements, that education shall be a repetition of civilization in little, that it shall be as much as possible a process of self-evolution, and that it shall be pleasurable.[47]

For Spencer self-evolution lacks the theological imperative of Pestalozzi and Froebel, instead recapitulating an anthropology commonly eliding children and savagery: 'every civilized man', he claims, 'passes through that phase of character exhibited by the barbarous race from which he is descended. As the child's features – flat nose, forward-opening nostrils, large lips, wide-apart eyes, absent frontal sinus, &c. – resemble for a time those of the savage, so, too, do his instincts. Hence the tendencies to cruelty, to thieving, to lying, so general among children'.[48] Spencer inverts this view in 'The Primitive Man – Emotional' in *Principles of Sociology*, claiming that 'the savage has the mind of a child with the passions of a man (or, rather, has adult passions which act in a childish manner) possesses a deeper meaning than appears . . . we may regard the co-ordination of them [emotions] in the child as analogous to the co-ordination in the primitive man'.[49] Though today an unpalatable analogy, the commonplace imputation of childhood with savagery not only served to intensify the liminality

46 Herbert Spencer, 'The Origin and Function of Music' (*Fraser's Magazine*, Oct. 1857), in *Literary Style and Music: Including Two Short Essays on Gracefulness and Beauty*, New York: Philosophical Library, 1951, 68.

47 Spencer, *Education: Intellectual, Moral, and Physical*, 89. See also Curtis M. Hinsley, *Savages and Scientists: The Smithsonian Institution and the Development of American Anthropology 1846–1910*, Washington, DC: Smithsonian Institution Press, 1981, 126–27.

48 Spencer, *Education: Intellectual, Moral, and Physical*, 122.

49 Herbert Spencer, *Principles of Sociology*, 3 vols. (1874–96), A System of Synthetic Philosophy 6–8, New York: D. Appleton and Co., vol. 1, 1883, 65.

of Victorian children but under the aegis of psychology fed general and musical pedagogy with increasingly scientific yet theoretically self-justifying anthropological designations. Spencerian recapitulation emboldened an idea eager to be first to raise its hand in class. Published in cutting-edge, evolution-facing International Scientific Series[50] Alexander Bain's *Education as a Science* (1879) begins with an equally forceful, recapitulationary proclamation at the foundation of Prussian pedagogical values: education is 'the harmonious and equable evolution of the human powers ... by a method based on the nature of the mind, every power of the soul to be unfolded, every crude principle of life stirred up and nourished, all one-sided culture avoided, and the impulses on which the strength and worth of men rest, carefully attended to'.[51] Yet according to Bain this recapitulationary declaration remains empty in the face of certain pressing evolutionary realities: 'the superior aptitude of individuals for some things rather than for others; the advantage to society of pre-eminent fitness for special functions, although gained by a one-sided development; the difficulty of reconciling the "whole man" with himself; the limit to the power of the educator, which imposes the necessity of selection according to relative importance'.[52] To overcome these pressures Bain grows a belief in the interrelatedness of emotional and intellectual, simple and complex psychological functions, known as a *relational* view of consciousness.

Overlooking its recapitulationary origins Rick Rylance describes Bain's relationalism as the idea 'that feeling and thought, simple and complex, even conscious and unconscious contents ... are mixed inextricably together'.[53] These psychological contexts entrain the child's liminality by organicizing his ontogeny and phylogeny into one continuously unbroken relationship from the simple to the complex. In education, therefore, 'It is through science that we take the best grasp of the method of unfolding a subject from the simple to the complex'.[54] Science for Bain is not what it is for Spencer, nor is education. Despite its recapitulationism Bain's *Education* is almost entirely intellectual in trajectory; Spencer's is much more sociologically orientated, encompassing moral and physical as well as

[50] Bernard Lightman, 'The International Scientific Series and the Communication of Darwinism', *Journal of Cambridge Studies* 5 (2010), 27–38.

[51] Alexander Bain, *Education as a Science*, The International Scientific Series 25, London: Kegan Paul, Trench, & Co., 1879/1886, 1.

[52] Bain, *Education as a Science*, 2.

[53] Rick Rylance, *Victorian Psychology and British Culture 1850–1880*, Oxford: Oxford University Press, 2000, 192.

[54] Bain, *Education as a Science*, 148.

intellectual education. Spencer, unlike Bain, is dictated partly by developmental ethics drawn from concepts of the greatest happiness (or justice) extended from Pestalozzi through James and John Stuart Mill. In Spencer's estimation individuals (and by analogy music) worth most to society – those that progress from the simple to the complex, from the known to the unknown – are the most fit and therefore most likely to survive. Robert Richards correctly describes this kind of evolution, and by extension education, leading to ever greater adaptation and happiness.[55] But even Spencer hesitates to posit whether evolutionary idealism can actually overcome socially inherent educational inequalities. And so in optimistically Victorian Britain, where according to Spencer all organic and inorganic matter responded to identical evolutionary principles, recapitulation continued to mould social, intellectual and ultimately musical educational mores in the face of pessimistically dissenting, yet empirically unproven, forms of scientific verification espoused by Baldwin, Romanes and many other scientists. Paradoxically the longevity of educational recapitulation – from Pestalozzi to Froebel to Spencer – benefited from scientific ignorance, as long as pedagogical practice remained practical and pedagogical theory remained theoretical. For Victorian musical culture Spencerian recapitulation eclipsed Darwinism for precisely the same reasons. Music education, like so much of the cultural landscape, exemplifies the non-Darwinian revolution and with it the lingering liminality of Victorian musical childhood.

Music Pedagogy

Singing Methods

If Victorian education begins in the liminality of the child music pedagogy was perfectly placed to prove it. In its earliest Pestalozzian incarnations music gave education a crucial teleological incentive: it claimed to help children develop morally if not spiritually. John Turner expresses this belief in his *Manual of Instruction in Vocal Music, Chiefly with a View to Psalmody* (1833), when through the words of the Rev. William Jones he reiterates one of Victorian Britain's most resonant recapitulationary tropes: 'Look out of Christendom into the kingdoms of China, Tartary, Turkey

[55] Robert J. Richards, *Darwin and the Emergence of Evolutionary Theories of Mind and Behavior*, Chicago and London: University of Chicago Press, 1987, 307.

and the regions of the southern world, and you will discover no music but what is beggarly and barbarous, fit only to amuse the ears of children and savages'.[56] Like Pestalozzi, Turner uses recapitulation to give music a missionary purpose, converting children from savages to civilized and material to spiritual, especially those of the morally impoverished working classes.[57] Despite its popularity, Turner's explicitly religious cause, inspired by the widely felt need to improve parochial psalmody,[58] was not shared universally by contemporaries. William Edward Hickson's *The Singing Master* (1836) invokes the same Pestalozzian spirit but without the paraphernalia of improving, religious severity. And significantly, Hickson contrives to indulge secular repertoire, considering it more amenable to the mentality of children than its more fearsome sacred counterpart.

Hickson's secular departure represents an important point in music pedagogy because it liberates recapitulation from its arguably more restrictive theological origins. It does this by amalgamating elements of Millsian happiness (happiness is both a sensual and intellectual pleasure) and natural interpretations of musical development:

> Let the teacher in this, as in all branches of instruction, follow nature; – she will not lead him wrong. A simple melody pleases at first, because it is understood without an effort. The ear can follow it, and the memory can retain it. By and by, the ear begins to tire of this very simplicity, and craves for melodies of a more elaborate construction.[59]

Hickson designates more elaborate constructions 'scientific' (i.e., intellectual), whereas simpler, more natural, compositions reflect a fundamentally material (i.e., emotional) construction. Thus because children are innately natural they are also intrinsically happy, and because they are intrinsically happy they should only sing music in a major mode:

> music, to please, must give expression to the feelings which govern the mind. In youth, the predominant feelings are those of hope and joy, arising from the flow of animal spirits. At a later period the mind has been saddened with disappointment . . . and solemn, or plaintive music, has then an inexpressible charm, because it awakens a kindred chord in our bosoms.[60]

[56] Rev. William Jones, cited in John Turner, *Manual of Instruction in Vocal Music* (1833), cited in John Potter, *Vocal Authority: Singing Style and Ideology*, Cambridge: Cambridge University Press, 1998, 79.

[57] Rainbow, *Land Without Music*, 30–31.

[58] Catherine Dale, *Music Analysis in Britain in the Nineteenth and Early Twentieth Centuries*, Aldershot: Ashgate, 2003, 12.

[59] William Edward Hickson, *The Singing Master*, 4th ed., London: Taylor & Walton, 1840, 4.

[60] Hickson, *The Singing Master*, 4.

True to his word Hickson banishes minor tunes from *The Singing Master*, except one single and dubiously modal tune in the key of F minor (but three flats).[61] Later editions would preserve the boycott, with even more unapologetically happy tunes, all in satisfyingly jolly major keys. Teleology from child to adult, animal to human, simple to complex, natural to scientific, major to minor, does not necessarily guarantee progression, however. Indeed children are by no means equally talented by nature or nurtured by environment. Rural children, for instance, are the savage major to town children's civilized minor:

> The town boys soon learn, because of their ears have been tutored by listening to airs performed by street musicians; but the boys of the village – those at least who spend their days tending sheep, or frightening away birds, and do not even live within the sound of a peal of bells – are seldom found to have the slightest notion of music, and can only be taught, with extreme difficulty, when upwards of twelve years of age.[62]

There may be a recognized disparity in the musical advantages of town and country but for Hickson, like Turner, all boys acquire meaning only as they are socialized; and only as they are increasingly socialized can they hope to achieve the morality which gives the much-vaunted happiness of developmental equality. For good Pestalozzians education recapitulates this process, revealing how pedagogy nurtures social structures mirrored and advanced within music and its contexts. As Pestalozzi taught, recapitulationary connections between music and morals reveal a developmental process imbued with social implications. Hickson's greater contemporary Sarah Glover (1786–1867) put this theory into practice with Norwich Sol-fa, the first systematically flexible method of teaching singing. Eldest daughter of the Curate of St Laurence's Church, Norwich, and originator of the most influential musical pedagogy of the nineteenth century, Glover begins from an axiomatically Pestalozzian starting point in which practice recapitulates theory: 'In teaching children music', she writes, 'I think it best to instruct them on the same principle as they are taught speech; viz. BY DEDUCING THEORY FROM PRACTICE, RATHER THAN PRACTICE FROM THEORY'.[63] To enable children to deduce theory from practice

[61] Hickson, *The Singing Master*, 186. [62] Hickson, *The Singing Master*, 1–2.

[63] Sarah Glover, *Scheme To Render Psalmody Congregational*, [n. p. Norwich?]: Hamilton and Jarrold, 1835, 66, cited in Rainbow, *The Land Without Music*, 50–51. The same paragraph does not appear in Sarah Glover, *Scheme for Rendering Psalmody Congregational, 1835; Together with the Sol-fa Tune Book, 1839*, Kilkenny, Ireland: Boethius Press, 1982.

Glover adopts the 'scientifically dangerous'[64] approach of simplifying the visual language of musical notation and removing as much unnecessary abstraction as possible. Out went the staff and along with it a multiplicity of non-accidental sharps and flats, clefs and other gratuitously complex characters.[65] In came solfeggio with helpfully moveable, rather than intransigently fixed, doh (or tonic key), and the ostensibly easier formula: Do Ra Me Fah Sol Lah Te (diatonic); Lah Te Doh Ra Me Fah Sol (minor); Lah Te Doh Ra Me Bah Ne (melodic minor); Doy Roy Moy Foy Soy Loy Toy (chromatic sharps); and Dow Row Mow Fow Sow Low Tow (chromatic flats), amongst many other increasingly complicated syllabifications.[66]

Glover's innovative system is more subversive – or perhaps progressive – than it looks, not because it superseded drier rote learning with child-centred teaching but because of the way Norwich Sol-fa intended to render psalmody congregational. Unusually Glover updated Guidonian hexachordal solmisation and allocated syllables to scale degrees rather than pitch. Doing so taught an implicitly moral lesson: just as all men are created equal in the eyes of God all pitches are essentially equal in form. It is their function – their purpose – which fills them with meaning, and their meaning which gives them individual identity. Accordingly, doh and all other syllables may be fixed in function but their form (their pitch) is capable of change, transformation, development, progression – even modulation. The equality of a moveable system has liberated their potential to progress. As such Glover's pedagogical reforms embody the quintessentially Victorian enterprise of 'improvement', or what theorist David Wayne Thomas calls 'cultivation'.[67] The vastly popular improvement coach Samuel Smiles would soon arrive at an identical mission statement: 'Every human being has a great mission to perform, noble faculties to cultivate, a vast destiny to accomplish. He should have the means of education, and of exerting freely all the powers of his godlike nature'.[68] Reiterating John

[64] Sarah Glover, *The History of the Norwich Solfa System*, 1844, cited in Rainbow, *The Land Without Music*, 46.

[65] Glover, *Scheme for Rendering Psalmody Congregational*, 16–17.

[66] B. B. E. Blum, 'Solmization in Nineteenth-Century American Sight-Singing Instruction', PhD diss., University of Michigan, 1968, 136.

[67] David Wayne Thomas, *Cultivating Victorians: Liberal Culture and the Aesthetic*, Philadelphia: University of Pennsylvania Press, 2004, 3.

[68] Samuel Smiles, 'The Education of the Working Classes: An Address given by D. Samuel Smiles to the Members and Friends of the Leeds Mutual Improvement Society', Leeds: Hobson, 1845, cited in Aileen Smiles, *Samuel Smiles and His Surroundings*, London: Robert Hale Ltd, 1956, 70–71.

Stuart Mill's dictum that the individual recapitulates the state – that 'The worth of the State, in the long run, is the worth of the individuals composing it'[69] – Smiles claims that 'National progress is the sum of individual industry, energy and uprightness, as national decay is of individual idleness, selfishness and vice'.[70] Likewise Glover feels people are changed by the morally democratizing force of music: 'Let singing become a branch of national education, not only in schools for the children of labourers and mechanics, but in academies for young ladies and gentlemen'.[71] As moral recapitulation music 'composes while it raises the spirits, refines the mind and under judicious regulations, is calculated to favour piety'.[72] Celia Waterhouse concludes that for Glover 'singing should be for all, as a part of national education. Everyone can sing ... Musical literacy, like linguistic literacy, should be available to all'.[73]

Though successful Glover had her competitors and one became a particular rival: John Hullah. Having become acquainted with the music critic Henry Chorley, Hullah became aware of the trendy music teacher Joseph Mainzer and his popular singing classes in Paris. Chorley and Hullah visited Paris in 1839, but having found his classes discontinued they sought other opportunities to observe Mainzer's system in action.[74] Mainzer based his system on the widely used primer *Manuel Musicale* (1841) by G. L. Bocquillon Wilhem. In 1841 Hullah would translate and adapt Wilhem for English usage, crucially – and later fatally – retaining its intractable use of fixed doh. James Kay, Hullah's sometime travelling companion, may have imagined that Wilhem's Method was Pestalozzian, observing that it allowed children to 'proceed from the simplest elements ... to those which are more difficult and complex'.[75] Rainbow corrects this misapprehension, claiming that Wilhem bears remarkably

[69] John Stuart Mill, 'On Liberty' (1859), http://archive.org/stream/onliberty34901gut/pg34901.txt, accessed 24/4/13.

[70] Samuel Smiles, 'Self-Help – National and Individual', *Self-Help; with Illustrations of Character and Conduct*, Boston: Ticknor and Fields, 1863, http://oll.libertyfund.org/title/297/36621/906086, accessed 24/4/13.

[71] Glover, *Scheme for Rendering Psalmody Congregational*, 7.

[72] Glover, *Scheme for Rendering Psalmody Congregational*, 14.

[73] Celia Waterhouse, 'An Appraisal of the Norwich Sol-Fa Method and Materials for Choir Training and Music Teaching, Devised by Sarah Anna Glover in 19th Century Norwich', www.britishkodalyacademy.org/public_downloads/Sarah%20Glover's%20Music%20Education%20Scheme.pdf, accessed 23/4/13.

[74] Rainbow, *Land Without Music*, 122.

[75] James Phillips Kay, *Minutes of the Committee of Council on Education: With Appendices. 1840–41*, London: William Clowes and Sons, 1841, 47.

little resemblance to Pestalozzi;[76] for all Hullah's charisma may have popularized Wilhem's system it failed to ameliorate its greatest weakness – fixed doh. Inexplicably Hullah fixated on fixed doh, and though for much of the early Victorian period he enjoyed a practical monopoly on singing methods his celebrity would have greater longevity than his system. After decades of criticism Hullah eventually conceded defeat and belatedly rectified the deficiency with *Time and Tune in the Elementary School* (1875). By that time, however, his principal competitor John Curwen (1847–1916) was positively flourishing and it was far too late to turn popular opinion against him.

Curwen is Sarah Glover's most obvious successor, adapting both the musical substance and moral spirit of Norwich Sol-fa into what would become Victorian Britain's most influential singing method. He graduated from University College, London, in 1838, and entered the ministry of the Congregational Church, where his first appointment revealed a natural ability for teaching children. As his *Memorials* (1882) attest he was widely read in contemporary pedagogical literature, including David Stow's *The Training System* (1840) and the writings of American Christian progressive Jacob Abbott, and was positively consumed the works of Pestalozzi and his disciples, beginning with Elizabeth Mayo's *Lessons on Objects* (1840). The 1894 edition to *How Gertrude Teaches Her Children* illustrates the extent of his devotion:

> There is no study in this country where his [Pestalozzi's] influence has been greater, or his method used so effectively as in the teaching of singing. The Rev. John Curwen, founder of the Tonic Solfa method, said at a discussion on Pestalozzi at the Education Society, that he came on purpose to testify how much and how deeply he was indebted to him; and Mrs J. S Curwen [John Curwen's daughter-in-law] tells me he was always ready to acknowledge it, and that before he attempted to teach singing he was familiar with Pestalozzi's method, and used it.[77]

In 1841 Curwen read Glover's *Scheme to Render Psalmody Congregational*, and convinced of its winning combination of religious and musical simplicity introduced the Norwich system into his own teaching. Tonic Solfa is rooted in a Pestalozzian educational theory based on eliciting knowledge (and ultimately moral, spiritual awareness) through sense impression.[78]

[76] Rainbow, *Land Without Music*, 117.

[77] Pestalozzi, *How Gertrude Teaches Her Children*, fn 3 (found on page 95), 231.

[78] For the fullest exploration of Curwen and morality, see Charles Edward McGuire, *Music and Victorian Philanthropy: The Tonic Sol-fa Movement*, Cambridge: Cambridge University Press, 2009.

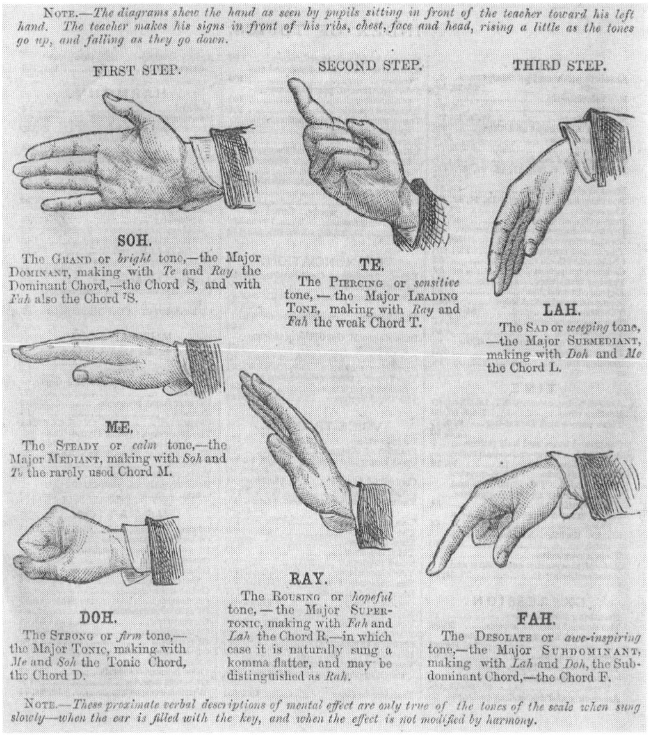

Figure 4.1 John Curwen, 'Mental Effects and Manual Signs of Tones in Key', *The Standard Course of Lessons and Exercises in the Tonic Sol-Fa Method of Teaching Music* (1876), iv

A 'musical sound', is not simply a sound but the sensed experience of a sound, and as a sensation it progresses towards musical note-hood as children progress in knowledge: 'Now, children, we are going to learn the art of singing in tune. What are we going to learn? First, then, you must remember that any musical sound is called a *note*. What is a musical sound called? This is a note. (*I hear you singing to the sound* ah *any note you please.*)'[79] Drawing upon Jue de Berneval's *Music Simplified* (1831) (Figure 4.1), Curwen allocates a different mental effect to each note corresponding to a degree of the scale, and in standard practice represents these in three columns, along with their corresponding hand signs. At their most basic level, these stress different emotional affects simulating perceptual sensitivity to each individual degree of the scale: 'These

[79] John Curwen, *Independent Magazine* (1842), 23, cited in Rainbow, *The Land Without Music*, 148.

proximate verbal descriptions of mental effect are only true of the tones of the scale when sung slowly – when the ear is filled with the key, and when the effect is not modified by harmony'.[80] Degrees of the scale combine to form significant harmonic and melodic hierarchies between doh-me-soh, ray-te, and fah-lah, denoted respectively first, second and third steps. As a linear progression, therefore, mental affects in major keys proceed from (1) Strong/firm-Steady/calm-Grand/bright, to (2) Rousing/hopeful-Piercing/sensitive, to (3) Desolate/awe-inspiring-Sad/weeping; in other words progression from (1) the happy (major) tonic, to (2) the emotionally indeterminate (yet minor) supermediant and unstable leading tone, to (3) the desolate (yet major) subdominant and the miserable (minor) submediant. Hand signs follow suit: (1) doh is a fist, me a horizontal, soh an open palm facing, fingers together; (2) ray the hand tilted upward at 45 degrees with fingers closed, te the hand twisted with the index finger pointing 45 degrees upward; and (3) fah a finger hanging 45 degrees downward limply from the rest of the hand, and lah the entire hand hanging loosely at 45 degrees.

Sustained by a teleological premise similar to Hickson (i.e., the experience of music progresses from major to minor, happy to sad), Curwen stresses the hierarchical nature of scale degrees, rather than the intervallic content of pitch relations:

> We are not anxious, at present, to teach the *absolute pitch* of sounds. Our first and chief work is to teach the relation of sounds in a tune to what is called the key-sound of that tune ... The pupil will learn to sing them by first learning to perceive their effects on the mind, and *not* by noticing their relative distances from each other.[81]

As with Pestalozzi exercises are organized to demonstrate this, revealing progression from mental effect to concrete idea in a cumulative, combinatorial, recapitulationary manner, shown in Figure 4.2.[82] According to Curwen

> The effect felt by the mind as it listens to these tones, arises first from their difference in pitch, one being higher or lower than the other, and secondly and chiefly from their *agreeing* well with each other – so that it is pleasant to hear them one immediately after the other, and pleasant to hear them sounded together. The science of sound shows how closely and beautifully these three tones are related to

[80] John Curwen, *The Standard Course of Lessons and Exercises in the Tonic Sol-Fa Method of Teaching Music*, London: Tonic Sol-Fa Agency, 1876, iv.

[81] Curwen, *Standard Course of Lessons*, 3.

[82] Curwen, *Standard Course of Lessons*, 4.

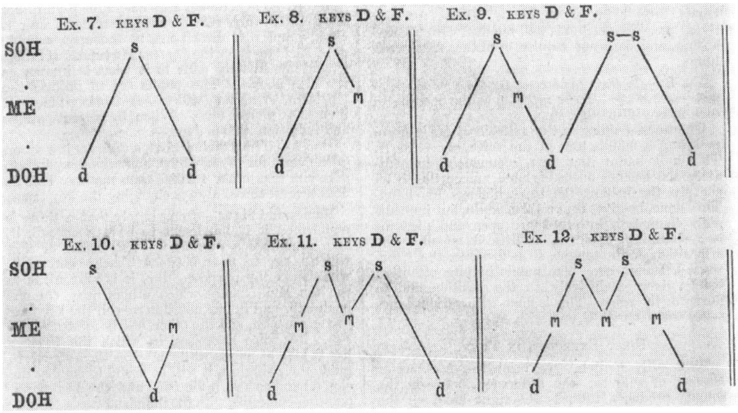

Figure 4.2 John Curwen, Exercises 7 to 12, *The Standard Course of Lessons and Exercises in the Tonic Sol-Fa Method of Teaching Music* (1876), 4

each other, in the number of their vibrations . . . When three tones are thus related, and sounded together, they are called a Chord. The pupils will be led to notice the different effect on their minds of the three tones of this Chord. As they form the Chord of the key tone, they are the bold, strong, pillar tones of the scale, on which the others lean, but they differ in the manner of their boldness, one being brighter, another stronger and more restful, another more peaceful, &c.[83]

As this suggests mental effect advances seamlessly by changes in performance. Speed, for example, previously and purposefully restrained, is encouraged in the third of six steps (chapters) making up *The Standard Course* of Tonic Sol-fa, effectively transforming or even modulating the affective nature of scale degrees. By progressing from slower to faster, minors can evolve into majors by recapitulating their affective origins in a particular mental effect:

Let the pupils sing any exercise containing *lah* and *fah* very slowly indeed, and notice how their mental effects are brought out. Then let them *sol-faa* [*sic*] the same piece as quickly as they can, keeping the time and observing the change. *Lah* and *fah* are now gay and abandoned instead of weeping and desolate in their effect, and the other tones undergo a similar modification. They will discover for themselves that great speed of movement makes the bold tones (**d m s**) sharper in their effect, though still firm; and makes the emotional tones (**r f l t**) more bright and lively, but leaves them still the emotional tones of the scale.[84]

On the surface Curwen's Tonic Sol-fa would seem to demonstrate recapitulationary similarities with the psychology of sensationalism, the theory

[83] Curwen, *Standard Course of Lessons*, 4.

[84] Curwen, *Standard Course of Lessons*, 28.

that 'ideas (the "thought" component of mental life) derive simply and directly from the sense experience'.[85] James Mill defines sensationalism as 'two classes of feelings; one, that which exists when the object of sense is present; another, that which exists after the object of sense has ceased to be present. The one class of feelings I call SENSATIONS; the other class of feelings I call IDEAS'.[86]

This is certainly the impression Curwen gives throughout his career. In the broadest sense mental effects are emotions springing from sensation – ideas which Victorian psychologist and educational theorist Alexander Bain defines as 'the coalescence of sensations and ideas'.[87] Bain goes on to categorize emotions by two different laws: (1) the law of harmony and conflict and (2) the law of relativity. In Curwen the emotions of mental effect such as Grand/bright-steady/calm-strong/firm (step one) and desolate/awe-inspiring-sad/weeping (step three) represent the extremes of harmony and conflict; rousing/hopeful-piercing/sensitive (step two), the law of relativity. These materialize in the aesthetic emotion of music:

> Some sounds are characterized as sweet; others are loud and strong, which within limits, is also an agreeable property. The pitch of sound is the basis of musical harmony. The waxing and waning of sounds is an effect particularly impressive. All these qualities are concerned in producing pleasure of the true artistic kind, and the additional circumstances of Harmony is alone wanting.[88]

For Bain and Curwen definitions of melody and harmony are similar, deriving from the same basic experiential sensations of sound elevated from the ear to the mind. Like Curwen, Bain describes their origins in Helmholtzian terms:

> A sound on one continuing pitch is more agreeable than on irregularly varying in pitch, or a mere noise. Now, what is there in such a sound? Mechanically, it consists of a series of vibrations at equal intervals of time, varying from twenty a second to upwards of seventy thousand a second. It appears, then, that even in this very minute scale of duration, equality of times between the shocks on the nerves is more agreeable than inequality. The same fact reappears on the larger scale, in what is properly known as Time in music, meaning the duration of the successive notes of a melody, which to be agreeable must be equal or proportional.[89]

85 Mark Rylance, *Victorian Psychology*, 59.

86 James Mill, *Analysis of the Phenomena of the Human Mind*, 2 vols., 2nd ed., London: Baldwin & Cradick, 1829, vol. 1, 52.

87 Alexander Bain, *The Emotions and the Will*, London: Longman's Green and Co., 2nd ed., 1865, 35.

88 Bain, *Emotions and the Will*, 216–17.

89 Bain, *Emotions and the Will*, 223.

Curwen would have been aware of Helmholtz's *Tonempfindungen* (1863) and John Tyndall's lectures, eventually published as *Sound* (1867), as his emphasis on particularizing sensation is described through '*the production of pure vocal klang*'.[90] Klang (short for Klangfarbe, clang-tint) runs like a current through the manuals of Tonic Sol-fa, signifying more than the 'compound' or 'musical' tone Helmholtz used to distinguish it from the 'simple' tone. As Ben Steege notes

> as a typical element of Victorian musical discourse, it draws attention to the particularity of aural sensation as opposed to the various notions of 'tone,' 'pitch' or 'note,' to which it formed and alterative. Whereas 'tone' carried the connotation of an a priori unified object of perception; 'pitch' seemed to imply a point along some physically measurable spectrum; and 'note' clearly pointed to the written sign for such an element, 'klang' could now be unambiguously attached to the raw sensory value of the musical sign in its specific difference from any other tone of the same pitch.[91]

According to Curwen klang is associated with 'organic impressions', or what he called musical 'statics', distinguishing it from their mental effect. Organic impression was 'not dependent on the aid of mental associations'[92] and could be produced by pleasure-giving single or harmonic relations. Steege detects haziness in Curwen's progression from organic impression to mental effect, or from sensation to emotion (or aesthetic emotion). Curwen himself does little to correct that impression: 'the chief contribution to mental effect comes from the relationships of the scale, whether its tones succeed one another in melody or co-exist in harmony. It is difficult to separate this from the other elements of organic passing over into mental impressions'.[93] For Curwen as for the English philosopher Isaac Taylor 'the precise point at which we pass the border from the world of Matter to the world of Mind escapes our keenest search'.[94] Curwen's progression from the world of matter to the world of mind may be hazy in psychological terms but in recapitulationary terms it makes perfect sense.

For Curwen a 'musical tone' is never merely 'a sound produced by vibrations and the air, which touch the drum of the ear in a rapid, regular

[90] Curwen, *Standard Course*, 3.

[91] Benjamin Steege, *Helmholtz and the Modern Listener*, Cambridge: Cambridge University Press, 2012, 202.

[92] John Curwen, *A Tract on Musical Statics: An Attempt to Show the Bearing of the Most Recent Discoveries in Acoustics on Chords, Discords, Transitions, Modulations, and Tuning, as Used by Modern Musicians*, London: Tonic Sol-Fa Agency, 1876, 13.

[93] Curwen, *Tract on Musical Statics*, 14.

[94] Curwen, *Tract on Musical Statics*, 13.

and continuous manner';[95] it is an organic impression recapitulating a mental effect recapitulating an idea (in Pestalozzi, 'The same laws of the physical mechanism that develop in me the sense-foundations of knowledge, are also the sense-means of facilitating my virtue';[96] and children progress 'from vague sense-impressions to clear ideas').[97] To succeed pedagogy must reflect and embody this principle:

> The true method of teaching is not by Dogmatic Rules of by the compact Statements of Science, but by leading the pupil along the path of Discovery – from the easy to the difficult, from the simple to the complex, from the common to the uncommon, from the known to the unknown – until he is able to see for himself the scientific principles, and to frame for himself the dogmatic rules, to which the facts have led him . . . *Music* must be taught first, and out of it must be developed the *theory* of music.[98]

In other words, practice must recapitulate theory in the same way matter recapitulates the mind. Pestalozzi locates this developmental process in a 'faculty'; song, for example, 'cannot be regarded as a means of rising from vague sense impressions to clear ideas, but rather as a faculty that must be developed',[99] and it is precisely that faculty which Curwen determined to create: 'His [the teacher's] object is to increase the *musical faculty* of each one of his pupils'.[100]

Were it not for their holistic attitude towards the relationship between mind and body, mind and matter, the idea of a musical faculty might situate Pestalozzi and Curwen in the realm of classic Victorian 'faculty psychology' in which faculties are arranged oppositionally or hierarchically in a Great Chain of Being. Higher more spiritual faculties like love, faith and morality govern or even recapitulate lower, more sensate faculties like appetite and desire. In arguing their brand of musical faculty both Pestalozzi and Curwen reflect a paradoxically more progressive, evolutionary attitude. An opponent of faculty psychology, mathematician and zealous anti-recapitulationist W. K. Clifford debunks the idea of higher and lower faculties; instead, mental development is 'process of simultaneous differentiation and integration which goes on in the paths of consciousness, between the mind and external things, between the mind and other

[95] John Curwen, *Musical Theory*, London: J. Curwen & Sons, Ltd, 1879, 1.
[96] Pestalozzi, *How Gertrude Teaches Her Children*, 181.
[97] Pestalozzi, *How Gertrude Teaches Her Children*, 111. [98] Curwen, *Musical Theory*, i.
[99] Pestalozzi, *How Gertrude Teaches Her Children*, 95.
[100] John Curwen, untitled lecture (1875), cited in John Spencer Curwen, *Memorials of John Curwen*, London: J. Curwen & Sons, 1882, 156.

minds'.[101] Unlike Clifford, however, Curwen believed that no theory should unrank recapitulationary stages coupled in teleological progression, especially those which have as their basis a moral, religious intention of the kind observed in Sarah Glover. For Glover, determined to render church psalmody congregational, one must 'teach, first, the simple and beautiful *thing*, MUSIC',[102] (the organic impression) and only afterward introduce the more complex matters (the mental effects) of visual representation and their attendant aesthetic concerns. Tonic and Norwich Sol-fa embodied that belief, in its 'simple formula of the Christian faith, and the pure principles of its morality being taught in the schools'.[103]

Piano Methods

By the end of the Victorian period this simple formula began to strain under changes in Victorian society and musical culture, and soon those changes would destabilize its hegemonic position. One problem concerned the traditional nature of a repertoire that focused on a core repertoire of hymns, songs and oratorios in the face of increasingly popular music, and not unrelatedly the nature of its religious message, moralizing to an increasingly secular society; another problem was the longer and more tonally intricate compositions being written in Britain and elsewhere; and yet another problem was the demand for increasingly instrumental repertoire.[104] The last of these problems – a rise in the popularity of instrumental music – presents an altogether different category of problem because Tonic Sol-fa never made claims to teach anything but singing. For instrumental teaching Victorians had to look elsewhere, and because of its diversity of instruments, pedagogical traditions and social constituencies Victorian instrumental teaching tended to lack the methodological uniformity that characterized the singing culture of Glover, Hullah and Curwen.

Piano pedagogy reflects this diversity, to some extent justifying Curwen's abiding fear of catechism-type teaching methods. As meticulously organized as they are incalculably dry, classic Victorian piano primers often seem to disregard the age of their audience, such as J. Alfred Johnston's

[101] William Kingdon Clifford, Leslie Stephen and Frederick Pollick (eds.), *Lectures and Essays by the Late William Kingdon Clifford*, London: Macmillan and Co., 1879, 101.
[102] John Curwen, cited in John Spencer Curwen, *Memorials*, 40.
[103] John Spencer Curwen, *Memorials*, 182.
[104] McGuire, *Music and Victorian Philanthropy*, 211–12.

The Art of Teaching Pianoforte Playing (no date), H. R. Palmer's *Piano Primer* (1885?) or Ernest Pauer's *The Art of Piano-Forte Playing* (1877). Mid-Victorian primers like the thirty-fifth edition of Burrowes's *The Piano-Forte Primer* (1849) often fare no better, treating the acquisition of knowledge as nothing more than the delivery of factual information. Burrowes opens by pronouncing that 'The principal object of every Teacher should be, to make his pupils thoroughly comprehend one question, before they proceed to another'.[105] There are exceptions, however, such as Anne Rodwell's *The Juvenile Pianist, or A Mirror of Music for Infant Minds* (1835), a charmingly fictionalized account of how to teach children piano. As literary as it is educational *The Juvenile Pianist* retells the story of a conversation between Mamma and her daughter Harriet. Even if the bulk of that learning experience is contrived to deliver relatively mundane information Rodwell brings the learning experience lovingly to life: 'Pray Mamma, am I to learn the Theory of Music?' Harriett asks. 'I hope so, my love', replies Mamma. 'I am very glad of that; for then I shall be able to compose a new lesson for myself'.[106] Not all authors catechise so endearingly, however, and even the most lovingly prepared primers can invert Curwen's axiomatic ordering of practice and theory. Unusually combining singing and piano-forte pedagogy *The Musical Guide* (1845) begins dishearteningly with the 'Theory of Music': 'To learn Music properly it is necessary, as with every other science, to begin with first principles – and to proceed by steps, till we have acquired a perfect knowledge of the theory, of which we must then make a practical use'.[107] A similar example is S. M. Wolfsheim's pedagogically arid primer *Kindergarten Child's Piano Instructor for Class and Private Teaching* (no date). Embarrassingly un-Froebelian Wolfsheim flaunts his imagined credentials: 'As the impression which the children receive in regard to the sense in Froebel's Kindergarten are further developed in its succeeding intermediate class, preparatory to school instruction proper, so the "Child's Piano-Instructor" is designed as an addition to the intermediate class of the Kindergarten; and, consequently, at the same time as a preparation to

[105] J. F. Burrowes, *The Piano-Forte Primer; containing the Rudiments of Music: Calculated either for Private Tuition, or Teaching in Classes*, London: by the author, 1849, iii–iv.

[106] Anne Rodwell, *The Juvenile Pianist, or A Mirror of Music for Infant Minds*, 2nd ed., London: Chapman and Hall, 1838, 13–14.

[107] No author, *The Musical Guide for Singing and the Piano-Forte: in which A Clear Method of Explaining the Various Clefs, Scales, and Keys is adopted; and A Variety of Information Necessary for the Musical Student is Given. With an Appendix on the Theory of Sound*, London: Cradock & Co., 1845, 3.

the piano method proper'.[108] As this suggests for all their promise few Victorian piano primers treat pedagogy as methodically or sympathetically as Curwen. Indeed, unlike their counterparts in singing most piano primers seem to validate collective scepticism towards Pestalozzi's recapitulationary influence.

The Child's Pianoforte Book (1882) by Henry Keatley Moore (1846–1937) is one prominent exception. Life-time councillor for the borough of Croydon, Keatley Moore helped found the Croydon Public Library system, instituted forward-looking open-access systems, organized public Shakespeare readings and held regular choral concerts which he conducted for twenty-nine years.[109] He was an amateur conductor, keyboard maker and an author dividing his time between musical and non-musical writings, including straight forwardly academic titles like *Concert Pitch* (1901) and *A Manual of Pianoforte Tuning* (1919); pedagogical publications centred around music for children, including 'Music in the Kindergarten' (1881), *The Child's Pianoforte Book* (1882), *The Child's Song and Game Book* (1889) and *The Nursery Song Book* (1906); and translations of Froebel, including *Autobiography of Friedrich Froebel* (1882), *Fröbel's Letters on the Kindergarten* (1891) and *Friedrich Froebel's Education by Development* (1899). As these latter suggest Keatley Moore was a devoted Froebelian, serving as ordinary member, Treasurer, and Examiner in Music to the Froebel Society and Vice-Chairman of the Croydon Kindergarten Company.[110] *Library World* focuses on Keatley Moore's significance for librarianship, while *The Times* captures his rich life as a musician:

Mr. Moore, who was chairman of Moore and Moore Limited, piano makers,[111] was twice Mayor of Croydon and for over forty years was a member of Croydon Corporation. After graduating at London University he advocated the provision of a degree for music. When the University of London decided to institute the

[108] S. M. Wolfsheim, *Kindergarten Child's Piano-Instructor*, New York: G. Schirmer, 1876, 4.

[109] 'Obituary: Henry Keatley Moore', *The New Library World* 40/2 (1937), 40.

[110] Tina Bruce, Anne Findlay, Jane Read and Mary Scarborough, *Recurring Themes in Education*, London: Paul Chapman Publishing Ltd, 1995, viii.

[111] Possibly founded as early as 1837, the firm traded in London under the names John and Henry Moore from 1878 to 1883, and Moore and Moore from 1884 to 1921, being taken over by Kemble's in 1933. They made pianos, organs and harmoniums, with some harmoniums designed to accommodate the intricacies of Indian tuning. 'Reed Organs in England: A Comprehensive Study of Reed Organs in England, Scotland and Wales', http://tardis.dl.ac.uk/FreeReed/English/organ_book/node14.html, accessed 29/4/13, and Reginald Massey and Jamila Massey, *The Music of India*, New Delhi: Abhinav Publications, 1993/1996, 74.

degree of Bachelor of Music he is stated to have been the first to take it. Some years ago he expressed the view that the piano as a drawing-room entertainer was a thing of the past, and that it might be that the time was coming when those who wished to take up music as a profession would be the teachers' only pupils. While prophesying that this might be the fate of his favourite instrument as a result of the popularity of the gramophone and the wireless, Alderman Moore thought that the greatly increased facilities for hearing recorded and broadcast music was a good thing, as there was now more love and appreciation of the best music than there had ever been. Mr. Moore was also one of the pioneers of the system of kindergarten schools in England.[112]

Commissioned by the Croydon Kindergarten and Preparatory School Company, *The Child's Pianoforte Book* is Keatley Moore's antidote to the pedagogical indifference he observed in other primers of the time: 'He found nothing already published which did not in some way or other run counter to the method of education adopted in the Company's School; and it became necessary, therefore, to attempt to produce something in harmony with their requirements'.[113] Keatley Moore begins with a four-part, constitutionally Froebelian edict 'To the teacher': (1) 'all teaching must proceed from the simple to the complex, from the concrete to the abstract, from the known to the unknown'; (2) 'The development of the race is reproduced in the development of the child'; (3) teachers should avoid 'anything dry and uninteresting'; and (4) 'knowledge to be gained, must be reproduced, or anew-created'.[114] Reading like a paraphrase of Spencer's *Education*[115] *The Child's Pianoforte Book* transforms music education into an organic whole, relating childhood and music in a recapitulationary programme from rudimentary five-finger exercises to advanced musical literacy: 'Constant recapitulation is imperatively necessary – even going back occasionally to the very beginning, that the groundwork may be constantly strengthened'.[116] The history of the musical alphabet conforms to expectation. Beginning and ending on A, it is rather sad and mournful:

the musicians of that day [Ancient Greece] were generally slaves, and slaves (unless they are very stupid slaves) are sad people, for they are always wanting to be free and bewailing their state of slavery . . . it is very sure that the children of those

[112] 'Obituary: Henry Keatley Moore', *The Times*, 13 Aug. 1937, issue 47762, col. D, 14.

[113] Henry Keatley Moore, *The Child's Pianoforte Book: Being a First Year's Course at the Pianoforte for The Home, The Kindergarten, and the School: With Upwards of Fifty Original Tunes and Songs*, London: W. Swan Sonnenschein & Co., 1882, 9.

[114] Keatley Moore, *The Child's Pianoforte Book*, 9–12.

[115] Spencer, *Education*, 65–71.

[116] Keatley Moore, *Child's Pianoforte Book*, 12.

ancient times were not so happy as you are – their fathers were very stern to them, and there were no Kindergartens then.[117]

Lucky for modern children that the scale has evolved, because its modern descendant in C major is 'bright and cheerful'.[118] By juxtaposing the ancient and modern *The Child's Pianoforte Book* replicates and reinforces two standard developmental tropes, simplicity and, not unrelated to it, degenerationism. Like the ancient musical alphabet the child is a simple savage enslaved by the depth of his natural emotions, but as he grows he becomes more civilized and complex – liberated, like the C major scale, by the intellect of his mind.

The fact that history begins in the minor and progresses to the major is arguably more difficult to explain, especially as Keatley Moore inverts the seemingly more common historical paradigm used by Curwen and his predecessor Hickson. An obvious reason is that the alphabet begins on A, not C; the first scale is therefore bound to be minor. Another explanation is degenerationism: the ancient musical alphabet is a remnant of the relative, more degenerated 'natural' minor of C major. Although at the time Keatley Moore published his primer degeneration had begun its own process of degenerationism,[119] it continued to explain the presence of savagery in the face of civilization. Field-leading anthropologist E. B. Tylor identifies degeneration in 'the history of particular tribes, [rather] than the history of Culture as a whole'[120] so it is possible that Keatley Moore is using degeneration to make a Froebelian point: so vital is the organism of 'life-union' that the 'whole life-power' can even sustain degeneration of individual parts. Greece illustrates a cautionary historical tale – even the greatest cultures pass away if they live in slavery. The C clef comes under similarly allegorical treatment. Beginning as a simple, honest letter C, 'some fanciful monk' thought he would improve its looks by tethering it to a double line, as if tied to a post. Charitably, later scribes would try to release it, only to find their efforts only partially rewarded; the C clef would be forever written as a long-legged spider wriggling free. The moral of the tale: development is not always progress – 'if it ain't broke, don't fix it'; the history of other clefs tells much the same story.

[117] Keatley Moore, *Child's Pianoforte Book*, 31.
[118] Keatley Moore, *Child's Pianoforte Book*, 32.
[119] George W. Stocking, *Victorian Anthropology*, New York: Free Press, 1987, 161.
[120] E. B. Tylor, *Researches into the Early History of Mankind and the Development of Civilization*, London: John Murray, 1870, 193.

Figure 4.3 Recapitulation in 'The Stairs', 'The See-Saw', 'The Swing' and 'The Skipping Rope', in Henry Keatley Moore, *The Child's Pianoforte Book: Being a First Year's Course at the Pianoforte for The Home, The Kindergarten, and the School: With Upwards of Fifty Original Tunes and Songs* (1882), 56

In strict Froebelian fashion, therefore, songs consistently recapitulate their lesson. As shown in Figure 4.3 Harry Payne goes up and down the 'The Stairs', and so does 'The See-Saw' and 'The Swing'; 'The Skipping Rope' skips. Not only does the right hand recapitulate in movement what the text expresses in words, the left hand moves in contrary motion, broadly mirroring the image of tune, contracting, expanding or skipping. Later songs extend and continue to recapitulate the use of imagery. Alternating sequences of steps and skips, 'The Butterfly' incorporates a five-finger exercise on G flitting around intervallically. 'Eels and Snakes' meander around a five-finger exercise on F, even slithering harmonically towards a pre-cadential secondary dominant. True to recapitulationary form, as illustrated in Figure 4.4 the song of 'The Musical Alphabet' begins in A minor (half-heartedly in the first inversion), and within two bars modulates, or drifts (without accidentals) back to C major. Palindromically, it returns properly to A minor for its first strong cadence. The second section begins as the first, tentatively in A minor (first inversion), edges towards

Figure 4.4 'The Musical Alphabet', in Henry Keatley Moore, *The Child's Pianoforte Book: Being a First Year's Course at the Pianoforte for The Home, The Kindergarten, and the School: With Upwards of Fifty Original Tunes and Songs* (1882), 28–29

C major, modulates to A minor, and then cadences in C major. The third and last section opens in the sub-dominant of C, applies a chromatic secondary dominant (V/ii in C) and then cadences in C major. Overall tonal structure is as follows: section 1: a – C – a; section 2: a – C – a – C; section 3: C – (d) – C – C. Apart from its modest variation technique, 'The Musical Alphabet' is a perfect example of recapitulation. The six-bar phrases recapitulate the opening three notes of the alphabet at their cadences, rising from A to C, both melodically in microcosm and tonally in macrocosm. The irrepressible movement from A minor to C major – the life-power – conforms to identical progression from savage to civilized, simple to complex, known to unknown.

Beautiful illustrations often accompany songs, revealing an equally recapitulationary mindset. Chapter and sectional head letters are correspondingly naturalistic, integrating letters in decorously conceived proliferations of flora and fauna. The head letter introducing the five-finger exercise, the root of all musical knowledge, shows a child surfacing from the delicate petals of a water lily, the flower's tendril-like roots dangling

TO THE CHILD.

I.

WE can all learn songs if people are kind enough to sing them over many times for us : also we can forget them easily. It would be very convenient if we could know a way of writing down the songs that we learn ; and so, when we began to forget them, we should be able to bring them back to our memory again. Beside this, if all men were to agree to write music in one way, we could read new music for ourselves; just as when we can read printing we can find stories for ourselves in story books without having to trouble our mother to tell us stories ; and this is very pleasant, especially for the busy mother who has other things to do, and who is often tired into the bargain.

Listen now to this melody.

[The teacher sings it several times, the child should not sing it.]

Figure 4.5 Head Letter illustration, in Henry Keatley Moore, *The Child's Pianoforte Book: Being a First Year's Course at the Pianoforte for The Home, The Kindergarten, and the School: With Upwards of Fifty Original Tunes and Songs* (1882), 15

beneath the shimmering watery depth.[121] Like the Great Chain of Being or Haeckel's wondrously arborial 'Pedigree of Man', in Figure 4.5 the child powers through the lower forms of nature, bursting into existence like a flower. 'The top' is another good example.[122] In Figures 4.6 and 4.7 two boys play with tops, one very small, perhaps two or three years old, the other five or six. The younger boy winds his top while the older, more spatially dominant child watches transfixed as it spins on his hand. Learning to spin is a developmental metaphor, recapitulating the younger and older stages of education: 'above all things play every note evenly; for a top which spins unevenly or wobbles on its point is good for nothing'.[123] The tune behaves in much the same way as 'The Musical Alphabet'. A four-note exercise literally whizzes in position, fixed in the earlier, simpler, more primitive key of E minor until it spins itself out into the more advanced, civilized key of G major.

[121] Keatley Moore, *Child's Pianoforte Book*, 15.
[122] Keatley Moore, *Child's Pianoforte Book*, 39.
[123] Keatley Moore, *Child's Pianoforte Book*, 39.

Figure 4.6 'The Top' illustration, in Henry Keatley Moore, *The Child's Pianoforte Book: Being a First Year's Course at the Pianoforte for The Home, The Kindergarten, and the School: With Upwards of Fifty Original Tunes and Songs* (1882), 39

Figure 4.7 'The Song of the Top', in Henry Keatley Moore, *The Child's Pianoforte Book: Being a First Year's Course at the Pianoforte for The Home, The Kindergarten, and the School: With Upwards of Fifty Original Tunes and Songs* (1882), 57

Like Froebel's *Mother and Nursery Songs* the totality of songs in *The Child's Pianoforte Book* mirrors development. Broadly speaking, songs progress through the book from experiential ('The Stairs', 'The See-Saw', 'The Swing', 'The Skipping Rope'); to perceptual ('The Echo', 'The Clock of St. Paul's Strikes Three!'); to observational ('The Frogs That Would a Hopping Go', 'The Song of the Top'); to naturalistic ('The Forest Pool, 'The Butterfly', 'Eels and Snakes', 'Floating with the Stream'); to play acting ('Poor Dolly', 'The March of the Princesses', 'The Sailors' Song', 'The Spinning Wheel'); and to music without words ('A Rippling Sea', 'The Pony's Canter', 'Mazurka', 'Dance Till You're Out of Breath', 'The Weathercock in a Storm', 'Patrick at the Fair', 'Let Us Dance on Loch Lomond', 'A Birthday Polka', 'Goodbye to the Old Book'). As the music advances it acquires more keys, modulatory devices, rhythmic complications, notational intricacies and other expected signs of greater complexity. The music also progresses towards wordlessness, transcendentalizing the teleological purpose of Froebelian education as children near their spiritual goal. Here, gradually shedding the worldly-ness of words, Keatley Moore exceeds even Froebel by liberating songs from their text, and freeing music to convey representation more directly to the emotion without the mediation of words. Unfettered by words, music sings beyond its textlessness as the fullest expression and embodiment of life's 'inner': according to Froebel as an emblem of the 'ultimate unity'[124] music, as art, recapitulates 'the pure representation of the inner'.[125]

If in piano pedagogy Keatley Moore represents the 'inner' recapitulationary ideals of Pestalozzi, Froebel and Spencer it was a popular contemporary who would do much the same for Curwen. Not long after Keatley Moore published *The Child's Pianoforte Book* Mrs Curwen (1845–1932), wife of John Curwen's son John Spencer Curwen, translated the ideals of Tonic Sol-fa into piano pedagogy with her hugely successful *Pianoforte Method* (1886). Reproduced in seemingly countless editions *Mrs Curwen's Pianoforte Method (The Child Pianist)* is, unlike Keatley Moore's, a 'practical course of the elements of music' rather than a primer including its own selection of repertoire. It starts as it means to continue with a veritable Curwenian constitution of 'A Few

[124] Friedrich Froebel, *The Education of Man*, trans. Josephine Jarvis, New York: A. Lovell & Co., 1826/1885, 154.

[125] Froebel, *The Education of Man*, 152.

Educational Maxims Showing the Principles on which the Method of the "Child Pianist" is founded':

1. Teach the easy before the difficult.
2. Teach the *thing* before the *sign*.
3. Teach one fact at a time, and the commonest fact first.
4. Leave out all exceptions and anomalies until the general rule is understood.
5. In training the mind, teach the concrete before the abstract.
6. In developing physical skill, teach the elemental before the compound, and do one thing at a time.
7. Proceed from the known to the unknown.
8. Let each lesson, as far as possible, rise out of that which goes before, and lead up to that which follows.
9. Call in the understanding to help the skill at every step.
10. Let the first impression be a correct one; leave no room for misunderstanding.
11. Never tell a pupil anything that you can help him to discover for himself.
12. Let the pupil, as soon as possible, derive some pleasure from his knowledge. Interest can only be kept up by a sense of growth in independent power.[126]

The Teacher's Guide to Mrs. Curwen's Pianoforte Method decants wisdom with an expectedly systematic approach. An introductory chapter covers the pedagogical groundwork and lessons follow, each comprising an 'Aim', 'Preparation', 'Method', 'Practice' and 'Home-work' (sometimes preceded by 'Ear exercise'). Like Pestalozzi and Froebel Mrs Curwen believes that education begins with life, and so just like Tonic Sol-fa 'from first to last, [music] is a thing of *hearing* . . . every musical fact should reach the mind through the ear [not the eye]'.[127] Learning 'Pulse', for example, is prepared by hearing a sequence of differences between the stately ticking of a hall clock, the moderate ticking of a mantle-clock, the impatient rapidity of a watch, and finally the pulse in one wrist. Indicatively, the process recapitulates pulse in each of the time pieces,

[126] Annie Jessy Curwen (Mrs Curwen), *The Teacher's Guide to Mrs. Curwen's Pianoforte Method. (The Child Pianist.) Being a Practical Course of the Elements of Music*, London: J. Curwen & Sons, Ltd, 1886, viii.

[127] Curwen (Mrs Curwen), *Teacher's Guide*, 8.

culminating in true Pestalozzian/Froebelian manner in the experience of the human body itself – the ontogeny of mechanism recapitulating the phylogeny of human organism. Preparation for learning the staff begins similarly in the ontogeny of *hearing*:

We know that tunes are made of sounds of different pitch. If we would read new tunes we must know how pitch-sounds are written. Little children have stories *told* them, the words are spoken. When they are bigger they like to read stories for themselves. When they look at the words in a book it is as if they *heard* them. So, if we learn to write musical sounds we can also read them – *i.e.*, hear the tune when look at it.[128]

Not all teaching is couched in such prosaically recapitulationary language; in fact most of Mrs Curwen is delivered in the more ordinary yet emphatically experiential prose of her father-in-law's *Standard Course* of Tonic Sol-fa. Inevitably Mrs Curwen introduces Tonic Sol-fa as an aid to the aural experience of music, replicating its main points as an appendix to the first grade of her method. While 'mental effects' are given merely passing reference in the *Pianoforte Method* they reappear in force in the pedagogically seminal treatise *Psychology Applied to Music Teaching* (1913), tellingly prefaced with forewords by educationist and philosopher S. S. Laurie (1829–1909) and Herbert Spencer. Steeped in an amalgam of Victorian and Edwardian psychology, philosophy and science of mind, Mrs Curwen delves into topics with chapters on 'mind and body', 'the human telegraph', 'the percept', 'the mental image', 'imagination proper', 'the concept', 'on the choice of language in teaching', 'ideas and their behaviour', 'how apperception masses are formed', 'attention and interest', 'habit and memory' and various aspects of method. Drawing not uncritically on Edward Titchener, Mrs Curwen extends John Curwen's recapitulationary philosophy into the realm of science. Musical perception occurs in three recapitulationary stages from pure to mixed and symbolic, and the clang or organic impression – what Mrs Curwen calls holistically 'sense impressions' and ultimately 'perception' – progress from 'outward impression, the nerve-fibre, and the message it carries'.[129] Through varying degrees and instances of reiteration perception imprints a mental image (or memory) which is then generally diffused into the construction of a functionally creative imagination. The imagination processes and separates perception (the concrete) from conception (the abstract) – or in Spencerian terms, the

[128] Curwen (Mrs Curwen), *Teacher's Guide*, 49.
[129] Annie Jessy Curwen, *Psychology Applied to Music Teaching*, London: Curwen, 1913, 46.

known from the unknown, practical from theoretical, simple from complex. The apex of this process generates a musical 'idea', such as pulse, accent and measure, which forms a potential capacity to influence behaviour and eventually musical and artistic creativity.

Conclusion

In so many ways her intellectual mentor, it is John Curwen's voice which speaks loudest through *Psychology Applied to Music Teaching*, despite its glossy reliance on contemporary psychologists of mind, such as James Ward, Conwy Lloyd Morgan, William James, Alexander Bain and John Adams, amongst others. Modelled aspirationally like a musical version of Adams's *Evolution of Educational Theory* (1912), *Psychology* is more testimony to the longevity of Curwen's Tonic Sol-fa than the untested newness of post-Victorian philosophy of mind. Indeed, *Psychology* forms the perfect recapitulation of Tonic Sol-fa, itself the perfect recapitulation of Froebel, itself the perfect recapitulation of Pestalozzi – more eruditely theorized and compellingly updated but still rooted in the confidently progressive yet morally redolent, language of John Curwen: 'A successful education method arranges all the necessary difficulties of the subject in such a progressive order that each one shall be a step toward the other, the steps from one to the other shall be sufficiently difficult to stimulate effort without being so difficult as to discourage; and the unnecessary difficulties it removes out of the way'.[130] Whether *Psychology Applied to Music Teaching* – or indeed recapitulationary psychology – removed unnecessary difficulties from teaching music is anyone's guess but one seemingly unnecessary complication it excised was any relation to morality. With its unapologetically scientific orientation *Psychology*, perhaps unintentionally, makes the moral imperative at the core of Tonic Sol-fa irrelevant, and along with it the purposive ethical background of Victorian piano and vocal pedagogy. Writing sometime afterward J. H. Elliot's article 'Music – and Morals' (1926) observes this process, self-consciously referencing H. R. Haweis's popular book *Music and Morals* (1871):[131]

> Yet apart from the fact that music and morals are both subjects upon which everyone is licensed to dogmatize, we have difficulty in tracing any very direct connection between the two . . . Not that music has no refining influence upon the

[130] John Curwen, cited in Annie Jessy Curwen, *Psychology Applied to Music Teaching*, 235.
[131] McGuire, *Music and Victorian Philanthropy*, 215.

> mind. It has . . . [a]n improvement of general 'taste', perhaps, and an increase of the all-round faculty for artistic selection; but certainly no keener recognition of the abstract morality which springs from the depths of the inner consciousness, and which is ignored at individual and social peril.[132]

Elliot's disagreement reflects more than just moral disillusionment. It also implies scepticism towards dependence upon recapitulationary constructions of man and society (the individual and social). As a consequence music, as an agent of morality, is no longer a privileged partner in the advancement of man but an equal influence among all the artistic forces of progress. The inner consciousness and abstract morality – the individual and society – no longer enjoy the recapitulationary certitudes of a previous age. Reflecting a more Darwinian vocabulary of understanding there is now a faculty for artistic 'selection'. As *Psychology Applied to Music Teaching* begins to suggest, music no longer recapitulates morality because as a scientific model recapitulation presents an increasingly unsustainable metaphor.

Mrs Curwen's contemporary, composer and music psychologist William Wallace, airs a similarly non-committal attitude towards recapitulation in the inconsistently, if opportunistically, Darwinian *Threshold of Music* (1908): 'what has led up to our present state of music has been an imperceptible preparation extending over centuries, culminating only in the last two hundred years or less, when men's minds became susceptible to impression in a new form, assimilating them, utilising them, and extending them as the intellectual process became more intricate and penetrating'.[133] For Wallace that process – education – is about evolutionary survival, not some greater sense of abstract morality: 'Self-education then, as now, meant self-protection'.[134] If the changing reputation of recapitulation helped music and morality separate and divorce in the early decades of the twentieth century in Victorian times it continued to underwrite a much more solidly married relationship. Vocal and piano pedagogy – indeed most musical pedagogy, even if purely technical – was brightly backlit by morality.[135] Morality is taught by example in Mrs Curwen's *Pianoforte Method* and Keatley Moore's *The Child's Pianoforte Book*, both generously indebted to the pedagogical philosophy of

[132] J. H Elliot, 'Music – and Morals', *Musical News and Herald* (19 June 1926), 531.
[133] William Wallace, *The Threshold of Music: An Inquiry into the Development of the Musical Sense*, London: Macmillan and Co., Ltd, 1908, 244.
[134] Wallace, *Threshold of Music*, 247.
[135] David Golby, *Instrumental Teaching in Nineteenth-Century Britain*, Aldershot: Ashgate, 2004, 27.

Pestalozzi, Froebel and Spencer. Both make implicitly similar recapitulationary assumptions and claims, consequently.

Like her father-in-law Mrs Curwen expects the child to conform to type, advancing from the ontogeny of organic impressions to phylogeny of mental effects in the fullness of educational development, and Keatley Moore openly believes that while children recapitulate the race music recapitulates its history. For all their similarities, however, the difference between the two pedagogies is not insignificant. Although Keatley Moore takes recapitulation at face value in some respects his book does more to represent it than Mrs Curwen, with songs that do what they say and illustrations that say what they mean. The contents banner is representative, a thick flight of swallows coursing over the pyramids below. The swallow, 'joyous prophet of the happy summer',[136] is a migratory bird, and like children recapitulates progression (flight) from the antiquity of the ancient world to the modernity of British civilization. Migration, like education, brings the joys of a development. The contents page of *The Child's Pianoforte Book* initiates the journey of musical, educational migration from the spring of childhood and humanity. Mrs Curwen may lack the purpose-built repertoire and lavish visual content of Keatley Moore but what she lacks in design she compensates in pedagogy, inheriting from Tonic Sol-fa an added pedagogical dimension missing from Keatley Moore – a vocal pedagogy. With it comes a graded system of teaching not just an instrument but a musical way of life, not merely a method but a musical mission with the highly developed, overtly scientific explanation of John Curwen's *Musical Statics*. For Curwen education is two-fold, both scientific and aesthetic, impression and effect: 'Musical Statics teach those truths in music which arise from the great laws of Nature, which are unchangeable and indubitable. Aesthetics teach principles which are founded on our own susceptibilities, and more or less subject to variation'.[137] Given their close connection, science recapitulates aesthetics for vocal pedagogy as well, especially pedagogy steeped in improvement-driven evangelical traditions like John Curwen's and Sarah Glover's. Glover determined to render 'psalmody truly congregational',[138] to help congregants 'cultivate their own voices as independent instruments and encourage others by their example to join in this part of public worship'.[139]

[136] Phil Robinson, *The Poets' Birds*, London: Chatto and Windus, 1883, 440.
[137] John Curwen, *Musical Statics*, x.
[138] Glover, *Scheme for Rendering Psalmody Congregational*, 25.
[139] Glover, *Scheme for Rendering Psalmody Congregational*, 30.

To do that Glover effectively analogises moveable doh as the recapitulationary relationship of individual and congregation. As a syllable doh stands for the moveable, modulatory nature of the individual; as a system, the congregational nature of larger compositional structures. Deducing theory from practice rather than practice from theory Glover democratizes and congregationalizes music through recapitulation. Even secularizing Edward Hickson is tinged by 'congregationalism', if by congregationalism one can also mean the betterment of 'society'. In all these cases the individual learner – the child – is never merely an individual but the ontogeny of a larger, improving social phylogeny. James Mill defines education as rendering 'the individual, as much as possible, an instrument of happiness, first to himself, and next to other beings'.[140] John Stuart Mill follows in his father's footsteps; the purpose of education is to prepare 'the culture which generation purposely gives to those who are to be its successors, in order to qualify them for at least keeping up, and, if possible, for raising, the improvement which has been attained'.[141] Spencer echoes the Mills, asking, 'Is it not that education of whatever kind, has for its proximate end to prepare a child for the business of life – to produce a citizen who, while he is well conducted, is also able to make his way in the world?'[142] In the broadest sense the same recapitulationary ideals of citizenship inform even the (intentionally more spiritual) pedagogy of Pestalozzi and Froebel. As a musical organism the child behaves much the same way as his more abstracted pedagogical ideal, evolving from the simple to complex, concrete to abstract, emotional to intellectual, secular to sacred, animal to human, human to divine and so on. But what music pedagogy illustrates emphatically is the inexhaustibly rich recapitulationary imagination of the Romantic mind.[143] Victorians saw music as they saw children and children as they saw music: a liminal unity and multiplicity, an ineluctable force to be developed if understood – a reality and an idea, an idea and an Ideal, a child and a Child, educated and Educated.

[140] James Mill, '*SECTION 1: Theory of the Human Mind. — Its Importance in the Doctrine of Education*', *Supplement to the Encyclopedia Britannica*, London: J. Innes, 1825, http://oll.libertyfund.org/title/1764/92873, accessed 8/5/2013.

[141] John Stuart Mill, 'Inaugural Address Delivered to the University of St Andrews, 1867', in *The Collected Works of John Stuart Mill*, ed. John M. Robson, vol. XXI – Essays on Equality, Law, and Education, Toronto: University of Toronto Press and London: Routledge and Kegan Paul, 1984, http://oll.libertyfund.org/title/255/21681, accessed 5/8/13.

[142] Spencer, *Education*, 99. [143] Berlin, *Roots of Romanticism*, 16–18.

5 | Music Biography

Introduction

Life writing, or biography, is 'our creative and non-fictional output devoted to recording and interpreting real lives'.[1] From the earliest times to the present day it has been one of the most popular and lucrative genres of literature. Victorians loved it and wrote biographies in phenomenal numbers. As Juliette Atkinson remarks, in 1880 alone the *Publishers' Circular* estimated publication of 363 biographies and histories.[2] All disciplines were involved, spanning the full range of arts and sciences. In the classic style of monograph music biography appears slow off the mark, beginning with Edward Holmes's *Mozart* (1847) but eventually crescendoing to a climax with several book series including the Great Composers and turn-of-the-century Master Musicians. Dictionaries and compilation biographies (volumes with biographies of more than one musician) follow a more or less parallel, if later, trajectory reaching an apogee in Grove's *Dictionary of Music and Musicians* (1879–90). Like music pedagogy, binding together all these different types of biographies is recapitulationism, itself rooted in the abiding evolutionary emblems of the Great Chain of Being – in theories of genius, the Great Man and the Fittest Survivor.

This chapter uses the genre of biography to trace and explain changes in recapitulationary theories within Victorian musical culture. Divided into two main sections, 'Victorian Biography' and 'Victorian Music Biography from Great Man to Fittest Survivor', the first section introduces basic background information on the genre and then delves into philosophical aspects of genius and their application in the literature of biography, emphasizing concepts of character and its manipulation under the influence of recapitulationary thought. The second section focuses on various

[1] Nigel Hamilton, *Biography: A Brief History*, Cambridge, MA, and London: Harvard University Press, 2007, 1.

[2] See 'English Literature in 1880', *Journal of the Statistical Society of London* 44 (March 1881), 96, cited in Juliette Atkinson, *Victorian Biography Reconsidered: A Study of Nineteenth-Century 'Hidden' Lives*, Oxford: Oxford University Press, 2010, 4.

sub-genres of music biography, beginning with a discussion of early nineteenth-century dictionaries of music. This leads to an exploration of recapitulationary thought in the work of Edward Holmes, C. Hubert H. Parry and others, tracing the process of biographical transformation from Carlylean Great Man to Spencerian Fittest Survivor. A conclusion speculates on the reasons why Victorian music biography failed collectively as a genre to refute the intractable criticism of England being a Land Without Music.

Biography

There is no single genre of Victorian biography. It might include life writing in pamphlets, tracts, tomes, religious biographies, lives of the saints, memorials sketches of philanthropists, tradesman local worthies, parents, family and friends. It might be authored, anonymous, pseudonymous, autobiographical, pseudoautobiographical, fiction based on fact, fact based on fiction, part-biography, part-autobiography, part-memoire, edited, selected, arranged and annotated.[3] Richard D. Altick describes it as 'an unstable compound';[4] David Amigoni speaks of its 'generic instability and hybridity';[5] and more recently Juliette Atkinson studies its 'many faces'.[6] Despite its diversity Atkinson divides Victorian biography into three phases, the 1830s and 1840s, and the middle and later parts of the century.[7] The earliest phase focuses on recently deceased men of letters such as Thomas Moore's *Life and Letters of Lord Byron* (1830), De Quincy's *Recollections of the Lake Poets* (1834–40) and John Gibson Lockhart's *Life of Scott* (1837–38). Stylistically these tended to reflect earlier Boswellian models in their relative candour and vivid personal portrayals. From Arthur Penrhyn Stanley's *Life of Arnold* (1844) biography became, arguably, more deferential creating a new kind of middle- and upper middle-class moral hero which helped prepare entry of mid-century's testimony to literary art, Carlyle's influential *Life of John Sterling* (1851).[8]

[3] Juliette Atkinson, *Victorian Biography Reconsidered*, 14–17.

[4] Richard D. Altick, *Lives and Letters: A History of Literary Biography in England and America*, New York: Alfred A. Knopf, 1965, 181.

[5] David Amigoni (ed.), *Life Writing and Victorian Culture*, Aldershot: Ashgate, 2006, 1.

[6] Atkinson, *Victorian Biography Reconsidered*, 14.

[7] Atkinson, *Victorian Biography Reconsidered*, 14–45.

[8] A. O. J. Cockshut, *Truth to Life: The Art of Biography in the Nineteenth Century*, London: Collins, 1974, 89–91.

Carlyle (and many others) responded to an increasingly pliant market in biographies of historical, literary and philosophical figures with works probing greatness in moral contexts like *Life of Friedrich Schiller* (1825), *Oliver Cromwell's Letters and Speeches, with Elucidations* (1845) and *History of Friedrich II of Prussia, called Frederick the Great* (1859). Carlyle's influence persists in the latest phase of Victorian biography in which great recently deceased men were again put to further account in seminal studies such as John Forster's *Life of Dickens* (1872–74) and George Otto Trevelyan's biography of his uncle, *Life and Letters of Lord Macaulay* (1876). By this time biographies offered 'a chance to assess the age they embodied'.[9] In some instances that assessment was ill-considered, as Froude's two volumes of *Thomas Carlyle* (1882 and 1884) proved. Froude eventually obtained Carlyle's consent to write a biography but drew upon material of the most personal nature – Carlyle's unkindness towards his wife and the implication of impotence, amongst other things. Froude was eviscerated in the press and a great biographer learnt a very damaging lesson about Victorian culture and mores: despite its more progressive tendencies conventional moral pieties could not yet be breached. The culmination of these three phases comes in collected works of greatness in the form of individual books or book series such as The English Men of Letters, The Twelve English Statesmen, Eminent Women, and in the field of music The Master Musicians and The Great Musicians. The most significant manifestation of this trend is Sidney Lee's *Dictionary of National Biography*, published between 1882 and 1901.

While theorists and historians may disagree about many aspects of biography on one point they are in complete agreement: the subject of any biography – musical or otherwise – must be exceptionally gifted *and* socially important. What defines that combination – in the broadest sense *genius* – was in the nineteenth century (and still is today) the subject of much debate, and its role in life writing is equally controversial. Victorians inherited their concept of genius through idealism traceable to Edward Young. Young considered true originality to be developmentally organic, spontaneous, preternatural, unprecedented and inspirational: 'Learning we thank, Genius we revere; That gives us pleasure, This gives us rapture; That informs, This inspires; and is itself inspired; For Genius is from Heaven, Learning from man. *This* sets us above the low, and illiterate; *That*, above the learned, and polite. Learning is borrowed knowledge;

[9] Atkinson, *Victorian Biography Reconsidered*, 20.

Genius is knowledge innate, and quite our own'.[10] For Young originals can only arise from genius, and genius is itself spontaneous and biological in origin:[11] 'An Original may be said to be of a *vegetable* nature; it rises spontaneously from the vital root of Genius; it *grows*, it is not *made*: *Imitations* are often a sort of *Manufacture* wrought up by those *Mechanics*, *Art*, and *Labour*, out of pre-existent materials not their own'.[12] Youngian concepts of organic genius appear with regularity from the middle part of the eighteenth century and would have great longevity into the nineteenth century. Music historian John Hawkins describes the French composer Marin Marais as bearing 'the pregnant marks of a fertile genius, united to an exquisite taste and judgment',[13] and some fifty years later the music lexicographer John Sainsbury would use not dissimilar terms. Mozart for Sainsbury was 'happily endowed' with genius – genius which was 'unfolded' and 'innate'; genius whose talent could do justice to even the simplest music 'as long as there was some traits of originality and genius in it'.[14] Inevitably, however, there were competitor models of genius and in some instances they were almost diametrically opposed to Young. Like Young, Alexander Gerard equates genius with the spontaneity of the imagination – for him (as for Coleridge after him) genius was 'an organic unifying power'[15] – but he also praises the very characteristics of invention Young derides in mechanism: 'Genius', he proposes, 'is properly the faculty of *invention*'.[16] Amongst music biographers Hawkins is particularly sensitive to this distinction, and coincidentally 'reiterates' it in the second edition of 1875[17] when he finds in Handel's organ playing 'the fullness of his harmony, the grandeur and dignity of his style, the copiousness

[10] Edward Young, *Conjectures on Original Composition*, London: A. Millar and R. and J. Dodesley, 1759, 36.

[11] Young, *Conjectures*, 34. [12] Young, *Conjectures*, 12.

[13] John Hawkins, *A General History of the Science and Practice of Music*, London: T. Payne, 5 vols., 1776, Vol. 5, 46.

[14] John S. Sainsbury, *Dictionary of Musicians from the Earliest Ages to the Present Time. Comprising the Most Important Biographical Contents of the Works of Gerber, Choron, and Fayolle, Count Orloff, Dr. Burney, Sir John Hawkins, &c. &c. together with Upwards of a Hundred Original Memoirs of the Most Eminent Living Musicians. And a Summary of the History of Music*, 2 vols., London: Printed for Sainsbury, 1824, vol. 2, 186; 195.

[15] Jonathan Bate, 'Shakespeare and Original Genius', in Penelope Murray (ed.), *Genius: The History of an Idea*, Oxford: Basil Blackwell, 1989, 93.

[16] Alexander Gerard, *An Essay on Genius*, London: W. Strahan and T. Cadell and Edinburgh: W. Creech, 1774, 8.

[17] John Hawkins, *A General History of the Science and Practice of Music*, new (2nd) edition, London: Novello, Ewer & Co. and New York: J. L. Peters, 2 vols., 1776/1875, Vol. 2, 912.

of his imagination, and the fertility of his invention'.[18] Other critics came forward: Samuel Johnson, for instance, complains that there is too much emphasis on 'unassisted genius and natural sagacity',[19] and Joshua Reynolds that too much is left to the 'the imaginary powers of native genius'.[20]

Despite criticism at home Young's organic genius would permeate thinking abroad in German idealism, and through Kant, Sulzer, Herder, Schelling and Goethe it would acquire a new biologically-infused focus on the imagination which would eventually influence Coleridge and the evolutionary framework of Victorian biography. Kant, for example, locates the very rules of beauty within genius so that organisms (such as the artist) and aesthetic objects (which the artist creates) can exemplify evolutionary purpose.[21] Genius cannot be imitated;[22] it is 'the talent (or natural gift) which gives the rule to Art. Since talent, as an innate productive faculty of the artist, belongs itself to Nature, we may express the matter thus: Genius is the innate mental disposition (*ingenium*), *through which* Nature gives the rule to Art'.[23] Schelling (whom Coleridge calls a 'great and original genius')[24] extends this formulation by collapsing the distinction between the genius and his art: both the genius and his art suppresses 'within a finite product an infinite opposition'.[25] Goethe tries to bring the genius down to earth by locating this opposition within time. Under Youngian laws of creativity works of genius (and eventually geniuses themselves) originate spontaneously in nature, yet precisely because they appear 'from nowhere' or 'out of the blue' works of genius present an ontological uncertainty, as if time played no part in their preparation. Goethe would

18 Hawkins, *A General History of the Science and Practice of Music*, Vol. 5, 413.

19 Samuel Johnson, [no title], *Rambler* 154 (7 Sept. 1751).

20 Sir Joshua Reynolds, *The Discourses of Sir Joshua Reynolds*, Discourse 6 (1774), London: T. Cadell, Jn., and W. Davies, 1801, 186.

21 Robert J. Richards, *The Romantic Conception of Life: Science and Philosophy in the Age of Goethe*, Chicago and London: University of Chicago Press, 2002, 70.

22 Simon Schaffer, 'Genius in Romantic Natural Philosophy', in Andrew Cunningham and Nicholas Jardine (eds.), *Romanticism and the Sciences*, Cambridge: Cambridge University Press, 1990, 89.

23 Immanuel Kant, *Kant's Critique of Judgement* (*Kritik der Urtheilskraft*, Berlin and Libau: Lagarde und Friederich, 1790), trans. J. H. Bernard, London: Macmillan and Co., 1914, 188.

24 Samuel Taylor Coleridge, *Biographia Literaria; Or, Biographical Sketches of My Literary Life and Opinions*, ed. James Engell and Walter Jackson Bate, 2 vols., London: Rest Fenner, 1817, vol. 1, 150.

25 Friedrich Wilhelm Joseph Schelling, *System of Transcendental Idealism* (1800), cited in Jacques Taminiaux, *Poetics, Speculation, and Judgment: The Shadow of the Work of Art from Kant to Phenomenology*, trans. Michael Gendre, Albany: State University of New York, 1993, 29.

solve the problem by referring to the role of invention through time, and locating works of genius within the developing storylines of civilization. Imbued with both past and future – not simply a 'spontaneous' present – works of genius communicate more than the timeless expression of nature (what William Jones describes as 'the voice of nature herself');[26] they reflect and embody the collective history of human invention.[27]

Coleridge would adopt this recapitulatory type of narrative and use it to illustrate how the creative imagination of genius works, especially in books loaded with evolutionary terms like 'permanence' and 'progress'.[28] There are two inter-related imaginations in Coleridge's estimation: a primary one which is the agent of finite perception, and a secondary one which he calls 'an echo … co-existing with the conscious will'.[29] The echo is effectively mechanical; the primary imagination, organic, and out of the imagination arises the form it creates: 'The form is mechanic when on any given material we impress a pre-determined form, not necessarily arising out of the properties of the material, as when to a mass of wet clay we give whatever shape we wish it to retain when hardened. The organic form, on the other hand, is innate; it shapes as it develops itself from within, and the fullness of its development is one and the same with the perfection of its outward form. Such is the life, such the form. Nature, the prime genial artist, inexhaustible in diverse powers, is equally inexhaustible in forms'.[30] This model would prove immensely attractive to Victorian music biographers influenced by biological development because it created narratives which recapitulate in life the very qualities possessed in the imaginative works of genius; together they reveal ultimately how the imagination of the genius resembles the universality of divine creation.[31] The same approach would reach its zenith in the biographical writings of C. Hubert H. Parry, whose determination to locate meaning in universality is an abiding feature

26 Sir William Jones, 'On the Arts, Commonly Called Imitative', in Sir William Jones, *Poems, Consisting Chiefly of Translations from the Asiatick Languages, to which are added Two Essays, I. On the Poetry of the Eastern Nations [and] II. On the Arts, Commonly Called Imitative*, Oxford: Clarendon Press, 1772, 207.

27 Michael Beddow, 'Goethe on Genius', in Murray, *Genius*, 106; and also Angus Nicholls, *Goethe's Concept of the Daemonic: After the Ancients*, Rochester, NY, and Woodbridge: Boydell & Brewer, 2006, 134.

28 Samuel Taylor Coleridge, 'On the Constitution of Church and States' (1830), in Coleridge, *The Complete Works of Samuel Taylor Coleridge*, 7 vols., New York: Harper & Brothers, 1853, vol. 6, ix.

29 Coleridge, *Biographia Literaria*, vol. 1, 296.

30 Coleridge, *Complete Works of Samuel Taylor Coleridge*, vol. 4, 55.

31 David Higgins, *Romantic Genius and the Literary Magazine: Biography, Celebrity, Politics*, London and New York: Routledge, 2005, 3.

of his biographical writing. In an evolution-inflected example Johann Sebastian Bach appears to exhibit an unerring ability to sift the universal wheat from the local or national chaff: 'The essential problem for the Teutonic composer was to segregate the adaptable from the unadaptable; to adopt that which was of universal application and to eliminate what was purely [in this case] Italian and local ... the greater minds instinctively reject the superfluous and the ephemeral, and take counsel of the permanent and the universal; and in this sense Bach always assimilated the features and methods from types of national art which were of value'.[32]

Universality, according to Coleridge, also distinguishes the great individual from the ordinary multitude which he represents, as well as a simple truth from a mere fact alone and 'some comprehensive truth'[33] from the brute realities of life. That same universality is also moderated by a genius's own historical context; according to John Wilson: 'He who, in the pride of his own age, believes himself independent of the ages to which he succeeds, shuts out from himself the highest influences under which it was given to his mind to live'.[34] Wilson removes genius from the intangible interiority of mind and relocates it in the exteriority of society; now public, genius comes under scrutiny for its adaptive, social and moral characteristics. Reflecting this orientation John Heraud defines genius in broadly Coleridgean terms, emphasizing its re-creative (rather than creative), almost redemptive nature: 'By abstraction from the world, and self-reflection, and self-improvement, and self-regeneration, genius will ultimately make man a partaker of the joys of heaven'.[35] Re-creation – in the form of self-reflection, self-improvement and self-regeneration – would be an abiding feature of Victorian music biography. Complaining about Wagner's criticism of Beethoven's orchestration, for instance, Frederick Crowest retorts with characteristically fulsome praise: 'it is absurd to imagine that the genius [Beethoven] would have left any known open door for improvement upon his masterly operations'.[36] Heraud takes a moral stand and

[32] C. Hubert H. Parry, *Johann Sebastian Bach: The Story of the Development of a Great Personality*, New York and London: G. P. Putnam's Sons and The Knickerbocker Press, 1910.

[33] Samuel Taylor Coleridge, 'A Prefatory Observation on Modern Biography', *The Friend* 21 (25 Jan. 1810), 338–39, cited in James L. Clifford (ed.), *Biography as an Art: Selected Criticism, 1560–1960*, London, New York and Toronto: Oxford University Press, 1962, 58.

[34] John Wilson, 'On the Analogy Between the Growth of Individual and National Genius', *Blackwood's Edinburgh Magazine* 6/34 (Jan. 1820), 380.

[35] John Abraham Heraud, 'On Poetic Genius, Considered as a Creative Power', *Fraser's Magazine* (Feb. 1830), 57.

[36] Frederick J. Crowest, *Beethoven*, London: J. M. Dent & Sons Ltd, and New York: E. P. Dutton, 1899/1913, 222.

denies genius to those who socially transgress their talent, such as the brilliant but morally bankrupt Bryon. For Heraud genius recapitulates more than God-given natural talent; its cultural context necessitates that it recapitulate the divine source of all ethical thought and being.

James Field Stanfield designates the purpose of biography similarly to 'obtain a deeper insight into the principles of the human mind, and to offer examples to practical observation and improvement'.[37] To do this the biographer must unite disparate incidents to 'advance one step further in a distinctive appreciation of general character – one degree more in the important science of cause and consequence'.[38] 'Science' is especially important for Stanfield, who distinguished in biography a twofold nature, both '*biology* and biography'[39] – the objective, natural growth of the biographer's subject, and recognition of biography's constructive subjectivity. The biographical subject is also contingent within society: 'Man absolutely solitary is a savage', he claims, '– in society a hero'.[40] Especially up to the 1830s advancing moralization catalyzed increasingly differentiated yet interrelated biographical interpretations of the 'real' and the 'ideal' nature of this kind of genius.[41] In *Ueber Lessing* (1801) for example, Schlegel speaks of Lessing in terms of 'Beziehung aufs Ganze' (relation to the whole), 'symbolischen Form' (symbolic form) and 'Individualität' (individuality).[42] 'In Seiner [Lessing's] Individualität lag seine Größe' (In his Individuality lies his Greatness), he claims.[43] This forms part of a larger movement in the development of 'Characteristik', a genre which attempts to reduce the multitudinous aspects of a life to its essentials.[44]

Carlyle would revisit this genre in his own essay 'Characteristics' (1831). For him the genius (or hero) is greater than the real and the ideal, representing their ethicized union in time and space – the ontogeny of the individual recapitulating the phylogeny of his age, 'the visible and

37 James Field Stanfield, *An Essay on the Study and Composition of Biography*, Sunderland: George Garbutt, 1813, 145.

38 Stanfield, *Essay on the Study and Composition of Biography*, 165.

39 Stanfield, *Essay on the Study and Composition of Biography*, 12.

40 Stanfield, *Essay on the Study and Composition of Biography*, 270.

41 Higgins, *Romantic Genius and the Literary Magazine*, 45.

42 Friedrich Schlegel, 'Ueber Lessing', in August Wilhelm Schlegel and Friedrich Schlegel (eds.), *Charackteristiken und Kritiken*, 2 vols., Königsberg: Friedrich Nicolovious, 1801, vol. 1, 266, 263 and 193.

43 Schlegel and Schlegel, *Charackteristiken und Kritiken*, vol. 1, 193.

44 Roger Paulin, 'Adding Stones to the Edifice: Patterns of German Biography', in Peter France and William St Clair (eds.), *Mapping Lives: The Uses of Biography*, Oxford: Oxford University Press, 2002, 110.

tangible representative of the total forces – intellectual, moral, historical – of an age or culture'.[45] By the end of the Victorian period this description would take on literally eucharistic properties. According to James Huneker, Wagner, for example, 'carried within his breast the precious eucharist of genius. In music he is the true Zeit Geist'.[46] Carlyle typifies this dynamic tension in his essay using the word 'unconsciousness' to support the idea that genius is hermetically 'ever a secret to itself': 'Unconsciousness is the sign of creation; Consciousness, at best, that of manufacture', he suggests.[47] The backdrop to Carlyle's apparent dismissal of the real is a profound sense of historical British self-degenerationism in which the return to an ideal itself signals evidence of a struggle to progress. In 'Characteristics' change is mechanistic and 'the product simply of *increased resources* which the old *methods* can no longer administer' but in 'Biography' (1832) it encodes a vital organic dimension, 'evolving more and more wisely its [Reality's] inexhaustible meanings'.[48] That organic, biographical dimension bursts into historical flower in Carlyle's lectures *On Heroes, Hero-Worship, and the Heroic in History* (1837–40) in which he codifies his recapitulationary theory of the Great Man: 'Universal History, the history of what man has accomplished in this world, is at bottom the History of Great Men who have worked here. They were the leaders of men, these great ones; the modellers, patterns, and in a wide sense creators, of whatsoever the general mass of men contrived to do or attain; all things that we see standing accomplished in the world are properly the outer material result, the practical realisation and embodiment, of Thoughts that dwelt in the Great Men sent into the world: the soul of the whole world's history, it may justly be considered, were the history of these'.[49]

Here the Great Man is triumphal 'creator' within 'That great mystery of Time'[50] – even perhaps a creator of time. The Great Man is called forth by time to become its voice. Luther, for example, is a 'creature of the Time'.[51] But Carlyle's time is degenerated into the barest semblance of itself, and the genius, therefore, retreats deeply into man's natural heavenly origins.

[45] Paulin, 'Adding Stones to the Edifice', 105.

[46] James Huneker, *Mezzotints in Modern Music: Brahms, Tschaïkovsky, Chopin, Richard Strauss, Liszt and Wagner*, New York: Charles Scibner's Sons, 1899, 292.

[47] Thomas Carlyle, 'Characteristics', review of Thomas Hope, *An Essay on the Origin and Prospects of Man*, 3 vols., London: John Murray, 1831, *The Edinburgh Review* 54 (Dec. 1831), 358, 363.

[48] Thomas Carlyle, 'Biography', *Fraser's Magazine* 5/27 (April 1832), 257.

[49] Thomas Carlyle, 'The Hero as Divinity' (1840), in Thomas Carlyle (ed.), *On Heroes, Hero-Worship and the Hero in History*, London: J. M. Dent & Sons, Ltd, 1840/1911, 239.

[50] Carlyle, 'Hero as Divinity', 246. [51] Carlyle, 'Hero as Divinity', 250.

He symbolically reminds us of our true human being living 'in the inward sphere of things, in the True, Divine and Eternal, which exists always, unseen to most, under the Temporary, Trivial ... His life ... is a piece of the everlasting heart of Nature herself'.[52] Yet the Great Man is also a 'Force', like time itself: 'Force, Force, everywhere Force', Carlyle cries, 'we ourselves a mysterious Force in the centre of that'.[53] As Force and Nature, as creation in and of time, the Great Man is the exemplar – the Hero – and as such the fittest representative of his age: 'then there is given a man fit to speak of this, to sing of this, to fight and work for this, in a great, victorious enduring manner; there is given a Hero, – the outward shape of who will depend on the time and the environment he finds himself in'.[54] The subject of Carlyle's *Life of John Sterling* is just this eucharstic force of nature: 'however the ground and form of it [his creed of life] might change, essentially it was the monition of his natal genius to this as it is to every brave man; the behest of all his clear insight into this Universe, the message of Heaven through him, which he could not suppress, but was inspired and compelled to utter in this world by such methods as he had'.[55] Early music psychologist William Wallace echoes Carlyle in his practically transcendental adoration of Wagner: 'The divine ictus that fell upon him struck into him a supreme belief in himself. He was from the outset inexorable in his conditions ... His mind was made up to be great and he became great. His faith in himself survived the wreckage of hopes and ambition; to him an obstacle was an incomprehensible thing'.[56]

Two final characteristics define the Victorian genius, musical or otherwise: reputation and concentration. For Coleridge men of the greatest genius exhibit an 'inward assurance of permanent fame' which they use to eschew 'immediate reputation'[57] but by the middle of the century Alexander Bain would claim otherwise: 'If genius is to be understood as the promulgation of ideas beyond the age, sanctioned by their complete recognition in succeeding ages, we ought not to refuse the name to the promoter of our new moral ideas, or more enlightened impulses towards our fellow-men'.[58] Morally

[52] Thomas Carlyle, 'The Hero as a Man of Letters' (1840), in Carlyle, *On Heroes*, 384.

[53] Carlyle, 'Hero as Divinity', 246.

[54] Thomas Carlyle, 'The Hero as Priest' (1840), in Carlyle, *On Heroes*, 346.

[55] Thomas Carlyle, *The Life of John Sterling*, Boston: Phillips, Sampson and Co., 1851, 68.

[56] William Wallace, *The Threshold of Music; An Inquiry Into the Development of the Musical Sense*, London: Macmillan and Co., Ltd, 1908, 184.

[57] Coleridge, *Biographia Literaria*, 25–26.

[58] Alexander Bain, *On the Study of Character including an Estimate of Phrenology*, London: Parker, Son & Bourn, 1861, 336.

disinterested thinkers like Francis Galton would enshrine this interpretation in hereditary science, giving genius an evolutionary slant. According to Galton genius is natural ability leading to eminence and enduring reputation: 'those qualities of intellect and disposition, which urge and qualify a man to perform acts that lead to reputation';[59] more specifically it reflects 'the opinion of contemporaries, revised by posterity – the favourable result of a critical analysis of each man's character, by many biographers . . . the reputation of a leader of opinion, an originator, of a man to whom the world deliberately acknowledges itself largely indebted'.[60] An important corollary to reputation is heightened powers of concentration. These depend upon 'the possession of the intellectual functions (discrimination, &c.) in an exceptionally perfect form',[61] and 'the power of sustaining mental conflict'.[62] William James develops this characteristic to include thinking in an 'unhabitual'[63] way which survives the regularity of normal perception. It is 'the possession of similar association',[64] a type of mind which has the extraordinary capacity to relate seemingly disparate objects yet from them arrive at most universally applicable analytical principles. The greatest genius – like Wallace and Huneker's Wagner, Parry's Bach or Crowest's Beethoven – is the one in whom the highest stage of intellectual functions survives transmission through the obstructively lower stages of intuitive development.

Music Biography

Unlike its counterparts in general literature English-language music biography (written originally in English) has few substantive antecedents before the early part of the nineteenth century. There is nothing in the eighteenth century akin to Johnson or Boswell apart from John Mainwaring's *Memoirs of the Life of the Late George Frederic Handel* (1760), translated into German by Johann Mattheson as *Georg Friedrich Händels*

[59] Francis Galton, *Hereditary Genius: An Inquiry into its Laws and Consequences*, London: Macmillan and Co., Ltd, 1869/1914, 33.

[60] Galton, *Hereditary Genius*, 33.

[61] James Sully, *Outlines of Psychology with Special Reference to The Theory of Education*, London: Longmans, Green & Co., 1884, 100.

[62] G. F. Stout, *Analytical Psychology*, 2 vols., London: Swan, Sonnenschein & Co., Ltd, and New York: Macmillan & Co., 1896, vol. 2, 162.

[63] William James, *The Principles of Psychology*, 2 vols., New York: Henry Holt and Co., 1890, vol. 2, 100.

[64] James, *Principles of Psychology*, vol. 2, 360.

Lebensbeschreibung, nebst einem Verzeichnisse seiner Ausübungswerke und deren Beurtheilung (1761) and recognized as the first English music biography.[65] A dribble of English translations appears in the early part of the nineteenth century, including Robert Brewin's *The Life of Haydn, in a Series of Letters Written at Vienna. Followed by The Life of Mozart, with Observations on Metastasio, and on the Present State of Music in France and Italy* (1817), a translation of Louis-Alexandre-César Bombet's *Lettres écrites de Vienne en Autriche: sur le célèbre compositeur Jh. Haydn, suivies d'une vie de Mozart, et de considérations sur métastase et l'état présent de la musique en France et en Italie* (1814); and the *Life of Bach: With a Critical View of his Compositions* (1820), a translation of Johann Forkel's *Über Johann Sebastian Bachs Leben, Kunst und Kunstwerke* (1802) attributed to Stephenson. Translations later proliferate as the genre progresses, with mid-century translations on Schubert, Beethoven, Bach, Handel, Mendelssohn, Mozart, Haydn, Chopin, Schumann, Brahms and Tchaikovsky.[66]

Although biography forms an integral component in the Victorian musical press[67] it is only from the middle of the nineteenth century that English biographical works of substance and depth begin to appear. The earliest is probably William Bingley's *Musical Biography; or, Memoirs of the Lives and Writings of the Most Eminent Musical Composers and Writers* (1814) but this could hardly be said to represent anything like the full-scale works more commonly associated with 'classic' full-length biographies. Like many biographies of this period that of Bingley is rather hybrid and combines genres of both biography and history in what Joseph Reed, Jr, designates 'historical polytheism'[68] – the subsumption of great lives under the greater movements of their times. Bingley exemplifies this by using wider sweeps of history to fill biographical gaps. His extended introduction is particularly troublesome for that reason, with its dearth of focussed biographical content, but the first chapter (on English composers) hits its stride with the controversial reformer John Marbeck (c. 1510–c. 1585). Marbeck's entry is indicative of Bingley's style – deferential, unaggrandizing, dispassionate and even subtle. Inferring his character from the

[65] Christopher Mark Wiley, 'Re-writing Composer's Lives: Critical Historiography and Musical Biography', PhD diss., Royal Holloway University of London, 2008, 12 and 39–40.

[66] Wiley, 'Re-writing Composer's Lives', 197.

[67] See for example: Therese Ellsworth and Susan Wollenberg (eds.), Introduction, *The Piano in Nineteenth-Century British Musical Culture: Instruments, Performers and Repertoire*, Aldershot: Ashgate, 2007, 10.

[68] Joseph W. Reed, Jr, *English Biography in the Early Nineteenth Century, 1801–1838*, New Haven and London: Yale University Press, 1966, 15.

recapitulationary nature of his work Bingley brings Marbeck back from the brink of martyrdom to become a celebrated scholar and composer of reforming High-Churchmanship. Seldom is Bingley uncritical, however. Like his life Marbeck's music has 'each an harmonical relation with the whole'[69] even if it is based upon previously discredited formulations of Rome. Thomas Tallis's 'astonishing effort of harmonical abilities', the 'Song in Forty Parts', is equally telling as a 'stupendous, though perhaps Gothic, specimen of human labour and intellect'.[70] And William Bird's genius was left unfulfilled: had he lived in another time 'his genius would have expanded itself in works of invention, taste and elegance'.[71] Purcell suffered from none of these limitations: 'The unlimited powers of his genius embraced every species of composition that was then known with equal felicity'.[72] Indeed so great is the totality of Purcell's genius that he encompasses both types of originality, invention *and* creation, 'whether he adhered to the elaborate and learned style of his predecessors . . . or (giving way to feeling and imagination) he adopted the new and more expressive style, of which he was himself one of the principal inventors'.[73]

Bingley's 'biographic spirit'[74] (to use Stanfield's term) may be historically polytheistic but it points to a biographical future conceptually unified in recapitulationary thought – when the seeds of recapitulation blossomed in the achievements of successive generations of composers. Bingley's contemporary Thomas Busby tells this recapitulationary story through the increasingly popular concept of improvement. *A General History of Music from the Earliest Times to the Present; Comprising the Lives of Eminent Composers and Musical Writers* (1819) claims that the history of music 'no less the child of fancy than of science' reveals an 'infinite susceptibility to improvement'.[75] Bingley and Busby's composers may reveal an infinite susceptibility to improvement, but some, like Handel, reveal more infinite susceptibility than others. Bingley's Handel evinced an 'extraordinary ear' and an 'extraordinary manner'[76] and from this would progress to compose melodies exposing 'the rich vein of genius that runs

[69] William Bingley, *Musical Biography: Memoirs of the Lives and Writings of the Most Eminent Musical Composers and Writers Who Have Flourished in the Different Countries of Europe During the Last Three Centuries*, 2 vols., 2nd ed., London: Henry Colburn, 1814/1834, vol. 1, 27.

[70] Bingley, *Musical Biography*, vol. 1, 34–35.

[71] Bingley, *Musical Biography*, vol. 1, 42.

[72] Bingley, *Musical Biography*, vol. 1, 206.

[73] Bingley, *Musical Biography*, vol. 1, 206–7.

[74] Stanfield, *Study and Composition of Biography*, 21.

[75] Thomas Busby, *A General History of Music from the Earliest Times to the Present; Comprising the Lives of Eminent Composers and Musical Writers*, 2 vols., London: G. and W. B. Whittaker and Simpkin and Marshall, 1819, vol. 2, 83.

[76] Bingley, *Musical Biography*, vol. 2, 125.

through them, inimitable'.[77] Busby's Handel positively swells in talent: 'How then shall the sober tints of historical painting do justice to the diversity and the splendor [*sic*] of his genius? . . . Every species of voice, and every kind of instrument, owed new charms to the fruits of his imagination'.[78] Bingley and Busby's contemporary John Sainsbury would subscribe to a similar biographical mindset in his *Dictionary of Musicians from the Earliest Ages to the Present Time. Comprising the Most Important Biographical Contents of the Works of Gerber, Choron, and Fayolle, Count Orloff, Dr. Burney, Sir John Hawkins, &c. &c. together with Upwards of a Hundred Original Memoirs of the Most Eminent Living Musicians. And a Summary of the History of Music* (1824). Handel's irrepressible natural talent 'could not be subdued'[79] by the restrictions imposed by his father (his environment); even he recognized his son 'to have been endowed by nature with very uncommon propensities and faculties'.[80] Like Bingley and Busby Sainsbury claims that Handel had 'a genius for composition original and inexhaustible'.[81] There are, however, geniuses of such stratospheric susceptibility to improvement that other geniuses pale in insignificance by comparison – Mozart, for example.

It is not surprising that the first full-length music biography is a *Life of Mozart* (1845), and needless to say Edward Holmes indulges the nature of his almost supranatural Youngian genius: at age six and without any recourse to the piano 'his power in mental music constantly increased, and he soon imagined effects, of which the original type existed only in his own brain'.[82] Three aspects of this description require comment: first, Mozart is extremely young but already boasts the fully formed ability of an adult; second, he needs no external stimulus (a piano) to create; and third and last, his mind produces an 'original' of which 'imagined effects' are the consequent. Mozart's genius is a paradigm of Coleridgean organic form – it is 'innate': it 'shapes as it develops itself from within' and 'the fullness of its development is one and the same with the perfection of its outward form'.[83] As a 'vegetable' (if it can be put that way) his genius was oblivious to the implications of its talent: 'The wonder that his talents created, and the applause that he received, had no ill effects; he remained a

[77] Bingley, *Musical Biography*, vol. 2, 163. [78] Busby, *General History of Music*, vol. 2, 365.
[79] Sainsbury, *Dictionary of Musicians*, vol. 1, 317.
[80] Sainsbury, *Dictionary of Musicians*, vol. 1, 317.
[81] Sainsbury, *Dictionary of Musicians*, vol. 1, 320.
[82] Edward Holmes, *The Life of Mozart*, London: J. M. Dent & Sons and New York: E. P. Dutton & Co., 1845, 9.
[83] Coleridge, *Complete Works of Samuel Taylor Coleridge*, 55.

simple and affectionate child'.[84] Mozart was also 'unconscious'. His first imagination – his unconscious – was in constantly heightened communion with his second imagination co-existing with his conscious will. The ramifications of this combination endow Mozart with not only extraordinary creativity but a perfect sonic memory as well – exemplified by his astonishing aural transcription of the Allegri Miserere, having heard it only once. With such boundless (and perfectly formed Coleridgean) talent at his disposal Mozart's progress from boyhood to youth accumulated a repository of creative ideas: 'his memory in after life became a perfect storehouse of melodies and subjects that had long been floating in his imagination, and which exquisite tact and judgment enable him instantly to apply'.[85] After a few years wowing audiences in Salzburg (1773–77), however, Mozart's luck would change and advantage would soon turn into adversity. Holmes prepares the transition by remarking on a decline in public attitude towards Mozart in his passage from boyhood to youth. Events begin to foreshadow Mozart's downfall, and fortune tellers, like Grimm and Haydn, predict the worst. Mozart responds naturally by intensifying his progress, composing pieces 'of which no model existed',[86] but while such originality fed on adversity he was constantly stalked by failure. Paris, following on the heels of Salzburg, was no better: 'some crooked destiny was at work to render all his labours fruitless'.[87]

True to the genre Holmes uses Mozart to establish a clear correlation between suffering and creativity, and between creativity and progress. Suffering becomes the Coleridgean primary imagination which catalyses the second imagination into creation; the 'echo' forges progress. Mozart is now a God and a force of nature, unstoppable like time itself: 'as well as borne out by its internal evidences of prodigal invention – of the exuberant mind rejoining and luxuriating in its new creations, and scarcely knowing when or where to stop'.[88] As a force Mozart's genius lies in his ineluctable progress and the extent and depth of its historical impact: 'Estimated by the universality of his power – the rapidity of his production, and its permanent influence on art, the models he created, and the constantly advancing march of his genius'.[89] As a model and creator of models Holmes's Mozart is Coleridge's comprehensive truth[90] but he is also Carlyle's hero writ large, not simply because of Mozart's own personal

[84] Holmes, *Life of Mozart*, 15.
[85] Holmes, *Life of Mozart*, 70.
[86] Holmes, *Life of Mozart*, 75.
[87] Holmes, *Life of Mozart*, 113.
[88] Holmes, *Life of Mozart*, 139.
[89] Holmes, *Life of Mozart*, 281.
[90] Coleridge, 'A Prefatory Observation on Modern Biography', 58.

heroism in the face of tragedy and despair, nor because of his posthumous reputation as universal hero, but because of his force of character and character's place in the order of genius. As well as history Holmes's *Mozart* is extended biographical 'Characteristik', depicting the composer's moral evolution alongside his musical talent. Mozart is both biology and biography to use Stanfield's words. For him and others – Carlyle, for example – the biographical humanizes the biological, and for them 'character' not only humanizes it universalises, and if it universalises it recapitulates. Only universality of character in combination with extraordinary individual talent affords the essential prerequisites for genius and heroism. The portrayal of character thus becomes an essential ingredient in the paradoxical nature of his biographical position, as someone failed in life yet sanctified in death – as someone who through suffering can recapitulate the totality of human moral, intellectual and historical endeavour. In the broadest possible sense, suffering builds character, and character is a force which denotes progress. Character is what gives the rule to talent, to paraphrase Kant, and only then does talent give the rule to art.

The poetics of Victorian biography frequently indulge this 'rule' and in Carlyle it assumes its characteristically monumental mid-Victorian historical dimension; as he says, 'The History of the World . . . was the Biography of Great Men'.[91] Music biographies follow suit but not for some time after Holmes's *Mozart*. It seems that it was not until the 1870s and 1880s – when British musicology had begun to professionalize through the advent of the Musical (then Royal Musical) Association in 1874 – that Victorian music biography began to take off in any serious way. After the phase of Bingley, Busby and Sainsbury music biography received its first real boost with the publication of George Grove's *Dictionary of Music and Musicians* (1879–90) in which potted biographies, or 'characteristik' – some more extended than others – appear in great number. Like their counterparts in general dictionaries and encyclopaedias they tend to be short, concise and objective. Longer articles expectedly appear in proportion to established greatness – Groves's own 'Beethoven' for example. Although Grove's entry on Beethoven serves purposes other than sheer estimation of talent and character (it includes a lengthy analysis of music as well) it is in many ways representative of late-century musical biography and its recapitulationary characterization of the Great Man. Despite his factual objectivity Grove, like Holmes, contrives a figure with Carlylean 'force' emanating from a

[91] Carlyle, *On Heroes*, 251.

vital, germinal creativity aligned with – communing with – nature: the 'simplicity which lay at the root of so many of his characteristic traits ... gave extraordinary force and freshness to much that he did and said'.[92] Elsewhere Beethoven 'uses the force' to overcome social adversity, showing what 'an immense power there must have been in his genius, and in the absolute simplicity of mind'.[93] He is 'full of trust in God, love to man, and real humility',[94] and fittingly for 'characteristik' ugly appearance belies beautiful character. His music reflects these environmentally advantageous characteristics. His first period (there are three) is described as having 'extraordinary force and mastery – though abounding also in beauty and sentiment'.[95] In the same period Beethoven progresses from being a great musician to being a great teacher, if not moralist, 'and in a manner which no one ever did before and possibly no one will ever do again, conveys lessons which by their intense suggestiveness have almost the force of moral teaching'.[96] And from being a great teacher Beethoven becomes a fully fledged Great Man in its fullest Carlylean meaning. Quoting Edward Dannreuther Grove venerates Beethoven, worshipping him as one who 'passes beyond the horizon of a mere singer and poet, and touches upon the domain of the seer and prophet; where, in unison with all genuine mystics and ethical teachers, he delivers a message of religious love and resignation, identification with the sufferings of all living creatures, deprecation of self, negation of personality, release from the world'.[97]

As with all Great Men their works (artistic, literary, political or religious) not only prove that the ontogeny of their originality recapitulates the phylogeny of their character, the development of their character recapitulates the phylogeny of musical progress as a whole. This is a common feature of many late Victorian compilation biographies (in which several composers are discussed in one book, one chapter per composer) and series biographies, including the Great Musicians (1881–90), Master Musicians (1899–1906) and early twentieth-century, post-Victorian series like Bell's Miniature Series of Musicians (1903–13) and Great Days with

92 George Grove, 'Beethoven', in George Grove (ed.), *A Dictionary of Music and Musicians (A.D. 1450–1880) by Eminent Writers, English and Foreign*, 4 vols. plus index, London: Macmillan and Co., 1879–90, vol. 1 (1879), 172.

93 Grove, *Dictionary of Music and Musicians*, vol. 1, 167.

94 Grove, *Dictionary of Music and Musicians*, vol. 1, 169.

95 Grove, 'Beethoven', in Grove, *Dictionary of Music and Musicians*, vol. 1, 202.

96 Grove, *Dictionary of Music and Musicians*, vol. 1, 202.

97 Edward Dannreuther, *Macmillan's Magazine* (July 1876), cited in Grove, 'Beethoven', in George Grove (ed.), *Dictionary of Music and Musicians*, 202.

Composers (1910–15). Series biographies all refer to the same methodological template beginning with genius in ancestry and moving to precocity in childhood, industriousness, strength through suffering, death as apotheosis, and religious motivations and aspirations – what Christopher Wiley calls the 'historical linearity upon which biography is ontologically dependent'.[98] Compositions mirror the personal circumstances of individual composers, fostering a clear correlation between life and stages of musical development, and in many instances they give lives suitably climactic shape by drawing attention to the drama of later works (as in Mozart and Brahms's *Requiems*, Mendelssohn's *Elijah*, Haydn's *Creation* or Tchaikovsky's Sixth Symphony). Dramatizing biographical moments to highlight artistic progress (in essence constructing the biography to conform to expectation) is nothing new to the genre of biography; Richard Holmes describes it as 'inventing the truth', subjectively prioritizing certain facts over others to proportion the meaning of events across the biography.[99] In Carlylean biography the tangibility of fact (the real) and the intangibility of fiction (the ideal) tussle endlessly, even if the best biographies give the illusion that the author is invisible and the life is speaking for itself.[100] Writing in *The Contemporary Review*, for example, Robert Goodbrand observes the dilemma in two mutually dependent streams of biography, one which apotheosizes a life and one which selectively disregards its inconsistencies. There is a more honest, balanced alternative, however; biography should reveal the fact that 'for each human life there is a separate and peculiar and absolutely individual ideal . . . the solitary and separate ideal for him – the ideal, or rather idea, on which and for which the subject of our biography was originally framed, and by his approximation to which he is to be judged'.[101]

More 'honest' Boswellian approaches to biography filter seamlessly into the portrayal of genius in single monographs or series. While overwhelmingly redeeming in character Mozart, for example, is often portrayed as volatile, profligate and pranksterish; Beethoven is rough and dismissive;

98 Christopher Wiley, '"A Relic of an Age Still Capable of a Romantic Outlook": Musical Biography and The Master Musicians Series, 1899–1906', *Comparative Criticism* 25 (2004), 184.

99 Richard Holmes, 'Biography: Inventing the Truth', in John Batchelor (ed.), *The Art of Literary Biography*, Oxford: Clarendon Press, 1995, 15.

100 Paula R. Backsheider, *Reflections on Biography*, Oxford: Oxford University Press, 1999, 3.

101 Robert Goodbrand, 'A Suggestion for a New Kind of Biography', *The Contemporary Review* 14 (1870), 20–28, cited in Ira Bruce Nadel (ed.), *Victorian Biography: A Collection of Essays from the Period*, New York and London: Garland Publishing, Inc., 1986, 25.

Wagner, imperious and condescending. In fact the more seemingly 'honest' Boswellian approach to biography would appear to erode confidence in the hero and more importantly disrupt his progress. Rather than progressing teleologically towards perfection he is dogged by egregious but perfectly explainable flaws in his own personality. Nature has bequeathed him a mixed blessing – creative genius and a destructive, degenerational character. This is taken to its extreme in Cesare Lombroso's *The Man of Genius* (1891), for example, in which genius is linked to madness. In its softer form in late Victorian music biographies character flaws are not generally as pronounced as that, and as a result have less influence on the nature of personal development. But when they are more threatening they can seriously thwart progress – Wagner's political forays a case in point. In cases like these, however, music invariably comes to the composer's rescue and absolves him of his personal sins. Music plays a redemptive role for the composer, rationalizing imperfections in the character of his genius and vaccinating him against disruptions in progress. It becomes his panacea, perfect in itself and inexorable in its teleological progress through time. Music expresses the universal ideals his own life fails to embody, and represents the higher ideal of his character. Despite the vicissitudes of his personal circumstances, for instance, Mozart's 'power developed steadily from first to last',[102] and despite the tragedies (some self-inflicted) of Beethoven's life his 'was a slowly unfolding genius, properly and naturally developed, that led up to that wondrous aggregation – Beethoven – the master'.[103] For Mozart, Beethoven and many other composers the 'ideal' of their music compensated for the 'real' of their lives, investing their music with a character not consistently evident in even the most exemplary life. Music biographies often exonerate composers, however, even if they register some degree of moral disapproval.

Whatever Mozart's personal deficiencies Eustace Breakspeare recognizes only the finer points of his composition-related character:

> 1. Intuitive (and inimitable) apperception of the subtle affinities and artistic tone-material. 2. Exhaustive expression and interpretation of the manifold and widely-contrasting types and modes of character and individuality as shown in life. 3. Ideal beauty of the formal means (*e.g.* Melody *par excellence*; exquisite arrangement, proportionment, and balance of the Thematic contents, and its parts, etc.)

[102] Eustace J. Breakspeare, *Mozart*, London: J. M. Dent & Sons, Ltd, and New York: E. P. Dutton & Co., 1902/1922, 93.

[103] Crowest, *Beethoven*, 121.

chosen for the clothing and conveyance of the artistic *ideas* – (so far, indeed, as with an art like music, distinctions of *idea* and *form* are at all possible).[104]

Character displacement (focussing on the character of a genius's work rather than his personal character) is a standard feature of Carlylean biography. George Bentley speaks of knowing 'two Carlyles', for example,[105] and in music and general biography alike it imbued recapitulation with a morally redemptive literary function. Recapitulation helped readers refine their moral spectrum by gauging ethical sincerity within their biographical subjects, and it bolstered their faith in progress by imputing universal principles of human development within the individual character of genius. In this way music – not the musician – became the voice of morality. The most central Victorian figure in the area or music and morals is arguably the Rev. H. R. Haweis, author of *Music and Morals* (1871).[106] Like John Hullah Haweis supposes music to be a relatively recent – and therefore highly developed – accretion to the history of art, four hundred years old to be exact. Haweis begins with a Victorian commonplace: music is the language of the emotions; but for him music does more than represent emotions – it *is* the emotions, expressing and commanding them. Emotions, he asserts, are 'coextensive with consciousness'.[107] In its most highly developed forms of civilization emotions intensify as their representation becomes increasingly abstract, Turner's paintings for example. Music, as the most abstract of all arts, is therefore the fittest art to convey and embody emotions, and the nineteenth century is its most advanced century.

Nineteenth-century music reflects 'the Spirit of the Age' and as shown by Kant, Hegel and Goethe it is 'at once an analytical Science and a pure Art-medium of Emotion ... Music is pre-eminently the art of the nineteenth century, because it is in a supreme manner responsive to the emotional wants, the mixed aspirations and the passionate self-consciousness of The Age'.[108] Put simply music recapitulates and to some extent regulates emotions. And those emotions must, if they are to be socially advantageous, result in a positive moral response in the listener: 'Everything, it may be said, music included, which excites an emotion not

104 Breakspeare, *Mozart*, 96.

105 George Bentley, 'Sincerity in Biography', *Temple Bar* 11/63 (1881), 329–36, in Nadel, *Victorian Biography*, 331.

106 See Elizabeth Blackmore, 'The Angelic Choir: Rethinking Female Voices in Anglican Sacred Music, c. 1889', MMus diss. Durham University, 2015.

107 Rev. H. R. Haweis, *Music and Morals*, London: Strahan and Co., 1871, 11.

108 Haweis, *Music and Morals*, 38.

destined to culminate in action, has a weakening and enervating effect upon character'.[109] This is the Victorian state of music 'in its highly-developed modern form'.[110] For Haweis the composer is deeply implicated in this progress, arguably shouldering the greatest part of its moral burden. It is after all the composer who originates musical expression, but like the reader of biography it is the listener of music (as well as its performer) who receives a composition's moral message. If according to Haweis moral musical composition is universal, then if universal it must also be more evolved: thus 'Civilized music must ultimately triumph over every other kind of music, because it is based upon natural principles discovered once and for ever, and capable of being universally applied and understood'.[111] Music's moral compass guides many Victorian music biographies. According to Reginald Poole a good example is the C sharp minor fugue in the first part of Bach's *Wohltemperierte Clavier*. Its theme provides almost homiletic instruction in its congregation of voices:

> It is the character-drawing of the several voices, and the nobility of them, that make their discourse sublime – three voices entirely contrasted and entirely blended – each time with a new and surprising effect, now of pomp, now of tenderest pathos – one a slow organ-voice, the next delicate and flowing, in the third vehement, striking hammer-blowers. The second and then the last gradually die away; the solemnity of the original theme communicates itself again to the whole web of thought, and the end is plaintive and restful.[112]

The representation of individual compositions as a totality of life is not uncommon in biographical writing for which moral character is essential. Stephen Stratton often uses the word 'dignity' to approve works in such a way. Mendelssohn's Italian Symphony lacks dignity, while the introduction and Coda to the Finale of the later (though published earlier) Scottish Symphony 'are full of dignity, the one that of pathos, the other of triumph'.[113] Dignity features regularly as a concern of much moralizing criticism. According to C. F. Abdy Williams the overture to Handel's *Messiah* was initially considered 'as not sufficiently dignified for its position'.[114] Later generations would revise their criticism as they grew to understand the work, proving that 'The "things of this world" seemed to

[109] Haweis, *Music and Morals*, 51. [110] Haweis, *Music and Morals*, 52.

[111] Haweis, *Music and Morals*, 129.

[112] Reginald Lane Poole, *Johann Sebastian Bach, 1685–1750*, London: Sampson Low, Marston & Co., Ltd, [n.d., 1880s], 115.

[113] Stephen S. Stratton, *Mendelssohn*, London: J. M. Dent & Sons, Ltd, 1901/1910, 205.

[114] C. F. Abdy Williams, *Handel*, London: J. M. Dent & Co., 1901, 213.

have no effect on the workings of his [Handel's] genius', and that, conversely, they very probably 'drove him to concentrate his powers more than ever on his lofty subject'.[115] For Abdy Williams morality – true human understanding – lay within the development of the work, note by note, piece by piece and section by section, exposing the real inner depth and universality of the composer and his composition. This defines the real recapitulationary genius of Handel, who as a composer evinced 'the loftiest possible ideal'.[116]

For Victorians morality claimed a long recapitulationary pedigree. George Romanes (and many other evolutionary psychologists of the time) is happy to confer upon animals and children emotional capacities including the experience of fear, surprise, affection, anger, benevolence, rage, shame and so on, but not those emotions which refer to religion, morality and the perception of the sublime. Those emotions are, it appears, uniquely human, representing a higher, more advanced and adult order of development. The intellect is the same in both animals and man up to the highest level an animal can attain. Beyond that level the human 'belongs to the very essence of evolution, considered as a process, when one order of existence passes on to higher grades of excellence, it does so upon the foundation already laid by the previous course of its progress'.[117] Romanes locates morality at the evolutionary junction of apes and man, on the cusp of human consciousness somewhere between the imagination and self-conscious thought and sensation and abstraction – in other words somewhere between the Carlylean real and ideal. The Victorian music biographer, equally keen to verify development through the acquisition of morality, positions music on a similar evolutionary cusp, poised between naturalness of creativity and the artifice of invention, between morally undifferentiated emotion and its more evolved counterpart in the intellect. The greatest compositions therefore produce the greatest moral recapitulation, Beethoven's *Eroica* for example. As soon as Beethoven heard news of Napoleon's accession to the throne he tore up the title page in a frenzied rage. H. A. Rudall responds:

> His idol had fallen; but not the hero of his imagination, whose struggles, triumph, and death are celebrated in this most heroic of tone poems. By the magic of genius, the external event first selected for his subject matter had been gradually idealized

[115] Williams, *Handel*, 213. [116] Haweis, *Music and Morals*, 211.
[117] George John Romanes, *Mental Evolution in Man: Origin of Human Faculty*, New York: D. Appleton and Co., 1889, 9.

and enlarged, and as the scope of his original intention widened, the theme, from being individual, grew to be representative of a type.[118]

It is probably Spencer who in his essay 'On the Origin and Function of Music' (1857) provides the methodological starting point for this kind of moral-musical recapitulation. As stated previously his version of musical development begins in the impassioned speech excited by the muscular activity of our ancestor's instinctual responses, and ends in the highly developed music of Western civilization. Music not only satisfies a pleasure principle, however, it also has the 'indirect effect of developing this language of the emotions ... so music, having its root in emotion language and gradually evolved from it, has ever been reacting upon and further advancing it'.[119] Framed by the intellect the emotions communicate and recapitulate morality, enabling 'the hearer not only to *understand* the state of mind they accompany, but to *partake* of that state. In short they are the chief media of *sympathy*'.[120] And unsurprisingly sympathy has an evolutionary purpose; advanced societies repress disadvantageous elements of character while favouring those which elicit benefit. For Spencer 'music must take rank as the highest of the fine arts – as the one which, more than any other, ministers to human welfare'.[121] Victorian music biographers would have roundly agreed, whether expressing their opinions in lengthy books or more modestly sized portraits. For them there is a direct connection between music and morality, morality and character, character and reputation, and ultimately reputation and genius. This is especially true of compilation biographies, the direct inheritor of 'characteristik'. In their shorter more concise biographical formats they offer readers more intensely revealing moral and intellectual insights. Books like *Famous Musical Composers, Being Biographies of Eminent Musicians* (1891) follow the same format as specialist music dictionary entries (length proportional to greatness), even if their purpose is more literary than musicological. Lydia T. Morris, for instance, aims 'to furnish some knowledge of their characters ... leaving the criticism of their music to abler persons'.[122] C. Hubert H. Parry's *Studies of Great Composers* (1887) is another good example of the compilation biography but unlike Morris Parry is an

[118] H. A. Rudall, *Beethoven*, London: Sampson, Low, Marston & Co., [n.d., 1880s], 87.

[119] Herbert Spencer, 'The Origin and Function of Music', *Fraser's Magazine* (Oct. 1857), in Herbert Spencer, *Literary Style and Music*, New York: Philosophical Library, 1951, 71–72.

[120] Spencer, 'Origin and Function of Music', 74.

[121] Spencer, 'Origin and Function of Music', 76.

[122] Lydia T. Morris, *Famous Musical Composers being Biographies of Eminent Musicians*, London: T. Fisher Unwin, 1891, 6–7.

accomplished musicologist and voracious reader of philosophy, science and literature. It is in Parry that we first observe the metamorphosis of biographical subjects from Carlylean Great Man into Spencerian Fittest Survivor.

Already in 1884 Parry had formulated his theory of musical history based on Spencerian refractions of German morphology. From Haeckel the principle of recapitulation (ontogeny recapitulates phylogeny); from Von Baer came the idea that all species progress from simplicity to complexity, from the homogenous to heterogeneous; and from Spencer, the concept of the survival of the fittest. These coalesce in an early article for the *Proceedings of the Musical Association* (1884) in which Parry argues that biographical hero worship has actually undermined our understanding of the Great Man. Far from enshrining the Great Man in the continuum of history the Carlylean cult of biographical divinity has given the impression that he appears spontaneously (like Young's vegetable genius) without the concourse of history to help inform his work. In other words the Carlylean construct of the Great Man emphasizes the individual at the expense of his historical influences, and rather than fully recapitulating the history of his art the Great Man has been removed from time:

> it has been customary to fix all attention upon isolated individuals to such an extent that they have seemed to have no connection with anything that went before them, and very often the most erroneous impression has been produced by the way in which their greatness has been estimated and described ... It inclines to that common weakness of looking upon an exceptionally able man as a divinely appointed prophet whose judgment is infallible, and to ignore the fact that the greatest men the world has produced were all of them products of the times that preceded them and of the circumstances in which they lived.[123]

Parry complements this concern with a belief in teleological progression from simplicity to complexity:

> to understand its qualities and characteristics, or to realise justly the light it throws upon the state of music in our own time, without tracing the conditions which led to it, and following the steps from the small and insignificant beginnings to the masterpieces which we regard as triumphs of our art ... The study of the steps from elementary simplicity up to our complex condition of art shows how progression after progressions became admissible by being made intelligible.[124]

[123] C. Hubert H. Parry, 'On Some Bearings on the Historical Method Upon Music', *Proceedings of the Musical Association* 11 (1884), 2–3.

[124] Parry, 'On Some Bearings', 3–4.

Set within a biographical context Parry's Great Man is no longer superhuman, but the product of slow successive evolutionary steps in history. He is supernatural rather than superhuman, recapitulating in one great individual the genetic inheritance of the time. Zarathustra-like the Great Man has been humanized, secularized, naturalized; he is no longer a Godlike hero but a Great Man bound to all the strictures of his own natural human heredity, and as a Great Man he can finally be understood instead of worshipped: 'the very greatness of some man when separated from their context is so dazzling that it almost numbs the judgment. But if we could study the successive steps, and follow progressive stages up to the great achievements, we should not only understand those achievements, but see how to direct our footsteps accordingly'.[125] Once understood the Great Man can also be studied more accurately, not just as a heroicized Great Man but as nature's Fittest Survivor, allowing historians

> to piece together a tolerably continuous record of the progress of musical art from the earliest times when its modern forms presented themselves . . . and we can judge how and why men were driven in particular directions by the limits of their opportunities or the circumstances of their lives, and why certain forms of art were prosperous at certain times in history, and different at others.[126]

Despite these methodological protestations Parry is inconsistent in approach, and in *Studies of Great Composers* Beethoven is framed in almost retrograde Carlylean terms. Accordingly, second-rate composers may reflect second-rate societies but occasionally a more elevated society experiences the incarnation of a Great Man: 'some exceptionally splendid genius appears, who is fully in sympathy with the best tendencies of his day'.[127] Beethoven becomes 'a prophet . . . a leader, instead of a mere illustrator, of contemporary emotion',[128] and his recapitulationary credentials are preserved entirely intact. He becomes the voice of his age; the mirror of his time and the embodiment of its social, political and artistic strivings. He serves

> not only as the representative of the very highest type of art of the new period, but also as a link between the new and the old world. For he accepted all that was best and purest in the art of his predecessors, and renewed and transformed it by the fervour and passion and sympathetic imagination of his naturally democratic disposition.[129]

[125] Parry, 'On Some Bearings', 7. [126] Parry, 'On Some Bearings', 2.
[127] C. Hubert H. Parry, *Studies of Great Composers*, London: George Routledge and Sons, Ltd, and J. Curwen and Sons, Ltd; and New York: E. P. Dutton and Co., [n.d., 1887], 156.
[128] Parry, *Studies of Great Composers*, 156. [129] Parry, *Studies of Great Composers*, 159.

Beethoven serves more than just a broadly recapitulationary purpose, and his character has implications beyond his art. He is a moral and ethical exemplar through which the progress of society can be observed, noted and determined. He is the fulfilment of a progressionary ideal, yet one rooted in time and place by personal, human and spiritual heredity. In *Studies of Great Composers* evolution has ethicized the Great Man into the Everyman. Yet not every man is a Great Man, and not every great man is an Everyman. In his art and character the truly Great Man survives the test of posterity; he is Galton's socially constructed genius, and Sully's more personalized one. Beethoven is the exemplar, the pattern, the modeller (to use Carlyle's terms): 'All over the civilised world it was felt [when Beethoven died] that a man of the most powerful character and unique genius was gone; and yet to the public of that day his music was not a tithe of what it is to musicians of the present'.[130] Beethoven survived; he is the Fittest Survivor.

Few composers rank higher amongst Parry's panoply of Carlylean musical heroes but amongst them the most important is undoubtedly Wagner. Reflecting a widespread if ideologically divisive figure[131] Wagner for Parry recapitulates musical history in a way that no other composer does (barring perhaps Beethoven and Bach). Repudiating earlier Coleridgean concepts of genius Parry criticises those who believe that genius 'sprang by inspiration from nowhere, and were the independent creations of their originators'.[132] It is not by chance that geniuses arise but through their unique ability to harness the past, transform the present and project the future. Wagner represents just this type of genius; indeed, 'The whole of its [opera's] history can be traced from its crudest beginnings up to the present day, and the earliest conception formed of the way to combine a drama or a theatrical poem with music is the same in principle with that of its latest great master, Wagner'.[133] Wagner is not only a Great Man, but also unquestionably a – possibly *the* – Fittest Survivor. He is incontestably the summation of operatic musical history and all of his music underscores this fact. Writing a few

[130] Parry, *Studies of Great Composers*, 194.

[131] See for example Richard Sessa, *Richard Wagner and the English*, Rutherford, NJ: Fairleigh Dickinson University Press, 1979; David C. Large and William Weber, *Wagnerism in European Culture and Politics*, Ithaca, NY, and London: Cornell University Press, 1984; Theresa Muir, 'Wagner in England: Four Writers Before Shaw', PhD diss., City University of New York, 1997; and Emma Sutton, *Aubrey Beardsley and British Wagnerism in the 1890s*, Oxford: Oxford University Press, 2002. See also Ruth A. Solie, *Music in Other Words: Victorian Conversations*, Berkeley, Los Angeles and London: University of California Press, 2004.

[132] Parry, *Studies of Great Composers*, 322.

[133] Parry, *Studies of Great Composers*, 323.

years later in *The Evolution of the Art of Music* (1893) Parry condenses musical history into the following account:

> The type of vague meandering melodies which formed the arias of Hasse and Porpora became far more definite in organisation in Mozart's hands . . . In the next generation the process of defining details progressed very fast in Beethoven's hands. Even in his first sonata the tendency to concentrate his thoughts into concise emphatic figures is noticeable . . . Schubert's accompaniments to songs are often made up of little *nuclei* . . . and the way in which he knits them together is a perfect counterpart, in little, of Wagner's ultimate method.[134]

As Great Man and Fittest Survivor – as Greatest Survivor – Wagner is a constant source of fascination (and repulsion) for Victorians.[135] In another compilation biography *Studies in Modern Music* (1893) historian and critic W. H. Hadow typifies this self-conflicted, analogue representation of Wagner, first by prefacing his chapter with a lengthy excerpt on music from Carlyle's 'Hero as Poet' and then by following it with a chapter headed 'A Struggle for Existence'. For Hadow Wagner is the recapitulationary apogee:

> it [opera] is to embody a national character and fulfil a national aspiration, if, in one word, it is to take rank as serious Art, then in the whole range of its record will be found no greater name than that of Wagner. He has clothed it with new life, he has taught it to deliver a new message, and the echoes of his voice will last, not only in his own work but in that of the days to come.[136]

Wagner clearly passes the test. Hadow could be accused of paying lip-service to evolutionism because he is also deeply rooted in Coleridgean concepts of creativity. For Hadow genius is something unconscious and almost passive in attitude, his original creation ring-fenced and sealed by its own uniqueness. In true Carlylean fashion music must also 'stand as the outcome of its composer's character, and primarily of his original imaginative power'.[137] But Hadow's evolutionary language soon creeps in when outlining his theories of musical criticism, the fourth and last principle of which is 'the Principle of Fitness': 'In every Art the style must to some

[134] C. Hubert H. Parry, *The Evolution of the Art of Music*, ed. H. C. Colles, New York: Greenwood Press, 1893/1968, 416.

[135] See Bennett Zon, 'From Great Man to Fittest Survivor: Reputation, Recapitulation and Survival in Victorian Concepts of Wagner's Genius', *Musicae Scientiae* (2009–10), 415–45.

[136] W. H. Hadow, *Studies in Modern Music: Hector Berlioz, Robert Schumann, Richard Wagner*, London: Seeley and Co., Ltd, 1911, 325–26.

[137] Hadow, *Studies in Modern Music*, 27.

extent be determined by the nature of the form employed'.[138] With fitness comes the characteristic armoury of late Victorian genius: concentration, reputation and endurance, but also heroic defiance of the ordinary: 'Genius indeed', Hadow decides, 'if it is to be successful, implies an enthusiasm and a concentration which almost inevitably warp the judgment to some extent'.[139] A defiant Wagner continues to fulfil the criteria of genius in late Victorian and turn-of-the-century compilation biographies, although his evolutionary credentials are often assailed, especially amongst anti-Wagnerian campaigners. Devoted Comtean John Frederick Rowbotham glimpses insouciantly into the lives of master musicians in *The Private Life of the Great Composers* (1892) without any of the theoretical trappings of his more ideologically restrictive *History of Music*. But Rowbotham's Wagner is no friend despite his genius, and thinly veiled attacks pervade his account. Wagner's theory of types – the belief that every hero represents an eternal type, and that eternal types are sinners – is dismissed as a rationalization of bad (implicitly Wagner's bad) behaviour and Wagner is rightly upbraided for loose morals, lack of moral fibre, moral reductionism and the decline of religion.[140] Unlike Parry's Wagner Rowbotham's Wagner failed the test of Greatness; in the end it was his much-contested character that let him down, not his music.

Pro-Wagnerians have an altogether different take on Wagner's moral legacy, which is often to disregard it entirely in favour of his music and/or aesthetic writings. Francis Hueffer's earlier encomium to Wagner's greatness, written while his object's celestial star was inexorably rising, offers a categorically biological model of biography. Arguably more aesthetics and history than biography – but biography nonetheless – *Richard Wagner and the Music of the Future* (1874) is an uncharacteristically un-moralizing compilation biography including chapters on Wagner, as well as Schubert, Schumann, Liszt and the lesser-known Robert Franz. In it – and on it, as it so happens – is one of Victorian Britain's clearest expressions of recapitulation and organic philosophy. Hueffer is uncompromisingly recapitulationary, emplotting his ontogenic Wagner with the very richest phylogenic musical history, from the beginning of musical time to its unrivalled apex in the nineteenth century. The message is clear: in his capacity as saviour Wagner will regenerate all national and individual life:

[138] Hadow, *Studies in Modern Music*, 48. [139] Hadow, *Studies in Modern Music*, 146.

[140] John Frederick Rowbotham, *The Private Life of the Great Composers*, London: Isbister and Co., Ltd, 1892, 336–38.

For the artist [Wagner] is to be the longed-for Messiah, to deliver future generations from the fetters of custom and prejudice, and I need not add that this artist is to speak to the nations from the hallowed boards of a truly popular stage, and that his words shall be accompanied by the inspiring strains of the divinest of the arts, *music*.[141]

Wagner is the Titan 'progressing in enormous strides towards Utopia'.[142] Overcoming the crush of personal setback he disperses his enemies and casts out his demons – indeed the weight of misery 'would have crushed a weaker man; Wagner's dramatic nature rose up in the contest, riding on the billows that were to submerge him'.[143] Wagner is the one true genius, the primary and secondary imagination, the real and the ideal, a Great Man, the Fittest Survivor, limited only by the imperfect powers of our human perception:

The bird's-eye view from the sublime heights of genius must needs ignore many of those minute differences and considerations with which we mortals of smaller stature are obliged to reckon. Besides, a great man must be accepted or renounced as a whole, like the sea or the sun in heaven. If you look at it through the lens of a dissecting criticism, its splendour will be diffused in a single ray of moderate lighting power; so the demigod will look very much like an ordinary mortal, not because he has ceased to be the *hero*, but because you are the *valet* of the proverb.[144]

Unusually for Victorian music biography, Hueffer helps his readers visualize this, extending his verbal metaphors to the cover of the book itself. Drawing upon a history of evolutionary trees, Hueffer positions Wagner in the celestial canopy of a great Haeckelian oak, its roots in Schubert, its branches reaching upward to Liszt, Schumann and Franz. Wagner ranges over them, centrally placed amidst the stars of the highest firmament. As embodied in Figures 5.1 and 5.2, Wagner has become Man, the apex of nature, recapitulating in ontogeny the phylogeny of musical creation.

Conclusion

While Wagner reaches an apogee in Victorian biography so too does recapitulationism. By the middle to later part of the Victorian period

[141] Francis Hueffer, *Richard Wagner and the Music of the Future: History and Aesthetics*, London: Chapman and Hall, 1874, 13.
[142] Hueffer, *Richard Wagner and the Music of the Future*, 70.
[143] Hueffer, *Richard Wagner and the Music of the Future*, 71.
[144] Hueffer, *Richard Wagner and the Music of the Future*, 72.

Figure 5.1 Front cover to Francis Hueffer, *Richard Wagner and the Music of the Future: History and Aesthetics* (1874)

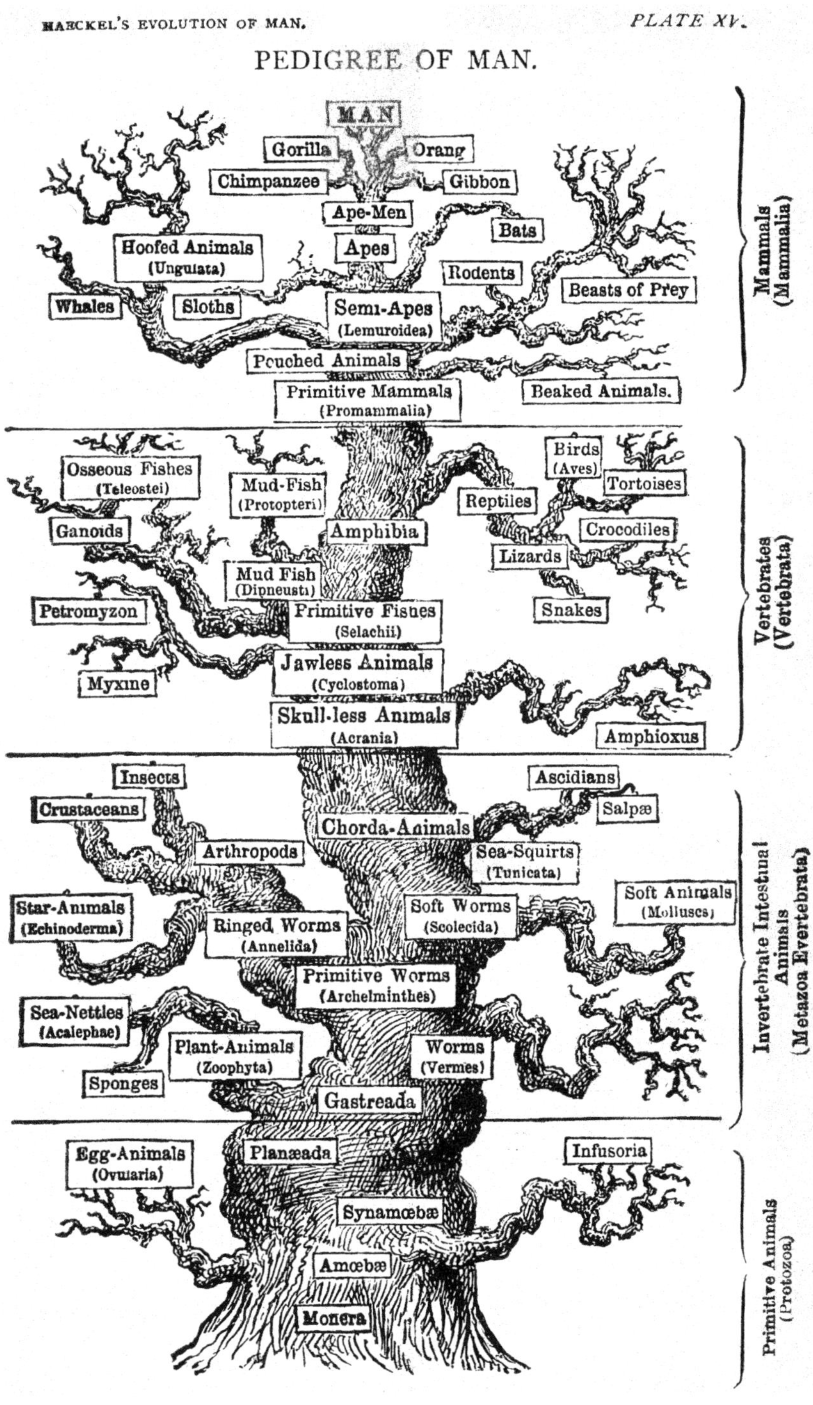

Figure 5.2 Ernst Haeckel, *The Evolution of Man* (1879), originally from *Anthropogenie* (1874)

Spencer's synthetic philosophy could not have become more representative of this relationship:

What Biography is to Anthropology, History is to Sociology, I mean, as commonly conceived. The kind of relation which the sayings and doings that make up the ordinary account of a man's life, bear to an account of his bodily and mental evolution, structural and functional, is like the kind of relation borne by that narrative of a nation's actions and fortunes its historian gives us, to a description of its institutions, regulative and operative, and the ways in which their structures and functions have gradually established themselves … in both cases it is with these structures and functions, in their origin, development and decline, that science is concerned.[145]

With Spencer's synthetic philosophy recapitulation – the biological relation of the individual to the whole – acquired not only philosophical but professionally accredited scientific endorsement. In biography this progression is observable in movements from early Youngian concepts of natural genius to Carlylean formations of the heroic Great Man, and from the Great Man to Spencerian concepts of the Fittest Survivor underwritten by trends in embryology and hereditary science. British music biography conforms to this pattern, creating Great British Men and transforming them into Fittest British Survivors through books, shorter essays and even more concise dictionary entries. But one crucial question regarding the genre of biography remains unanswered. If as Spencer says biography is to history as anthropology is to sociology why then does Victorian music biography seem to lack the same apologetical purpose we shall see driving the genre of music history in the next chapter? In the broadest sense, why does Victorian music biography – both individually and collectively – fail to convince detractors of Victorian musical culture? Why does it seem to entrench rather than refute Britain's longest-running criticism of being an inferior musical culture without great composers – of being a – no, *the* – Land Without Music?

Of the twelve books covered in Crowest's Master Musicians series there is not one native Englishman. There are ten Austro-Germanic composers (Bach, Beethoven, Brahms, Handel, Haydn, Mendelssohn, Mozart, Schubert, Schumann and Wagner), one Russian (Tchaikovsky) and one Pole (Chopin). In Francis Hueffer's earlier Great Musicians series there are ten Austro-Germanic composers (Bach, Beethoven, Handel, Haydn, Mendelssohn, Mozart, Schubert, Schumann, Wagner and Weber), two Italians (Rossini and Cherubini), one Englishman (Purcell) and one broad study of English church music. Even slightly later series, produced a few years

[145] Herbert Spencer, *The Study of Sociology*, London and New York: D. Appleton, 1873, 58.

after the death of Victoria, such as Bell's Miniature Series of Musicians (c. 1903 to ?) and Days with the Great Composers (c. 1910 to ?), favour foreign composers almost exclusively. Henry Saxe Wyndham's *Sullivan* (1903), from Bell's series, is the only single English composer to appear in either series. Influential dictionaries like Bingley, Busby, Sainsbury and Grove include strong coverage of British composers, as does Stratton and Brown's much-vaunted *British Musical Biography* (1897) and even compilation portrait books such as John Warriner's late nineteenth-century *National Portrait Gallery of British Musicians* (1896). But while these definitely place British musical culture on the biographical map none of them could be said to possess or advance any determined apologetical agenda. Cummings's *Purcell*, for instance, is laudatory but ultimately disappointing, Purcell was 'born the favoured child of the muses, destined to raise the musical fame of England to a height it had never before attained'[146] but his afterlife extends only to the early nineteenth century. Amongst late and post-Victorian biographies there are homages to Sullivan, such as A. H. Lawrence's *Sir Arthur Sullivan: Life Story, Letters and Reminiscences* (1899), and posthumous biographies like Benjamin Findon's *Sir Arthur Sullivan: His Life and Music* (1904) but there appear to be very few solidly Victorian biographies of English musicians like Charles Barrett's *English Church Composers* (1882) or William Alexander Willeby's *Masters of English Music* (1893).

Barrett does nothing more than use biography as an excuse to organize musical periods in the same way that Kiesewetter does in his *History of the Modern Music of Western Europe* (1834, translated into English, 1848). Like Kiesewetter's his aim is mainly historical. Willeby's compilation biography is rather different, apparently being the first of its kind to bring together in one book essays on exclusively British – and living – composers, including Sullivan, Alexander Mackenzie, Frederic Cowen, C. Hubert H. Parry and Charles Villiers Stanford. But even Willeby can claim no wider apologetical rationale. While establishing critical mass, his Great Men in-the-making are not treated collectively but organized into discreet chapters. So why as a genre did Victorian music biography fail to play apologetical ball and refute the criticism of the Land Without Music? There are several possible reasons. One might have to do with a readership which by necessity expected its reading to reflect the musical experience of a national concert hall dominated by foreign (mostly German) composers. Another

146 William H. Cummings, *Henry Purcell, 1658–1695*, London: Sampson Low, Marston & Co., Ltd, [n.d., 1880s?], 7.

and not unrelated reason may be the authors' own not dissimilar musical preferences and prejudices. Many biographers try to offset this by playing up the significance of English forays in cultivating foreign genius, English cultural influence abroad and England's creatively conducive working environment. A more compelling reason may come down to the simple fact that individual lives, set in relative historical isolation, have only limited capacity to express the totality of national history. This is arguably the basic flaw with recapitulation and its application within Victorian biography. Through their creative talent and moral character, their individual genius and social fitness, individual composers embody traits of national identity only when gauged against the longer span of historical time in which those traits have evolved into something identifiably national.

In some respects this advantages the compilation biography over the monograph because the compilation biography assumes by its very nature a more historically implicit biographical format often spanning several centuries through its chronologically based sequence of individual portraits. Series biographies do much the same in macrocosm but on a scale predictably too large to consume as easily, and seldom in any recognizably coherent order. Biography might recapitulate History according to Spencer but the ontogeny of the compilation biography does not necessarily recapitulate the phylogeny of the book series; neither does the individual necessarily recapitulate the whole. In fact the recapitulationary metaphor breaks down because it cannot be proven and like most things that cannot be proven it makes for a bad defence against an idea like the Land Without Music. The breakdown is not something unique to Victorian biography as a genre. It was symptomatic of larger unease with a science which, despite Spencer's proclamation, had begun to lose confidence years before he accepted and promulgated its precepts. In his synthetic philosophy Spencer accepted Haeckel and casually disregarded Von Baer's early nineteenth-century rejection of recapitulation, but many scientists – Darwin especially – would side with von Baer.

To them recapitulation was unproven and unable to be proven because it lacked scientific verifiability. For Charles Lyell there is 'no support whatever to the notion of a gradual transmutation of one species into another, least of all, of the passage, in the course of many generations, from an animal of a more simple to one of a more complex structure'.[147] Robert

[147] Charles Lyell, *Principles of Geology: Being an Inquiry How the Former Changes of the Earth's Surface are Referable to Causes Now in Operation*, 2 vols., Philadelphia: J. Kay, Jr, & Brother, Pittsburgh (1st American ed. from the 5th and last London ed.), 1837, 526.

Chambers echoes this in *Vestiges of the Natural History of Creation* (1844) when he sides with von Baer and insists that the growth of the individual (embryo) is always and only the growth of the individual. It emphatically does not recapitulate higher forms: 'It has been seen that, in the reproduction of the higher animals, the new being passes through stages in which it is successively fish-like and reptile-like. But the resemblance is not to the adult fish or the adult reptile, but to the fish and reptile at a certain point in their foetal progress'.[148] Recapitulation may have translated the Great Composer into the Fittest Musician, but it could not – without the increasingly scientific methodology of history – transform the Fittest Musician into a Great – or even *the* – Greatest Musical Nation. On its own, without reference to history, the literary analogy of recapitulation remained always and only what it is – a poetic and literary device. Perhaps the first editor of the *Dictionary of National Biography* (from 1885 to 1891) Leslie Stephen is right, therefore, when he claims for all biography that as a genre it is effectively an ontogeny for which there is no literary phylogeny:

> True biography is no handmaid of ethical instruction. Its purpose is not that of history. It does not exist to serve biological or anthropological science. Any assistance that biography renders these three great interests – ethical, historical and scientific – should be accidental; such aid is neither essential nor obligatory. Biography rules a domain of its own; it is autonomous – an attribute with which it is not always credited.[149]

[148] Robert Chambers, *Vestiges of the Natural History of Creation*, London: J. Churchill, 1844, 212.
[149] Leslie Stephen, *Principles of Biography*, Cambridge: Cambridge University Press, 1911, 18.

6 | Music History

Introduction

Shaping history was a preoccupation in Victorian culture and developmentalism was amongst its chief historiographical tools. Reflecting this preoccupation historian Frederick Harrison claims that 'The great truth which marks the science of our time is the sense of unity in the course of civilisation, and of organic evolution in its gradual growth'.[1] Developmentalism – the supposition that history progressed towards perfection – breathed life into British historical narratives, driving them inexorably forward from a past perfectly emplotted to explain the present. Victorian music histories are entirely characteristic in this regard but where general histories projected their past optimistically to their triumphal present, music histories drove their glorious past towards a present almost universally recognized as bleak. As intimated at the end of the previous chapter Victorian music history was burdened by the deeply resented criticism of being the Land Without Music, a country ostensibly with no great composers and no significant musical culture. Victorian music histories fight back, however, by exploiting forms of developmentalism like cyclical and progressive developmentalism and recapitulation; together they supply the rich apologetical brief examined in this chapter. An introductory section on general histories elaborates these concepts in broad terms. The role of these in forming the apologetics of Victorian music history – especially in their refutation of the Land Without Music – forms the second part. Here the role of genius is explored in fuller depth, providing at the same time an overview of Victorian music histories. Section three investigates changing models of originality in nineteenth-century Britain, with emphasis on shifting paradigms of creation and invention. The implications of their intersection with developmental models are explored through portrayals of individual and national musical genius. A conclusion

[1] Frederick Harrison, *The Meaning of History and Other Historical Pieces*, London: Macmillan and Co., Ltd, 1916, 123.

brings it all together, summarizing main points and offering an alternative reading and explanation of the Land Without Music.

History

Victorian history is commonly referred to as the 'Whig interpretation' of British history. Using often triumphal political hindsight to endow the constitution with germinal British powers, Whig history emplotted the past to justify the present. Herbert Butterfield defines it as 'the tendency in many historians to write on the side of Protestants and Whigs, to praise revolutions provided they have been successful, to emphasize certain principles of progress in the past and to produce a story which is the ratification if not the glorification of the present'.[2] Subsequently downgraded amongst historians to nothing more than a self-justifying narrative of imperial progress, that history illustrates the strength of currency narrative held in the Victorian historical mindset. As Michael Bentley says, 'Narratives varied in their force but narrative itself never did: it held a monopoly on how to write about the past and did not even begin to weaken before 1890'.[3] Those narratives being what they were, Whig histories inevitably spoke the language of progress patterning 'all of history as moral lesson',[4] and as such history became imbued with the developmental values of the time. Complemented by manifest achievements in colonial, industrial, technological, economic and social progress by the end of the Victorian period those values positively radiated developmental self-fulfilment. Indeed, a sense of privileged predestinationism marks even the most cautiously progressive historical thinkers. An unswervingly ateleological Darwinism eventually helped stem the flow of Victorian developmental arrogance, but not until Spencerian developmentalism had lured the middle and later generations of historians into a seemingly inescapable teleological cul-de-sac. Ultimately, the horrors of the First World War would put pay to any triumphally teleological approach but not after breathing life into countless general and subject-specific histories. Victorian music histories reflect this profile in microcosm.

[2] Herbert Butterfield, Preface, *The Whig Interpretation of History*, London: G. Bell and Sons, 1931.

[3] Michael Bentley, *Modernizing England's Past: English Historiography in the Age of Modernism, 1870–1970*, Cambridge: Cambridge University Press, 2005, 6.

[4] Robert F. Berkhofer, Jr, *Beyond the Great Story: History as Text and Discourse*, Cambridge, MA, and London: Harvard University Press, 1995, 125.

Whig history is often more variegated – and self-conflicted – than it might at first seem,[5] stressing developmental purpose and unpredetermined organic evolutionism in turn. Historian Thomas Macaulay typifies this state of blissfully untroubled co-existence. On the one hand he is unreconstructedly developmental in approach: 'For the history of our country during the last hundred and sixty years is eminently the history of physical, of moral, and of intellectual improvement';[6] on the other hand he presents a proto-evolutionary stance foregrounding the necessity of adaptation:

> The fact is, that a good government, like a good coat, is that which fits the body for which it is designed. A man who, upon abstract principles, pronounces a constitution to be good, without an exact knowledge of the people who are to be governed by it, judges as absurdly as a tailor who should measure the Belvidere Apollo for the clothes of all his customers.[7]

Macaulay's brand of developmental inconsistency brought to historiography a view of 'change within continuity';[8] for him 'The History of England, in particular, is the history of a government constantly giving way, sometimes peaceably, sometimes after a violent struggle, but constantly giving way before a nation which has been constantly advancing'.[9] Constant advance was not only the preserve of collective constitutional history but the force driving key figures within its narrative. Thus Macaulay and many other Whig historians indulge the deeply human – and at times superhuman – individuality of his key players. Whig musicologists follow suit, foregrounding people even in the most trenchantly theorized historical narratives. Though he writes mainly on the history of formal structures in music C. Hubert H. Parry embodies this approach in *The Evolution of the Art of Music* (1893/96), often suspending the theoretical flow of his ideas to recount highly personalized stories of composers. Gluck is a particular fascination for him, although there are many more: 'Gluck deserves great homage as a man of the rarest genius. But he deserves fully

[5] Michael Bentley, *Modern Historiography: An Introduction*, London and New York: Routledge, 1999, 65.

[6] Thomas Babington Macaulay, *History of England from the Accession of James II*, 5 vols., London: Longman, Green, Brown and Longmans, 1849, vol. 1, 3.

[7] Thomas Babington Macaulay, *The Works of Lord Macaulay Complete*, ed. Lady Trevelyan, London: Longman, Green, and Co., 1866, vol. 7, 687–88.

[8] A. Dwight Culler, *The Victorian Mirror of History*, New Haven and London: Yale University Press, 1985, 38.

[9] Thomas Babington Macaulay, *Speeches of the Right Honorable T. B. Macaulay, M.P.*, 2 vols., Leipzig: Bernhard Tauchnitz, 1853, vol. 1, 92.

as much again for the splendid sincerity with which he refused to put up with the shams which the rest of the world found quite good enough to amuse them, and made men wake up to realise that opera was worth reforming'.[10]

Inevitably Whig humanism was strongly backlit by Carlylean hero worship, casting players in compellingly driven narratives. Sceptical of Enlightenment principles of universal human progress Carlyle sought to invigorate history with identifiably real people. Seeming to emphasize men's passions rather than reason[11] his figures supply a debatably richer narrative project than Macaulay; indeed his formal commitment to character-based narrative remains unassailable, if problematizing: 'Cut us off from Narrative, how would the stream of conversation, even among the wisest, languish into detached handfuls, and among the foolish utterly evaporate! Thus, as we do nothing but enact History, we say little but recite it . . . what is all Knowledge too but recorded Experience'.[12] While narratives of passion may seem more realistic than histories of reason without purposeful narrative direction histories often default to Enlightenment reason.[13] Carlyle overcomes – or perhaps complicates – this problem by collapsing history into two seemingly antithetical evolutionary methodologies: one, which he calls the 'Chaos of Being'; and the other, recapitulation embodied in the identity of the hero.

Disingenuously titled, the 'Chaos of Being' actually describes the formal developmental structure underlying Carlyle's historiographical hermeneutic, chronicling the fraught relationship between the representation of synchronic events within a diachronic narrative. Drawn from the morphological language of his Continental contemporaries[14] the 'Chaos of Being' records a developmental principle operating upon and within randomness. It is a principle which Whig music historians would also have understood implicitly, Henry Tipper's *Music and Its Growth* (1898) a good example.

[10] C. Hubert H. Parry, *The Evolution of the Art of Music*, ed. H. C. Colles, New York: Greenwood Press, 1893/1968, 277.

[11] J. R. Hale, *The Evolution of British Historiography: From Bacon to Namier*, London and Melbourne: Macmillan, 1967, 41.

[12] Thomas Carlyle, 'On History', *Fraser's Magazine* 2, no. 10 (Nov. 1830), 413–18, cited in Chris R. Vanden Bossche, *Thomas Carlyle, Histories*, Berkeley and London: University of California Press, 2002, 4.

[13] Hayden White, *The Historical Imagination in Nineteenth-Century Europe*, Baltimore, MD, and London: Johns Hopkins University Press, 1973/1975, 47.

[14] See Rosemary Ashton, *The German Idea: Four English Writers and the Reception of German Thought, 1800–1860*, Cambridge: Cambridge University Press, 1980, and C. F. Harrold, *Carlyle and German Thought, 1819–1834*, 1st AMS ed., New York: AMS, 1934/1978.

Tipper, like Parry, turns unsurprisingly to self-determined Gluck to convey that perfect sense of development within chaos. The Renaissance introduced a time of massive, sustained change in society – 'acute minds scanned with closest scrutiny the origin of all things. No institution was sacred; no creed of necessity inviolate'[15] – and against that backdrop Gluck appeared like a beacon of truth, reforming opera and the hypocrisy it had become.

Christopher Parker characterizes this as classical idealism 'with an admixture of Scots Calvinist predestination'.[16] According to Carlyle:

> The most gifted man can observe, still more can record, only the *series* of his impressions: his observation, therefore, to say nothing of its other imperfections, must be *successive*, while the things done were often *simultaneous*; the things done were not a series, but a group. It is not acted, as it is in written History: actual events are nowise so simply related to each other as parent and offspring are; every single event is the offspring not of one, but of all other events, prior or contemporaneous, and it will be in its turn combined with others to give birth to new; it is an ever-living, ever-working Chaos of Being, wherein shape after shape bodies itself forth from innumerable elements.[17]

Carlyle's use of the term 'bodies' is significant because it is individual bodies which populate historical narrative with developmental purpose, and here again the language of German idealism is abundant. George Levine describes it as Carlyle's pursuit of a 'best self . . . a secular pilgrim's progress'.[18] In Schiller's terms, 'Every human being . . . carries within him, potentially and prescriptively, an ideal man, the archetype of a human being, and it is in his life's task to be, through all his changing manifestations, in harmony with the unchanging unity of this ideal. The archetype . . . is to be discerned more or less clearly in every individual'.[19] Putting theory into practice Carlyle claims that 'History is the essence of innumerable Biographies. But if one Biography, nay our own Biography, study and recapitulate it as we may, remains in so many points unintelligible

15 Henry Tipper, *The Growth and Influence of Music in Relation to Civilization*, London: Elliot Stock, 1898, 160.

16 Christopher Parker, *The Idea of History from Coleridge to Collingwood*, Aldershot: Ashgate, 2000, 37.

17 Carlyle, 'On History', in Vanden Bossche, *Thomas Carlyle, Histories*, 7.

18 George Levine, *Dying to Know: Scientific Epistemology and Narrative in Victorian England*, Chicago and London: University of Chicago Press, 2002, 69.

19 Friedrich Schiller, *On the Aesthetic Education of Man in a Series of Letters, Über die äesthetische Erziehung des Menschen*, trans. Elisabeth M. Wilkinson and L. A. Willoughby, Oxford: Oxford University Press, 1967, letter 2, 17.

to us, how much more must these millions, the very facts of which, to say nothing of the purport of them, we know not, and cannot know!'[20] For Carlyle the 'Chaos of Being' is regulated by recapitulation and as such every individual, no matter how historically insignificant, becomes a kind of hero: 'Every living man is a visible mystery: he walks between two Eternities and two Infinitudes'.[21] Believing, more abstractly, that 'the Present is the living sum-total of the whole Past'[22] he soon elevates that mystery to the realm of divinity, asking how 'an individual Delineation [of a Whole] may be 'informed within the Infinite''.[23] From the stuff of human divinity individual greatness – the hero – is born, and with it history moves into action. Amongst Carlyle's most popular works the *French Revolution* (1837) typifies this, with each individual action propelled 'by the same relentless inevitability possessed by the Revolution as a whole'.[24] These are the kinds of heroes that bring order to chaos and elicit imitation in Victorian historiography; whether revolutionaries, parliamentarians or great composers they were the models, patterns and creators of what Victorians sought to attain.[25]

Immersed in a culture of recapitulation Victorian historical narrators faced another methodological quandary, this time over alternating notions of cyclical and progressive development.[26] Cyclical history came in a variety of predetermined, self-completing forms, each taking into account its own rise and fall. In contrast progressionists inflected their histories with upward trajectories, idling at the most advanced stages of development (as man in nature, and Britain in world culture). Both generally accepted positivist Comtean stages of development but situated their scalar position within them differently, from savagery to civilization, childhood to maturity, or Spring to Winter. It is not surprising, for example, that (discussed in Chapter 2) John Frederick Rowbotham equates the drum,

[20] Carlyle, 'On History', in Vanden Bossche, *Thomas Carlyle, Histories*, 5.

[21] Thomas Carlyle, *Two Note Books of Thomas Carlyle, from 23rd March, 1822, to 16th May, 1832*, ed. Charles Eliot Norton, New York: The Grolier Club, 1898, 136.

[22] Thomas Carlyle, 'Characteristics', review of Thomas Hope, *An Essay on the Origin and Prospects of Man*, 3 vols., London: John Murray, 1831, *The Edinburgh Review* 54 (Dec. 1831), 380.

[23] Carlyle, *Two Note Books of Thomas Carlyle*, 187.

[24] Rosemary Jann, *The Art and Science of Victorian History*, Columbus: Ohio State University Press, 1985, 59.

[25] Thomas Carlyle, 'The Hero as Divinity' (1840), in Thomas Carlyle, *On Heroes, Hero-Worship, and the Heroic in History*, London: J. M. Dent and Sons, 1840/1911, 239.

[26] Peter J. Bowler, *The Invention of Progress: The Victorians and the Past*, Oxford: Basil Blackwell, 1989, 9.

pipe and lyre stages in the history of music to the stone, bronze and iron in archaeology and the theological, metaphysical and positive in Comtean historiography.[27] Nor perhaps is it surprising that Rowbotham should end at the music of the Crusades, a crucial juncture in the metanarrative of musical cycles: 'Thus ... we may leave the Moorish Music, nor need we pursue its decline, which was soon to come. For by this time its work was done, and its influence, though divested of its magnificence, had ere this been disseminated through Europe'.[28] Cyclical history essentially politicised recapitulation as a historiographical tool, imbuing individual and national evolution with ripening ideological implications. It is usually progressionist, not cyclical historians who ask what would be considered today as profoundly uncomfortable questions about race, colour, class and breeding. Peter Bowler calls this the 'darker side of developmentalism'[29] and from it arose the inimical theory of human development – degenerationism, an abiding concern amongst both general and music historians. C. Hubert H. Parry summarises it for music history when he complains that 'the tendency to revert to primitive conditions is frequently to be met with even in the most advanced stages of art; and an antagonism, which it is one of the problems of the art to overcome, is persistent throughout its history'.[30]

Comtean positivism and its bubbling threat of degenerationism were taken to an extreme by the heavily derided Henry Thomas Buckle, author of the *History of Civilization in England* (1857–61). Buckle was accused of shoe-horning historical events into calculably predetermined outcomes and structuring his history without any regard for the individual nature of human experience: 'Rejecting, then, the metaphysical dogma of free will, and the theological dogma of predestined events, we are driven to the conclusion that the actions of men, being determined solely by their antecedents, must have a character of uniformity, under precisely the same circumstances, always issue in the same results'.[31] With science forcibly conscripted into Buckle's methodological army individuality languished amidst a sea of putatively interrelated statistics and facts. Critics were coruscating, for as many reasons as there were critics. Lord Acton gristles: 'The true historian takes the individual for his centre ... Mr. Buckle's

27 John Fredrick Rowbotham, *A History of Music*, 3 vols., London: Trübner & Co., 1885, vol. 1, xx.
28 Rowbotham, *A History of Music*, vol. 3, 579. 29 Bowler, *Invention of Progress*, 11.
30 Parry, *The Evolution of the Art of Music*, 1893/1968, 12.
31 Henry Thomas Buckle, *History of Civilization in England*, 2 vols., London: J. W. Parker and Son, 1857 and 1861, vol. 1, 18.

history, if he *could* write a history according to this programme, would be the reverse of all this; he would merge the individual in the company, the person in the body, wishes, passions, character, conscience, all would be abstracted'.[32] Macaulay considered him incoherent, and Darwin called it 'wonderfully clever and original' yet doubted his generalizations were 'worth anything'.[33] Given their similarities, one cannot help wonder whether C. Hubert H. Parry had read Buckle with an eye to organizing the structure of the *Evolution of the Art of Music*. Despite the occasional foray into story-telling it is an unremittingly cyclical tract with progressive concerns on the design of music through history. Earlier, more primitive stages were organized around human feeling but after a deep crisis music veered away from expression to return to it after the advent of absolute music and the symphony.[34]

While individuality may have languished in Buckle (and to a lesser extent in Parry) it consumed vast swathes of Victorian historiography, and with the help of recapitulation great men came to embody great nations. The critically besieged author James Anthony Froude follows in Carlyle's recapitulationary wake with books like the *History of England from the Fall of Wolsey to the Defeat of the Spanish Armada* (1856–70) and *Thomas Carlyle: A History of the First Forty Years of His Life, 1795–1835* (1882 and 1884), sadly mired in controversy. Froude is often portrayed as a self-conflicted, half-hearted demi-Whig, uncertain and uneasy about English national progress[35] yet he lavished attention on individuals; for him 'the object of history is to discover and make visible illustrious characters, and pay them ungrudging honour'.[36] Where for Macaulay the individual 'was merely a passive receptor of the current of progress, for Froude he was the instrument of Providence'.[37] Couched in such terms of progress Froude's individual forms an irrevocable part of an organically unfolding process sympathetic to evolutionary interpretation: 'unlike Macaulay's, his cycles insured diversity, not uniformity . . . his conception of change . . . was more evolutionary than revolutionary'.[38] Like Carlyle

[32] Lord Acton, in W. H. McNeill (ed), *Essays in the Liberal Interpretation of History by Lord Acton: Selected Papers*, Chicago: University of Chicago Press, 1967, 17.

[33] John Kenyon, *The History Men: The Historical Profession in England since the Renaissance*, London: Weidenfeld and Nicolson, 1983, 113.

[34] Parry, *The Evolution of the Art of Music*, 1893/1968, 18.

[35] J. W. Burrow, *A Liberal Descent: Victorian Historians and the English Past*, Cambridge: Cambridge University Press, 1981, 281.

[36] James Anthony Froude, 'Inaugural Lecture', *Longman's Magazine* 21 (Dec. 1892), 162.

[37] Jann, *Art and Science of Victorian History*, 114.

[38] Jann, *Art and Science of Victorian History*, 114–115.

Froude spoke the language of cyclists and through the agency of key individuals like Henry VIII and Elizabeth I he let institutions he deemed increasingly moribund wane in the absence of utility (the Catholic Church during the English Reformation, for example). With such ineluctable forces of character behind it the Reformation could only have 'spread the Anglo-Saxon race over the globe, and imprinted the English genius and character on the constitution of mankind'.[39] Indeed the English Reformation provides Froude with a seemingly perfect forum for nationalizing individual British genius. Yet Froude's ostensibly impartial characterizations were not ideologically disinterested, nor did they necessarily or uniformly promote confidence in national progress. For Froude 'the course of history was divided against itself. The energy of the new religion propels him forward, the decay of the social system drags him back'.[40] John Richard Green claims that 'though of great merit, [it] is disfigured by a love of paradox'.[41] The spectre of degeneration rears its ugly head once more.

While amongst historians like Froude English individuals came to recapitulate an increasingly democratized national identity Victorian historiography and developmentalism come closer to meeting their evolutionary destiny. In some cases that destiny carried a cautionary, degenerational message for nations dismissive of, and unresponsive to, their own national moral obligations. In the genre of music history that warning is signalled most loudly and unrelentingly by Henry Davey (discussed later in this chapter). Davey is obsessed with the implications of cultural centralization in London, and his *History of English Music* (1895) stands as one of the most forceful historiographical polemics of its age. According to Davey Britain has only one nation to blame for receiving the criticism of being a Land Without Music: Britain. Davey's history is not alone in registering fear of degeneration, and in the increasingly 'naturalized' historiographical world progress is often not assured by human right but by divine privilege. Committed imperialist and author of *The Expansion of England: Two Courses of Lectures* (1883) John Seeley believes in a cyclical organicist theory of history in which God revealed his purpose through nations and national leaders;[42] as such

39 James Anthony Froude, *The Divorce of Catherine of Aragon*, London: Longmans, 1891, 18.
40 Jann, *Art and Science of Victorian History*, 118.
41 John Richard Green, *A Short History of the English People*, London: Macmillan and Co., Ltd, 1874/1899, 331.
42 Christopher Parker, *The English Historical Tradition Since 1850*, Edinburgh: John Donald Publishers Ltd, 1990, 67.

'the true Bible of every nation is its national history'.[43] In principle Davey (and many other music historians) would agree:

> The future historian who then sums up this tale of the nation's achievements will not begin with musical matters, nor even with the discovery of the laws of gravitation and evolution, nor with the conquest of India, and the empire of the sea, but with the creation of *King Lear*, *Macbeth*, *Hamlet*, and *Othello*. Yet let him not put much below these the invention of musical composition by John Dunstable.[44]

Perhaps more than any other historian Seeley encapsulates the Hegelian, recapitulationary nation-state 'as a self-realizing idea, coming, as a state, to consciousness and realization of its national individuality ... [in which] the self needs the not-self to recognize its own identity'.[45] This same reflexively aspirational or degenerational principle occupies the work of Edward A. Freeman, acid critic of Buckle and author of books including a *History of the Norman Conquest* (1867–76) and *The Chief Periods of European History* (1886). Freeman recognizes the ebb and flow of events, and what he calls the 'strange cycles of history'[46] which appear to return nations ineluctably to earlier, less advanced periods in their historical development. Though some deem Freeman to be amongst optimistically progressivist historians[47] – like Davey in Victorian music history – the presence of a degenerational, cyclical realism cannot be ignored. Indeed, although Whig history was largely 'optimistic' that realism points to an English historiography on the cusp of self-recognition and change.

As Davey and Seeley's works suggest a combination of degenerational pessimism and progressive evolutionary optimism forms an important strand in both musical and general Victorian historiography. Author of the magisterial *Constitutional History of England in its Origin and Development* (1874–78) William Stubbs is emblematic, facing Hegelian logic with evolutionary rhetoric. Indicatively for Stubbs 'the continuity of life,

43 John Seeley, cited in Deborah Wormell, *Sir John Seeley and the Uses of History*, Cambridge: Cambridge University Press, 1980, 131.

44 Henry Davey, *History of English Music*, London: J. Curwen and Sons, 1895, 76.

45 John Burrow, 'Historicism and Social Evolutionism', in Benedikt Stuchtey and Peter Wende (eds.), *British and German Historiography 1750–1950: Traditions, Perceptions, and Transfers*, Oxford: Oxford University Press, 2000, 256.

46 Edward A. Freeman, *The Chief Periods of European History: Six Lectures read in the University of Oxford in Trinity Term 1885*, London: Macmillan, 1886, 176.

47 Rosemary Mitchell, *Picturing the Past: English History in Text and Image, 1830–1870*, Oxford: Oxford University Press, 2000, 114.

and the continuity of national purpose, never fails'.[48] Indeed, Stubbs weaves Darwinian evolution even into the very fabric of his title. For him early constitutional history is 'a process of natural selection . . . in constant working . . . exemplified in the gradual formation of the seven kingdoms and in their final union under Wessex'.[49] And as a whole the English constitutional evinced progressive tendencies: 'the continuous development of representative institutions from the first elementary stage, in which they are employed for local purposes . . . to that in which the national parliament appears'.[50] Like Parry Stubbs peppers his work with Darwinian terminology, frequently picturing structures manifesting life 'from order to organization to organism'[51] yet disingenuously for some[52] teleology subsists in the very projection of his history:

> Under the monotonous details there is going on a process of hardening and sharpening, a second almost imperceptible stage of definition which, when life is infused into the mechanism, will have no small effect in determining the ways in which that new life will work. In the record of asserted privilege may be traced the flashes of consciousness that show the forms of national action to be no mere forms, and illustrate the continuity of a sense of earlier greatness and of an instinctive looking towards a greater destiny.[53]

In many respects Stubbs, like Davey, epitomises the ideological and methodological tension inherent within Whig historiography as a whole: national identity crowned by a confidence fearful of degeneration; scientifically alert albeit selective; recapitulationary and organic yet constitutionally teleological. It is this kind of historiography against which British music histories would be written. Musical histories would tell the same story but under different historical pressures, for unlike their self-confident counterparts in Victorian constitutional history they had to use evolution to convey an apologetical agenda: to dispute and refute the long-standing accusation of being a Land Without Music. Unlike ascendant political historiography Victorian music historiography was on the defensive.

48 William Stubbs, *The Constitutional History of England in its Origin and Development*, 3 vols., Oxford: Oxford University Press, 1874–78, vol. 1 (1874), 638.

49 Stubbs, *Constitutional History of* England, vol. 1, 170–71.

50 Stubbs, *Constitutional History of England*, vol. 1, 544.

51 James Campbell, 'Stubbs, Maitland, and Constitutional History', in Stuchtey and Wende, *British and German Historiography 1750–1950*, 103.

52 Bentley, *Modernizing England's Past*, 32–44.

53 Stubbs, *Constitutional History of England*, vol. 3 (1878), 2.

Music History

The Land Without Music is a common epithet bestowed upon a generally unrecognizing Victorian musical public. Although the phrase was coined by Oskar Adolf Hermann Schmitz to entitle *Das Land ohne Musik* (1914) it had subsisted in the anti-British imagination for almost two centuries (and possibly earlier)[54] and continues today to be a source of grim fascination. By the early part of the nineteenth century it formed a frustratingly unyielding maxim, exciting British defensiveness across a range of popular British and Continental musical and general magazines. *The Quarterly Musical Magazine and Review* claims 'It has been frequently asserted . . . that the English are by no means a musical people'.[55] *The Harmonicon* gave internationally famous French musicologist François-Joseph Fétis a platform to discuss the problem in 'Letters on the State of Music in London' (1829). Rhetorically, he asks why 'An English composer beholds neither the glory nor profit in the effects of his labour; who, then, shall induce him to write? . . . We need not wonder if in London we find only arrangers, who esteem their labours no more than the public'.[56] Robert Schumann echoed Fétis's concerns in *Neue Zeitschrift für Musik*, decrying the commonplace attitude that an English composer is no composer.[57] *The British Critic* (1840) says that 'dull [English] ears and tuneless voices are now a proverb on the continent',[58] and *The Musical World* 'England is not a musical nation'.[59] The following year *The Musical World* continues discussion dismissing that 'old-fashioned and still fashionable twaddle in high quarters – that the English cannot be a musical people'.[60] Over time several theories have been put forward to explain the invective including general faults in British character, a decline in British musical genius, the oppressive hegemony of Purcell and Handel, over-centralization of London musical culture, poor education, the Continental values of a foreign monarchy and

[54] Percy M. Young, *A History of British Music*, London: Ernest Benn Ltd, 1967, 281.

[55] *The Quarterly Musical Magazine and Review* 4 (1822), 134.

[56] François-Joseph Fétis, 'Letters on the State of Music in London', *The Harmonicon* 7/11 (Nov. 1829), 276.

[57] Robert Schumann, 'Englischer Komponist, kein Komponist', *Neue Zeitschrift für Musik* (24 Feb. 1837), cited in Nicholas Temperley, 'Xenophilia in British Musical History', in Bennett Zon (ed.), *Nineteenth-Century Music Studies*, 3 vols., Aldershot: Ashgate, 1999, vol. 1, 4.

[58] 'Art. IV. – *The Psalter, or Psalms of David, pointed for Chanting. 24mo. London: Burns. 1840*' (review), *British Critic* (Oct. 1840), 374.

[59] *The Musical World* 14 (1840), 262.

[60] *The Musical World* 15 (1841), 155.

xenophilia.[61] More recently these strands have come together in theories exploring the Land Without Music as a function of British self-criticism involving anti-urban socialist utopianism[62] and evolutionary concepts of genius and the hero.[63]

Like their journalistic counterparts Victorian musical histories also express condemnation of the Land Without Music but unlike popular magazines their apologetical grammar is substantively different. Owing in part to their overarching chronological frameworks Victorian music histories – the same as corresponding works in British general history – deliver extended narratives structurally incompatible with the more concentrated bursts of reply populating contemporary journals. Like constitutional histories Victorian music histories tell a story first and through it make their ideological point so their apologetical intention is woven into the narrative drapery itself, often subtly inscribed with representations of historical progress. Evolution – or development more specifically – plays a significant role in the signification of British national resentment over the Land Without Music for it naturally precipitates discussion of cause and effect – and apology is defenceless without that discussion. Cause and effect are threateningly teleological words, however, in the philosophically predatory habitat of evolutionary England. Like so many words constructed from context rather than consensus cause and effect imply determination, and determination the action of some higher form of predetermined design or Will. That Will – inimical to later evolutionary thought yet abundant in the Victorian developmental conscience – runs like a seam through contemporary histories of British music. Having been unfairly maligned, British music historians used their progress of history to trace the origins and development of that cultural wrong. Taking up positions against foreign and domestic criticism of the Land Without Music, they readied themselves for battle.

Full, complete and extended accounts of British music history found expression in two places, general histories of music published in Britain and/or America and histories of British music published in Britain, America or both. In the (slightly 'long') Victorian period general histories of

[61] Bennett Zon, 'Histories of British Music and the Land Without Music: National Identity and the Idea of the Hero', in Emma Hornby and David Maw (eds.), *Essays on the History of English Music: Style, Performance, Historiography*, Woodbridge: Boydell & Brewer, 2010, 311–12.

[62] Bennett Zon, '"Loathsome London": Ruskin, Morris, and Henry Davey's *History of English Music* (1895)', *Victorian Literature and Culture* 37/2 (2009), 359–75.

[63] Zon, 'Histories of British Music and the Land Without Music', *Essays on the History of English Music*, 311–24.

music including reference to British music considerably outnumber histories of British music by a proportion of roughly seven to one. General histories of music are helpful in contextualizing British attitudes towards its musical history in relation to European (and sometimes American) culture. Histories of British music can also provide intensely focussed comparisons with general histories. Inevitably general histories betray their pedigree in Burney and Hawkins yet were quick to establish independent ideological identities. Fubini aptly describes Burney as 'aiming at a reassessment of music within the ambit of civilization and culture'[64] and 'stereotyping certain Enlightenment concepts of progress'.[65] Those concepts are expressed in the preface to *A General History of Music from the Earliest Ages to the Present Period* (1776–89):

Music, indeed, like vegetation, flourishes differently in different climates; and in proportion to the culture and encouragement it receives; yet, to love such music as our ears are accustomed to, is an instinct so generally subsisting in our nature, that it appears less wonderful it should have been in the highest estimation at all times, and in every place, than it should hitherto never have had its progressive improvements and revolutions deduced through a regular history, by any English writer.[66]

Contrasting with Burney's buoyantly organic, progressive teleology is the work of his contemporary and arch-competitor John Hawkins, author of *A General History of the Science and Practice of Music* (1776). Unlike the constitutionally optimistic and modern-facing Burney the curmudgeonly Hawkins casts music reaching its height in the polyphony and counterpoint of the sixteenth and seventeenth centuries, and from that time suffering unstoppable decline and debasement, especially since the days of Handel. Often portrayed as the ying to Burney's yang, Hawkins is actually misunderstood. The misunderstanding arises by confusing his temperament for his method. In method both Hawkins and Burney subscribe to identical approaches but Hawkins is a cyclist and Burney a progressionist. In theory at least Hawkins believes in progress and improvement:

a weightier argument in favour of modern music, at least so far as regards the improvements in theory and practice that necessarily result from the investigation

[64] Zon, 'Histories of British Music and the Land Without Music', 231.

[65] Enrico Fubini, *The History of Music Aesthetics*, trans. Michael Hatwell, Houndsmill: Macmillan Press Ltd, 1964/1991, 229, 230.

[66] Charles Burney, Preface, *General History of Music from the Earliest Ages to the Present Period, to which is prefixed, a Dissertation on the Music of the Ancients*, 4 vols., London: Printed for the Author, 1776, vol. 1, [vii].

of new principles and the discovery of new combinations, may be drawn from the natural course and order of things, which is ever towards perfection . . . so that of music it may be said, that the discoveries of one age have served but as a foundation for improvements in the next.[67]

In then more recent practice improvement is not guaranteed, however: 'why is it to be assumed of music that it is continually improving, or that every innovation in it must be for the better?'[68] Simply put Hawkins is a card-carrying degenerationist, and as such a self-styled or even accidental mouthpiece for the Land Without Music. Against the teleological background of Burney and Hawkins's monumental if in criticism diametrically opposed achievements nineteenth-century British music historiography began to take shape with new approaches to the Land Without Music. Perhaps in vain the earliest nineteenth-century general music history, Thomas Busby's *A General History of Music* (1819) seen in the previous chapter, tries to eschew and overcome their influence. The more condensed format (two small volumes), designed to entertain as much as educate, trims the fat while updating and correcting. Proportionally Busby emphasizes recent musical culture but for all his methodological manoeuvring there remains an inescapable teleology invisibly guiding the individual lives of his narrative towards perfection, culminating in the lives of great modern (English)men. In fact the lives of great men increasingly refract the chronological narrative as Busby heads towards the present, gradually exposing a fundamentally recapitulatory mindset. If anything it is that methodological mindset which distinguishes him from his predecessors, not his own self-acknowledged intellectual debts: his history 'chiefly devotes its pages to the narration of interesting events, and the investigation and eulogium due to sound science and extraordinary talents'.[69] Crucial to the following excerpt is Busby's closing sentence in which Purcell the individual triumphally recapitulates England the nation:

If in the general history of a science a single master, for the full estimation of his merits, can claim the devotion of an entire chapter, it is such a master as Purcell; if any talents can sustain and excuse the historian who, for the purpose of their just

[67] John Hawkins, *A General History of the Science and Practice of Music*, new (2nd) edition, London: Novello, Ewer & Co.; and New York: J. L. Peters, 1776/1875, xxiv.

[68] Hawkins, *A General History of the Science and Practice of Music*, xxxv.

[69] Thomas Busby, *A General History of Music, from the Earliest Times to the Present; Comprising the Lives of Eminent Composers and Musical Writers. With the Whole Accompanied with Notes and Observations, Critical and Illustrative*, 2 vols., London: G. And W. B. Whittaker, 1819, vol. 1, vi.

delineation, quits for a while the broad path of his progress, and descends into the limited track of a biographer, they are such talents as those of our English Orpheus. This musician shone not more by the greatness than the diversity, by the diversity than the originality, of his imagination . . . His career was preceded by the transit of great and conspicuous lights; but his superior radiance eclipsed and survived their glories . . . of Purcell we fondly discourse, pride ourselves in the brilliancy and the expanse of his genius, and are content to identify with his the musical pretensions of our country.[70]

As explained in the last chapter Busby's emphasis on lives – and genius in particular – is significant in representing a growing biographical trend in general historiography as much as it is a developmental methodology intentionally harnessed to refute the Land Without Music, especially, it would seem, regarding the advent of nineteenth-century music dictionaries. But while early compendiums like William Bingley's *Musical Biography* (1814) and Sainsbury's *Dictionary of Musicians* (1824) reflect and develop this trend as dictionaries neither had the inherent capacity to narrate cumulative, linear histories. Nor as dictionaries were they uniformly successful in repudiating the Land Without Music. In foregrounding British musicians Sainsbury's *Dictionary*, for example, was according to Leanne Langley as important as it was poorly executed.[71] The developmental baton continued to be passed amongst histories, and from the 1830s those histories develop increasingly developmental programmes. Like all good natural theology William Stafford's *History of Music* (1830) and William Hogarth's *Musical History* (1835) both return musical history to its origins in nature, itself designed by God. According to Stafford 'Natural causes may sufficiently account for its origin, without referring to a miracle for the event'[72] and for Hogarth 'Music, though now a very complex and difficult art, is in truth, a gift of the Author of Nature'.[73]

In many respects natural theology is the perfect foil to a developmentalism based on the recapitulatory divinity of genius. Refusing to relinquish purposeful – or intelligent – design, natural theology became secularized under Romantic developmentalism, inverting concepts of musical origins

70 Busby, *General History of Music*, vol. 2, 249.

71 Leanne Langley, 'Sainsbury's Dictionary, the Royal Academy of Music, and the Rhetoric of Patriotism', in Christina Bashford and Leanne Langley (eds.), *Music and British Culture, 1785–1914: Essays in Honour of Cyril Ehrlich*, Oxford: Oxford University Press, 2000, 85.

72 William C. Stafford, *A History of Music*, Edinburgh: Constable and Co.; and London: Hurst, Chance, and Co., 1830, 1.

73 William Hogarth, *Musical History, Biography, and Criticism: Being A General Survey of Music, from the Earliest Period to the Present Time*, London: John W. Parker, 1835, 1.

in divinity with a concept of divinity (genius) in musical origins (God). Emphasis on an incarnated presence rather than an aloof absence gave spiritual reality to human greatness (as we shall see in the next chapter). Genius was therefore created by God not invented by man, and history was co-opted to deliver this message of transformation. Like their contemporaries Stafford and Hogarth contrive this to be perceived on both a personal and national level, defending themselves against the Land Without Music by claiming British greatness *ab initio*. Stafford maintains, 'There are few countries in the world, in which music is more extensively cultivated, or more generally admired than in England; nor is this passion for the art one of late date in our island. The aboriginal inhabitants, the ancient Britons, were passionately fond of music; and their bards . . . were held in the highest estimation'.[74] Having germinated from a musical people, however, modern Britons struggle to realize their potential: 'we should love to see the native genius of our own country more warmly cherished,'[75] Stafford laments.

While in Stafford and Hogarth individual British talent remains sporadic both portray recent British musical fortunes with the generally optimistic tone of a progressivist. Although dismayed over historically indifferent support for domestic talent – and conversely awed by respective levels of Continental support – Hogarth notes that the 'improvement of music, in all its branches, is much promoted by numberless societies, clubs, and other associations of a private nature, both in the metropolis and the provinces'.[76] This pragmatically upbeat tone also defines what appears to be the first sustained Victorian history of British music, the anonymously authored *Music in England, Ireland, Wales, and Scotland; from the Earliest Ages to the Present Time* (1845). Where Stafford and Hogarth dissimulate their apologetical purpose, however, *Music in England, Ireland, Wales, and Scotland* takes it as its starting point:

> We are told that the English are not a musical people; an assertion which never had truth for its basis; and is less than ever warranted in the days in which we live. In every age of our history, music has been popular; it has also formed a prominent feature in the religious rites, the public festivals, and the domestic recreations of the people; and those who would trace its origin and progress amongst us, must revert to that period, when arts were young; and when 'the noble savage', if he did not run 'wild in woods', was very far removed from that civilization which

[74] Stafford, *History of Music*, 283.
[75] Stafford, *History of Music*, 386.
[76] Hogarth, *Musical History*, 428.

characterises the modern European. The Britons, – a branch of the Celtic Gauls, – were, like their progenitors, distinguished for their love of music.[77]

In many respects this reads very similarly to Stafford or Hogarth although gone is any lingering whiff of divine musical origins. Instead there is only a driving narrative of personal and national genius – the 'wild and enthusiastic genius' of the bards; the genius of British audiences appreciative of Handel; and the history's 'humble record of British genius', itself dedicated to 'the brightest impersonation of that genius', Clara Novello.[78] Presented as a 'sketch of the Progress of Music', *Music in England, Ireland, Wales and Scotland* treads steadily in its developmental progress, departing only briefly from its chronology to explore (national, but not art) music of the Celtic fringe. As this shows by 1845, not long after Carlyle published *On Heroes* (1841), progressive, recapitulationary and teleological history had become fully intertwined with the Land Without Music.

Between 1845 and the publication of Henry Davey's *History of English Music* in 1895 the historiographical demography of apology changes, as there appear to be no British histories of British music covering the full expanse of historical time; nor are there any others until 1907, when Ernest Walker published his immensely popular *History of Music in England*. Even the prolific Frederick Crowest, author of *Phases of Musical England* (1881), a *Dictionary of British Musicians* (1895) and *The Story of British Music* (1896), fails to provide as complete picture as would Davey, or later Walker. From the 1840s British music histories, whether general or more nationally localized, begin to limit their chronological boundaries. Some histories continued taking in the widest possible sweeps of time, such as G. A. Macfarren's *Musical History* (1885), William Rockstro's *A General History of Music* (1886) and C. Hubert H. Parry's *Art of Music* (1893), but fragmentation – or selective segmentation – is increasingly evident as Victorian music historiography progresses. John Hullah writes *The History of Modern Music* (1862 and 1875), following by *The Third or Transition Period of Musical History* (1876); William Chappell produces only one volume of the *History of Music* from Egypt to the fall of Rome (1874); H. G. Bonavia Hunt publishes *A Concise History of Music from the Commencement of the Christian Era to the Present Time* (1878); John Frederick Rowbotham issues his sizeable three-volume set of *A History of Music* (1885–87) covering the period from the beginning of time to that of

[77] *Music in England, Ireland, Wales, and Scotland; from the Earliest Ages to the Present Time*, London: Cradock and Co., 1845, 1.

[78] *Music in England, Ireland, Wales, and Scotland*, 39, 60, 60.

the troubadours; and Henry Tipper stops at Beethoven in *Music and Its Growth* (1898). Perhaps inevitably British histories of British music also begin to focus on the nineteenth century, such as Charles Willeby's *Masters of English Music* (1893) and J. A. Fuller Maitland's *English Music in the Nineteenth Century* (1902). British music historians were not alone in their British musical interests and a number of important works were also produced by foreigners publishing simultaneously in America and England or by foreign expatriates living and publishing in Britain. Amongst these are Frederic Ritter's *History of Music* (1870), D. W. Henderson's *Story of Music* (1889), F. Weber's *Popular History of Music From the Earliest Times* (1891) and significantly, Ritter's *Music in England* (1883).

Uniting these seemingly myriad sources are again two distinct yet complementary features: firstly, a developmental methodology producing either positive (progressionist) or potentially negative (cyclist) historical results; and secondly, a symbiotically responsive formula of apology. Emboldened and invigorated by Carlyle's biographical model from the 1840s music histories usually intensify the relationship between these features with the addition of more clearly articulated emphases on development and degeneration (in varying proportion), often limiting the chronological framework to magnify focus on key individuals. Hullah's histories are teleological only from the eleventh century; before then Western music (as well as Oriental music in modern times) is unknown and static. For Hullah 'music is a *new* art'.[79] 'There is', moreover, 'not the slightest evidence to show that anything analogous to it existed among the nations of antiquity'.[80] Without the benefit of 'aboriginal' British origins this chronological constraint delivers considerably less thrust to the historical narrative and to the force behind Hullah's apology. It also signals a certain developmental indecision. 'The connexion between one age and another', he proffers, 'is no more to be broken off suddenly, with impunity, in art than in anything else. Change, whether by aggregation, rejection, or growth, is, no doubt, a condition of life; but it is itself subject to another condition – that it be gradual . . . The music of the future may be safely left to grow out of the music of the present, as the music of the present has grown out of the music of the past'.[81] As a consequence Hullah's histories are (some might say) satisfyingly unpredictable. Composers like Purcell and Handel still register adoration – if not imitation – but their

[79] John Hullah, *The History of Modern Music: A Course of Lectures Delivered at the Royal Institution of Great Britain*, London: Parker, Son, and Bourn, 1862, 4.

[80] Hullah, *History of Modern Music*, 114.

[81] Hullah, *History of Modern Music*, 258.

apologetical imperative is in many respects developmentally undermined. The paradoxically titled book *The Transition Period* attempts to rectify this deficiency in microcosm with two chronologically sequential lectures devoted to England, and as expected of any self-respecting history of English music the first lecture begins with the genre's constitutional apology: 'Whatever claims – and they are very strong claims – we English may have to be called a musical people, it cannot be pretended that we have exercised much – some would say, any – influence on Continental music'.[82] Here the narrative thrust is restored, beginning iconically with the putative inventor of counterpoint John Dunstable, eventually connecting to Purcell and culminating in the double-edged sword of Handel, a foreign expatriate many of whose works are largely unknown in his own German homeland. For Hullah the Land Without Music is due to Britain's increasingly national insularity and as a result its history marks a degenerative decline in home-grown talent.

As degeneration the Land Without Music continues to occupy an apologetical pride of place in mid- to late Victorian histories of British music and general histories including reference to Britain. William Chappell moans about it in *The History of Music* (1874), and his developmental credentials are entirely intact. For him the rot sets in when a society (England) fails to cultivate music from infancy; conversely, 'The most brilliant examples of development are among those who heard music from the cradle. Haydn, Mozart, and Beethoven, were all sons of musicians'.[83] Like Chappell Frederic Ritter (about whom, more later) speaks the language of cyclical degeneration, tracing Britain's loss of hegemony to a historical decline in individuality. Fundamentally agreeing with Haweis's well-known claim that 'The English are not a musical people'[84] – moreover, that there are only great composers *in*, rather than *of*, England – Ritter's history of British national music climaxes with Purcell: 'The history of musical development in England after the epoch of Purcell is, on the whole, only a narrative of the doings of great and small foreign musicians who have visited London up to our time'.[85] Undergirding this sentiment throughout his work is a recapitulationary philosophy of individuality rooted in Herderian concepts of national genius: 'The individual art

82 John Hullah, *The Third or Transition Period of Music History: A Course of Lectures Delivered at the Royal Institution of Great Britain*, London: Longmans, Green, and Co., 1876, 181.

83 William Chappell, *The History of Music*, London: Chappell and Co., 1874, xlviii.

84 Rev. H. R. Haweis, *Music and Morals*, London: Strahan & Co., 1871, 491, and cited in Frédéric Louis Ritter, *Music in England*, New York: Charles Scribner's Sons, 1883, 4.

85 Ritter, *Music in England*, 83.

character of a nation – if the gift of inventiveness be innate to its emotional existence – will never gain general importance and aesthetical meaning if the mind is satisfied by mere slavish imitation of that which, on general principles, is recognized as excellent in the artistic labours and effort of other and foreign nations'.[86] Unexpectedly, in *The History of Music* Ritter sings this philosophical chorus about German musical culture, but in *Music in England* it is clear that time's degenerationist arrow points straight to the Land Without Music.

Questions of English musical degeneration litter works of an even more avowedly evolutionary kind. Consummate Spencerian and author of *The (Evolution of the) Art of Music*, C. Hubert H. Parry denies English musical development after Purcell but not for the usual reasons. Instead, he seems to describe Purcell in almost polygenist terms, as if arising from an altogether different stock of human talent:

> Almost completely outside the direct course of musical evolution stands the unique and highly individual genius of Purcell. The sources of his artistic generalisations can be traced, as is inevitable even with the most pre-eminently 'inspired' of composers; but isolation was entailed by the peculiarly characteristic line he adopted, and the fact that almost all the genuine vitality dropped straight out of the English art directly he died; while none of his remarkable English achievements penetrated so far afield as to have any sort of influence upon the course of musical progress on the Continent.[87]

Why Parry echoes the apologetical content of Chappell, Ritter, Haweis and others without subscribing to its developmental form requires some unpacking. Not unusually for his time, as an evolutionist (in a Spencerian way) he complements progression with recapitulation without questioning their compatibility. As we have heard before, historians can

> piece together a tolerably continuous record of the progress of musical art from the earliest times when its modern forms presented themselves . . . [and] can judge how and why men were driven in particular directions by the limits of their opportunities or the circumstances of their lives, and why certain forms of art were prosperous at certain times in history, and different ones at others.[88]

[86] Frederic Louis Ritter, *The History of Music, from the Christian Era to the Present Time*, London; William Reeves, 1870/1892, 257.

[87] Parry, *The Evolution of the Art of Music*, 1893/1968, 181.

[88] C. Hubert H. Parry, 'On Some Bearings of the Historical Method Upon Music', *Proceedings of the Musical Association* 11, 1884, 2.

But recapitulation does not harmonize with progression in a polygenist world in which there are different human origins, and thus different and potentially unconnected streams of history. Were Purcell to symbolise an undeveloped potential in an unprogressing British music history it would place the British in effectively the same category as modern-day savage culture – inexplicably unevolved. For Parry savage music is aboriginal music and as such bears the seeds of its own future. Despite his genius Purcell is a seed which fell on stony ground (like modern savage music continues to do), and according to Parry British musical history continues to answer for it.

So why, therefore, isolate Purcell from the rest of music history if he was Britain's only chance to repudiate the Land Without Music? One reason may involve confirming the validity of his English passport. Only if Purcell is truly English and not (an imitation of) French or Italian can he be truly individual, as Ritter and many others claim. And only if he is an individual can he be a genius and therefore seed later generations of British greatness. Another related reason may have to do with the Spencerian origins of Parry's evolutionism. By parachuting Purcell into history and airlifting him from it Parry does not have to sacrifice his fundamentally progressionist (rather than cyclist) approach to history. But then that would make Purcell the exception that proves the rule of the Land Without Music. A final, linked reason: Parry intentionally isolates Purcell to elevate and magnify his individuality, and by temporarily suspending Purcell in history there is no developmental degeneration – no Land Without Music. Parry is, in other words, selective in his degenerationism.

Historians have every right to be selective in their methodology of course and Parry's developmentalism is remarkably inconsistent, or synthetic to use Spencer's term. But then so is much general history of the time, especially in the increasingly critical methodological world of post-Darwinian Britain. Perhaps unsurprisingly degeneration continues to answer many questions in Henry Davey's *History of English Music*, Victorian England's most substantial history of musical Britain. I say 'history of musical Britain' because Davey's book is not simply a history of music but a social desideratum designed to rouse Britain from its long musical slumber. It is also, despite its date and pronounced ideological imperative, a book which continues to be consulted as a work of accurate, frequently valuable scholarship.[89] What differentiates Davey from other

[89] Bennett Zon, '"Loathsome London": Ruskin, Morris, and Henry Davey's *History of English Music* (1895)', *Victorian Literature and Culture* 37 (2009), 361.

developmentalists like Parry is the depth of his social critique and the recipe he uses for successfully overcoming the Land Without Music. Steeped in the philosophy of Ruskin and Morris Davey recommends a substantial – if utopian – diet of socialist reform to be implemented systematically; for Britain is not unmusical, only its social structures are. British music has not degenerated, just the British. Fuelled by the optimism of totally unreconstructed nationalism the *History of English Music* sets Britain on the developmental road to recovery by reacquainting readers with its true historical greatness, and then using that history to justify the need for social change. Davey begins by locating England unabashedly at the inception of music: 'The art of musical composition is an English invention . . . A History devoted solely to English music is therefore necessary, to throw light on the origin of the whole art of composition'.[90] But equally, 'the history of English music is longer than that of any other nation . . . but in the 18th and 19th centuries it is of little interest except as regards the performances of foreigners' works; and a General History of Music might after 1700 omit the compositions of Englishmen almost entirely'.[91]

Like many of his predecessors Davey plays music historical connect-the-dot from Dunstable to Purcell and Handel, pausing to observe and consider individual English greatness in the period of Henry VIII to Elizabeth I and to a lesser extent in the seventeenth century. True to form, by the eighteenth century things begin to look pretty bleak for British music, and from the increasingly piled-up historical wreckage Davey begins to coax out his thesis: the Land Without Music is due to 'the centralisation of England in London, which gives no chance to performers or composers elsewhere'.[92] The principle defects in British musical culture arise from the over-centralisation of culture in London: the absence of Grand Opera in provincial cities and towns, the lack of great compositions, ignorance of English musical history, impoverishment of aesthetic standards, irresponsibly amateur performance and musical editing, unawareness of national folk song, and the vulgarities of street music.[93] Beginning his tale of woe with the reign of James I Davey's down-beat degenerationism belies his progressionism, however, and he certainly cannot be accused of being a cyclist when at the same time proffering a rather involved recovery plan. No, there is in Davey a factor which continues to motivate his optimistic developmentalism in the face of manifest decline. Like all musical

[90] Davey, *History of English Music*, 1–2.
[91] Davey, *History of English Music*, 2–3.
[92] Davey, *History of English Music*, 366.
[93] Davey, *History of English Music*, 495–97.

progressionists Davey ultimately recognizes the Land Without Music but not for reasons conventionally given or understood. Davey and many other musical apologists believe that the English subscribe to a different, English notion of originality not shared by their foreign or domestic critics – a notion based on distinguishing creativity from invention; indeed, that notion is at the very root of British musical identity, and Britain's answer to the Land Without Music: as Victorian music histories show us, the British are at heart inventors, not creators.

Creation, Invention and Musical Genius

As markers for or against originality, creation and invention have a long-established pedigree in British culture going back to Edward Young's *Conjectures on Original Composition* (1759) in which an original is created and an imitation is manufactured.[94] Some years later Coleridge would revive Young's theory, dividing originality into roughly the same two strands, imagination (creativity) and fancy (invention).[95] Carlyle would do much the same, musing in apophatic terms on Coleridge but simply renaming creation unconsciousness and invention consciousness or 'manufacture'.[96] By mid-century that uniformity began to wither on the British vine, and what M. H. Abrams calls 'unconscious invention' starts to creep in and counter-balance 'creation'.[97] While its counterpart in German philosophy retained that dialectical relationship – 'obsessed, by the turbulent dark of inception and instauration'[98] as George Steiner suggests – British philosophy was beginning to change. In 1818 Keats could proclaim 'That which is creative must create itself'[99] but by 1825 William Hazlett was defining Walter Scott as 'pure invention'.[100] Further signs of that change can be detected from the 1850s, when

[94] Edward Young, *Conjectures on Original Composition*, London: A. Millar and R. and J. Dodesley, 1759, 12.

[95] Samuel Taylor Coleridge, *Biographia Literaria; or, Biographical Sketches of My Literary Life and Opinions*, 2 vols., London: Rest Fenner, 1817, vol. 1, 295–96.

[96] Carlyle, 'Characteristics', 363.

[97] M. H. Abrams, *The Mirror and the Lamp: Romantic Theory and the Critical Tradition*, Oxford: Oxford University Press, 1953, 213.

[98] George Steiner, *Grammars of Creation*, London: Faber and Faber, 2001, 93–94.

[99] John Keats, Letter to James Hessey (8 Oct. 1818), John Keats, *Letters of John Keats*, ed. Edward Rollins Hyder, 2 vols., Cambridge: Cambridge University Press, 1958, vol. 1, 374.

[100] William Hazlitt, *The Complete Works of William Hazlitt*, ed. P. P. Howe, 21 vols., J. M. Dent and Sons, 1930–34, vol. 6, 60.

psychology began to emerge as an experimental field independent from the philosophy of mind.[101] Under the aegis of this fledgling discipline, connections began to be modelled between individual unconscious and its relationship to the organic growth of the species. Shuttleworth and Taylor describe individual identity in genealogical terms that would later become conversant with the theories of gestalt and recapitulation.[102] Samuel Butler encapsulates this in *Life and Habit* (1878) when he suggests that 'every individual person is a compound creature',[103] and G. H. Lewes, husband of George Eliot and keen evolutionist, writes of the individual in relation to 'the General Mind … Common Sense, Collective Consciousness, Thought (*Das Denken*), Reason, Spirit of the Age'.[104]

Music historians would follow this example when, for instance, Huneker describes Wagner as Zeit Geist or when Henry Davey elides the time of Dunstable and the time of Sullivan (or even Elgar): when 'the time of Dunstable will seem practically contemporaneous with the time of Sullivan [or Elgar in the 1921 revised edition]'.[105] Like so many of his contemporaries Davey elevates the recapitulationary analogy to the level of society; similarly for G. H. Lewes 'Man apart from Society is simply an animal organism; restore him to his real position as a social unit, and the problem changes. It is in the development of Civilisation that we trace the real development of Humanity. The soul of man has thus a double root, a double history'.[106] Man's now metaphorically double roots in both himself and his species project an increasingly compound understanding of human origins, where previously his more unified nature was assured. As a result origins receded into the mists of time, when the age of man and the universe was indeterminably lengthened. Time seemed to be moving backward; in *Prehistoric Times* (1865), for example, John Lubbock time-travels from the Bronze Age to caves, intent on capturing that 'vague and overpowering sense of antiquity' which proves that 'man has existed on

101 L. S. Hearnshaw, *A Short History of British Psychology 1840–1940*, London: Methuen, 1964, v. See also Rick Rylance, *Victorian Psychology and British Culture 1850–1880*, Oxford: Oxford University Press, 2000, 5–17.

102 Sally Shuttleworth and Jenny Bourne Taylor, *Embodied Selves: An Anthology of Psychological Texts 1830–1890*, Oxford: Oxford University Press, 1998, xiii.

103 Samuel Butler, *Life and Habit*, London: Trübner and Co., 1878, 104.

104 G. H. Lewes, *The Study of Philosophy*, London: Trübner & Co., 1879, 159.

105 Davey, *History of English Music*, 23 (in the 1921 edition, p. 19)

106 G. H. Lewes, *Problems of Life and Mind*, London: Trübner & Co., 1874–79, cited in Shuttleworth and Bourne Taylor, *Embodied Selves*, 90.

the earth for a much longer was until recently supposed'.[107] While mid- to late nineteenth-century Britain struggled to recalibrate its origins – helped very largely by Darwin's *Origin of Species* – concepts of inheritance and development continued to flow into Victorian culture. Butler argued that the individual germ of memory stores the infinite repository of its inherited memories,[108] and for Lewes the creation of personality relies upon recognition of a 'Collective Experience combining with inherited tendencies in the formation of Experience for the individual'.[109]

Literary works of all kinds would reflect this crucial recalibration between the individual and collective experience of history. George Eliot describes her assessment in her essay 'Value in Originality': 'Great and precious origination must always be comparatively rare, and can only exist on condition of a wide massive uniformity'.[110] Translating this into her novel *Impressions of Theofrastus* (1879) Eliot claims that 'originality can only arise out of similarity'.[111] Aesthetician and critic Walter Pater holds an almost identical opinion objecting to originality in its 'created' form, instead favouring compound 're-' words like regenerate, restore, renew and refine.[112] In *Appreciations* (1888) he expresses it more broadly: 'The material in which he [the artist] works is no more a creation of his own than the sculptor's marble. Product of a myriad various minds and contending tongues, compact of obscure and minute association, language has its own abundant and often recondite laws'.[113] It is in this atmosphere of change that literature and music poignantly intersect in Victorian culture, and here that we begin to understand and comprehend the reasons for the Land Without Music.

Steeped in the German romantic aesthetics of creation, in which imitation is considered anathema, the Coleridgean model of musical originality centred on the composer's autonomy. In subtly adjusted forms traces of that model subsist in contemporary music histories throughout the nineteenth century. But even from the 1830s it is apparent that the status of the

107 John Lubbock, *Prehistoric Times, as Illustrated by Ancient Remains, and the Manners and Customs of Modern Savages*, 2nd ed., London: Williams and Norgate, 1869, 410.

108 Butler, *Life and Habit, passim.* 109 Lewes, *Study of Philosophy*, 171.

110 George Eliot, 'Value in Originality' [mid-1870s], in George Eliot, *The Works of George Eliot: Essays and Leaves from a Note-Book*, Edinburgh and London: William Blackwood and Sons, 1883, 303, cited in Robert Macfarlane, *Original Copy: Plagiarism and Originality in Nineteenth-Century Literature*, Oxford: Oxford University Press, 2007, 107.

111 Eliot, 'Value in Originality', cited in Macfarlane, *Original Copy*, 107.

112 Macfarlane, *Original Copy*, 180.

113 Walter Pater, *Appreciations, With an Essay on Style*, Rockville, MD: ARC Manor, 1888/2008, 11.

ontologically static, autonomously created work waned as the more actively evolutionary terminology of invention waxed. Extolling the virtues of Beethoven Stafford quotes the Editor of the *Quarterly Musical Review*: 'his peculiar beauties may be encountered as follows: – originality of invention, uncommon passages, a very energetic manner, imitative passages almost unnumerable, and abstruse scientific modulation'.[114] For Stafford Beethoven was 'strikingly original'.[115] Hogarth reiterates this claim but finds (late) Beethoven frustratingly hermetic, as if retreating into the chrysalis of a more autonomous, 'created' model for which there is little to the more contingent, constructed 'invention'. Nonplussed by Beethoven's non-developmental progress Hogarth paraphrases Dugald Stewart's Kantian reflections on the imagination:

> The imagination is defined by metaphysicians as the faculty which enables us to create new forms, by throwing the parts of existing objects into new combinations; but, in order that the new creation may be comprehensible, all its parts must be previously familiar to the mind . . . Originality in the arts consists in the novelty of the combinations into which the artist throws known materials.[116]

Abandoning preformational materials – abandoning invention and indulging creation – may heighten individuality, but it obscures meaning and universality. Late Beethoven, no matter how intriguing, is a lesson learnt. Purcell was unfortunately (or fortunately) spared that dilemma, dying early in his mid-thirties and at the height of his powers. Purcell's originality balances creation and invention: 'He studied the genius of the Italian music . . . and he endeavoured to give his own music corresponding qualities. It thus arises, that Purcell's music . . . is perfectly original, and more truly and essentially English than that of any composer who has appeared before or since'.[117] Clearly drawing upon both Stafford and Hogarth the author of *Music in England, Ireland, Wales and Scotland* effectively reiterates this formula of originality, claiming that 'to name Purcell, is a sufficient answer to those who assert that England has no national music'.[118]

As these examples amply illustrate by the 1830s the earlier, increasingly deprecated, aesthetics of imitation (of fancy) was being gradually replaced by the more purposeful, constructive and adaptive terminology of invention. Concurrently the changing language of creative imagination transposed the

[114] Stafford, *History of Music*, 243. [115] Stafford, *History of Music*, 244.

[116] Hogarth, *Musical History*, 348–49. See Dugald Stewart, *The Collected Works of Dugald Stewart*, ed. Sir William Hamilton, 2 vols., Edinburgh: Thomas Constable, and Boston: Little, Brown and Co., 1854, vol. 2, 431.

[117] Hogarth, *Musical History*, 99. [118] *Music in England, Ireland, Wales and Scotland*, 33.

universality of nature into the individuality of human expression. In literature and in music invention gradually came to stand for the greater mass of which creation was its individual expression. And so in recapitulationary terms creation became the ontogeny to invention's phylogeny: creation recapitulates invention. The great composer (like Beethoven) could, therefore, no longer remain aloof from influence – like a musical anchorite. No, in order to progress he needed to engage with the stuff of musical history. Hullah, for example, captures the young Beethoven battling over fancy and the imagination: 'There might be little of the modesty of genius, if genius is ever modest; but there was plenty of that good sense which, however often it may be dissociated from mere *fancy*, is a necessary complement of high *imagination*'.[119] 'Like all great geniuses', Beethoven 'worked out his own purposes in his own way' (creation), and yet as his symphonies 'get later, [they reveal] a continually enlarging, plan, in comparison with those of any foregoing master . . . The richness of his invention in detail is inexhaustible'.[120]

This model of originality, frequently realized through Beethoven, sanctioned a new benchmark for British music histories and Purcell came to typify the new aesthetic regimen. Intriguingly Ritter is conflicted over Purcell, selectively reinterpreting the Coleridgean model: although his compositions

> give proof of his great mastery over the melodic and harmonic material, yet, in the interest of a free and original inventiveness, it is to be regretted that he thought it necessary to adhere to it so often. His fancy and imagination seemed at times, when a more independent and a bolder progression was required, to be kept down by voluntary but unnecessary fetters.[121]

This is arguably Ritter's gripe with any country. A nation simply fails to produce musical genius when its musical culture is impeded, proven in England by the baleful time leading up to Purcell, when 'national traits belonging to his people had not had time to crystallize into new original art-expression, based upon aesthetic forms called forth by the spirit of a new time'.[122] Ritter's use of the phrase 'spirit of a new time' speaks very clearly of the relationship of individual genius and the recapitulationary gestalt or *zeitgeist*. It is therefore axiomatic that although 'a genius stands above his time yet he must first learn how to master that which his time presents at its best: before he is able to mould his own original thoughts, he

119 Hullah, *History of Modern Music*, 224. 120 Hullah, *History of Modern Music*, 234.
121 Ritter, *Music in England*, 76. 122 Ritter, *Music in England*, 76–77.

must learn to cast that mould from the material he finds around him'.[123] Elevated to the level of national standing those materials help forge and develop a country's identity. Echoing such Herderian concepts of national genius Ritter cites a passage from the *English Cathedral Service*, itself reflecting and extending Carlyle's recapitulationary claims for biography. While for Carlyle 'History is the essence of innumerable Biographies'[124] the *English Cathedral Service* purports that 'Every work of genius forms a part of that foundation on which a nation's character, station, and form are created'.[125] For Ritter himself: 'The individual art character of a nation – if the gift of artistic inventiveness be innate to its emotional existence – will never gain general importance and aesthetical meaning if the mind is satisfied by mere slavish imitation of that which, on general principles, is recognized as excellent in the artistic labours and efforts of other and foreign nations'.[126]

Ritter's hostility to naked, indiscriminate imitation offsets the more evolutionary nature of his conception, because development is not a function of repetition but invention, which is itself a function of selection and adaptation. And for creation to encapsulate invention, in the same way that the individual embodies the species and ontology recapitulates phylogeny, originality must – at its most basic level – speak the language of invention. It must reflect the sum of which a composition represents one of its parts. It must, in some way or other, reflect Carlyle's maxim, that 'the Present is the living sum-total of the whole Past'.[127] So for British historians eager to disprove the Land Without Music changes in the aesthetics of originality provided an alternative developmental profile. Where previously creation, as autonomous, natural imagination, had governed originality as an antithesis to imitation, invention acquired the momentum of progress. Invention provided historians with a Whig interpretation of musical (and literary) history, investing within history a past which justified a present. The individual became Froude's organic 'instrument of Providence'[128] – and stood for Green's increasingly democratized national identity. For Victorian music historians, therefore, the first goal was to (re)construct the British past to fit the British present, but unlike their progressionist counterparts in constitutional history they had to make

[123] Ritter, *Music in England*, 77.

[124] Carlyle, 'On History', in Vanden Bossche, *Thomas Carlyle,* Histories, 5.

[125] Carlyle, 'On History', 43, from Edward Taylor, *The English Cathedral Service: Its Glory, – Its Decline, and Its Designed Extinction*, London: Simpkin, Marshall and Co., 1845, 66.

[126] Ritter, *History of Music*, 257.

[127] Carlyle, 'Characteristics', 380.

[128] Jann, *Art and Science of Victorian History*, 114.

the square peg of an illustrious past fit the annoyingly round hole of a depressingly arid present. Invention could help if only British music historians could square their musical ancestry with contemporary musical culture.

Enter Henry Davey, author of Victorian Britain's foremost history of English music. For him 'The art of musical composition is an English invention'[129] and like so many before him who misread the fifteenth-century scholar Tinctoris, it was John Dunstable who invented it. Dunstable was 'the originator of the whole art of composition'[130] and breathed life into composition, freeing its vocal and instrumental parts from the grievous oppression of homophony. Dunstable democratized its parts, releasing them to express their individuality while at the same time contributing to the essential sum of polyphonic composition. Dunstable gave musical parts their independence like Englishmen their constitution. And it was through Dunstable that Davey freed British musical history from the Land Without Music, not simply by accepting Sebald Heyden's (mis)translation of Tinctoris, but by investing within Dunstable the developmental fullness of English originality. By ascribing to Dunstable germinal characteristics Davey invented England's musical past, but also created its present:

> In examining the origin of an art . . . historians should continually bear in mind the resources at the original inventor's command. Before his time there has been a mechanical application of undeveloped resources, for useful purposes; these resources in time are improved, and then a man appears who uses them artistically. The *art* is now created; the resources are further improved and enlarged, the creative artists use them better and better until the culminating point is reached, and the natural decline follows.[131]

Using invention to create art Davey inverts Coleridge and conjugates his developmental verb, and that verb is degeneration, or cyclism. Degenerational Davey, like Freeman and Hawkins, believed in 'strange cycles of history'[132] and the winnowing coda of a nation's musical life. But as a cyclist a nation's natural rise and fall does not end in permanent musical oblivion but begins again. Borrowing Stubbs's evolutionary optimism Davey constructs a history which illustrates 'the continuity of a sense of earlier greatness and of an instinctive looking towards a greater destiny'.[133]

[129] Davey, *History of English Music*, 1. [130] Davey, *History of English Music*, 52.
[131] Davey, *History of English Music*, 52. [132] Freeman, *Chief Periods*, 176.
[133] Stubbs, *Constitutional History of England*, vol. 3, 2.

Dunstable is that earlier greatness – he was 'the genius ... [who] had thrown back the veil'[134] – and the greater destiny can be realized provided the problematical present is re-invented. Or even re-created. Intriguingly Davey changes tack when he brings his book to its conclusion. By the end of the book, when British music desperately needs to start again and overcome the Land Without Music, Dunstable becomes a creator, rather than inventor: 'No light thing is it, that John Dunstable certainly created vocal composition ... Let us ever reverence [him] ... who dowered the whole world with a glorious new art'.[135]

That Dunstable was in the end converted from an inventor to creator serves political, not simply aesthetic, purposes. Arguing through Dunstable that Britain subscribed to a developmental model of originality based on invention rather than creation Davey deprives the Land Without Music of its thesis in the same way the Land Without Music deprives Britain of its musical inheritance. Reinstating that inheritance, however, requires a new invention of which Dunstable creates the basis. Although unoptimistic – there is 'little hope that a genius will arise in the future'[136] – Davey is smugly buoyed by the synchronicity of Germany's fall and Britain's imminent rise: 'The Teutonic nations have apparently said all they have to say in music, Germany seems worked-out now, as England was 200 years ago; but the highest pinnacles of musical composition have been reared by Germans, and the foundations were laid by Englishmen'.[137] The synchronic nature of one nation's rise and fall of another fits comfortably into Davey's developmental conception, especially when those nations are respectively Britain and Germany. The Land Without Music is after all a German philosophical import, and like Germany's music is 'worked-out'. The many defects Davey identities in modern Victorian musical culture do not equate to a Land Without Music but to a Music Without a Land. Everywhere in England a musical nation is evident but dogged by the foreign alien criticism of the Land Without Music that nation has come to deprecate its own native genius. Like Parry's Purcell modern Britain's musical genius is a seed fallen on stony ground. But as the case of Dunstable's originality proves that ground can recover and revive, so where Germany has become the Land *Without* Music Britain will become the Land *With* Music.

[134] Davey, *History of English Music*, 498.
[135] Davey, *History of English Music*, 499.
[136] Davey, *History of English Music*, 499.
[137] Davey, *History of English Music*, 499.

Conclusion

It must have galled Davey to observe that Kiesewetter fails to mention Dunstable in *The History of the Modern Music of Western Europe* (1834, translated 1848), a fact not missed in Crowest's historically incomplete *Story of British Music* (1896). Crowest, who was unaware of his contemporary's book, is less ideologically drawn than Davey, and inadvertently bursts Davey's bubble by trying to right the historical wrong of Dunstable's position as inventor of counterpoint or composition. For killjoy Crowest 'Dunstable could not have *invented* counterpoint however, inasmuch as several works upon it were written before he was born'.[138] Later composers like Thomas Morley were dragooned into criticism as well, complaining about Dunstable's detrimental habit of separating syllables by rests. Unlike Davey Crowest merely tells a story in narrative form bereft of ideological necessity. Crowest's book tells the story of the Land *With* Music, English music up to the Tudor period in which Dunstable stands for but one of its many British greats. His genius could afford to be 'modest', to use Hullah's term,[139] because history had not yet become apologetical. The Land With Music sustained a rich supply of home-grown talent and the Land Without Music was some time away in the future. That said, British music was never entirely free from vicissitudes and some worryingly long hiatuses appeared between the third and fourth periods of Crowest's history. Auguring rough times ahead the 150 years between the famous thirteenth-century round 'Sumer is icumen in' and Dunstable marks a curiously fallow time when 'history is silent upon the point'.[140] Looking backward and forward Crowest speaks presciently:

> This early English School fluctuated both in its measure of excellence and succession of masters, but this was due solely to the disturbed conditions of the times. The English musical spirit was always there, and could not be exterminated, whatever happened. No other country possessed so much dormant, slumbering musical vitality, even when national matters here were at their worst. Yet all that we have had has been ignored by foreign, as well as by several English, music historians![141]

Like all music historians of the time Crowest's concern appears to become more intensely apologetical as his topic approaches the present,

[138] Frederick J. Crowest, *The Story of British Music (From the Earliest Times to the Tudor Period)*, London: Richard Bentley and Son, 1896, 345.
[139] Hullah, *History of Modern Music*, 224. [140] Crowest, *Story of British Music*, 381.
[141] Crowest, *Story of British Music*, 383.

such is the need to combat the Land Without Music. But is this trajectory at its most basic level merely a survival mechanism adapted into the very psyche of a battered Victorian musical culture, or does it represent a genuine change in the way Victorian musicologists conceptualize musical progress? In the late Victorian period, with empire at its height, historical progressionism was *de rigueur*, summarized in the type of history called the Whig interpretation. That history – strong, confident, unassailable – created a past leading ineluctably to a culturally triumphant present. But that triumphant present was not shared across its culture, and music, perhaps most amongst the arts, was demeaned as the butt of long-standing abuse. According to the decree of the Land Without Music Britain had no musical empire. The absence of musical empire in that sense inverts the presence of constitutional empire manufactured by Whig historians. It eschews progression and conceives of a thwarted, degenerated present; it proposes a cyclical response to a largely progressional historical culture. Ironically (or not), just as Britain started edging slowly towards the brink of losing its empire music began to resurface. Music assumed the role of progression and general history became arguably more cyclical. Darwinism helped of course, glazing developmentalism with the sheen of evolutionary uncertainty. The transition is obvious in Stubbs, for example, whose mixture of Spencerian progression and Darwinian evolutionism was, later, very heavily criticised. The next history of English music, comfortably post-Victorian, is Ernest Walker's comprehensive *History of Music in England* (1907). It would continue and develop the apologetical tradition with a restrained, yet compellingly sustained thesis: 'Between Dunstable and the present day there are nearly five centuries of continuous creative work . . . No doubt we have slept the sleep of the dull too often and too long; but now we are awake again'.[142]

What also helped awaken British musical culture from its deep slumber were changes to prevailing notions of originality and the way those changes impinged upon notions of recapitulation and genius. Supported by increasingly popular evolutionary language (not least that of Darwin), originality evolved from a spontaneous, autonomous and natural creative act to a more historically contingent, hereditary process. The Coleridgean model of originality superseded, invention came to embody the new historical, genetic consciousness expressed by Butler, Lewes and Eliot. The Carlylean hero or genius, moulded from the earlier stuff of Coleridge, was adapted to

[142] Ernest Walker, *A History of Music in England*, Oxford: Clarendon Press, 1907, 356–57.

this process. The genius, the hero – the artist – of Pater recapitulates his materials, the product of 'myriad various minds and contending tongues'.[143] In Victorian music histories several composers represent British genius; perhaps none better than Henry Purcell. But while Purcell is often portrayed as a truly inventive artist his creativity is often criticised for being imitative. In Purcell fancy and the imagination remain at odds. In Dunstable, however, Britain finds its true genius, for it is Dunstable who consistently lays the foundation for English musical art. It is Dunstable who would encapsulate Victorian concepts of true originality and it is Dunstable who recapitulates the sum of British musical history. Dunstable speaks for the nation, its glorious past, its troubled present and its symmetrically glorious future.

143 Pater, *Appreciations*, 11.

7 | Music Theology

Introduction

If Victorian science and religion were genuinely at war nothing could typify their positions better than debates between the relative and absolute. 'Modern thought', Walter Pater proclaims, 'is distinguished from ancient by its cultivation of the "relative" spirit in place of the "absolute" ... To the modern spirit nothing is, or can be rightly known, except relatively and under conditions'.[1] Frederick Harrison concurs: 'The doctrine of the Relativity of Knowledge ... has been accepted by so many schools of thought for their basis as almost to have passed into the sphere of subjects which are little liable to question'.[2] St George Mivart would question the doctrine of relativity, however; according to the biologist and Catholic apologist there is nothing but permanent law and order established through the rigid laws of nature.[3] 'We can know absolute truth', Mivart insists; 'truth really exists.'[4] Catholic theologian John Henry Newman adopts a more conciliatory tone. Rigid absolutism is to be avoided because it undermines human individuality:

> we are ever measuring and sounding, framing cross classes and cross divisions, and thereby rising from particulars to generals, that is from images to notions. In processes of this kind we regard things, not as they are in themselves, but mainly as they stand in relation to each other ... He [man] is attenuated into an aspect, or relegated to his place in a classification. Thus his appellation is made to suggest, not the real being which he is in this or that specimen of himself, but a definition ... he is made the logarithm of his true self.[5]

[1] Walter Pater, 'Coleridge's Writings', *Westminster Review*, New Series 29 (Jan. 1866), 107.

[2] Frederick Harrison, 'The Subjective Synthesis', *The Fortnightly Review* 8 New Series (1 July–1 Dec. 1870), 184.

[3] St George Mivart, *The Groundwork of Science: A Study of Epistemology*, New York: G. P. Putnam's Sons; London: Bliss, Sands & Co., 1989, 283.

[4] St George Mivart, *On Truth: A Systematic Inquiry*, London: Kegan Paul, Trench & Co., 1889, 134, 238.

[5] John Henry Newman, *An Essay in Aid of A Grammar of Assent*, London: Burns, Oates, & Co., 1870, 28–29.

This chapter tries to make sense of the relationship between the relative and absolute in Victorian musical culture. It does this by probing the way notions of the relative and absolute affected Victorian concepts of spiritual progress, and how spiritual progress was itself affected by the Christological theology underlying some of the most extreme apologetics of the English plainchant revival. A general introduction examines the theological – and in particular, Christological – implications of Darwinian and non-Darwinian evolution on Victorian notions of spiritual progress. The section following explores their application, venturing to explain the epistemological barriers of representing and embodying Christ through music in oratorios and hymns. A third section focuses in on the way incarnational theologies structure spiritual progression in the plainchant revival, examining the Anglican and, in greater scrutiny, the Roman Catholic revivals through a case study of the contentious theological world of plain song extremist the Rev. Henry Formby. Lastly, a conclusion returns to reflect on spiritual progress through further considerations of Newman and his understanding of the relative and absolute in Victorian culture.

Theology

Explaining the Victorian – and Romantic – obsession with the relative and the absolute only invites contradiction. Christopher Herbert describes it confusingly as 'Universal Relativity', where differential signifiers 'act as proxies for the real things' obeying 'two great principles that seem contradictory but figure in the logic and poetics and in the ethical allegory of relativity as one and the same: Unity and Proliferation. Relativity means that all things are rigorously bound up together in a single individual world; it means also that this world is not one after all, but uncontrollably multiple'.[6] Herbert takes his cue from his namesake Herbert Spencer who in *Principles of Sociology* maintains that within the organization of any society or organism 'changes in the parts are mutually determined, and the changed actions of the parts are mutually dependent'.[7] The same principle applies to the evolution of music from speech: 'the law that alike in

[6] Christopher Herbert, *Victorian Relativity: Radical Thought and Scientific Discourse*, Chicago and London: University of Chicago Press, 2001, 49–50.

[7] Herbert Spencer, *Principles of Sociology*, 3 vols. (1874–96), A System of Synthetic Philosophy 6–8, New York: D. Appleton and Co., vol. 1, 1883, 469.

occupations, sciences, arts, the divisions that had a common root, but continual divergence have become distinct, and are now being separately developed, are not truly independent, but severally act and react on each other to their mutual advancement'.[8] Evolutionary Spencer declares everything materialistically relative and states that 'organisms in their totality are mutually dependent, and in that sense integrated',[9] but he also admits that Relativity is impossible without the presence of some kind of Universality – a multiplessness or multiplicity transcendentally beyond comprehension, an absolute unknown or unknowable: 'the Unknowable – the Complex – is a cause or "manifestation of an unlimited and incomprehensible Power"'.[10] For the Rev. George Ladd Spencer's 'Unknowable is known to be a Power; and it must be a great Power, for the *Universe* – that is, all manifestations of power – is manifest to us. But power, inconceivably great – enough to accomplish all things done and even more – has been by Theists from time immemorial held to be an attribute of God'.[11] For Spencer antithetical spirit and matter are to 'be regarded as but a sign of the Unknown Reality which underlies both',[12] and it is music which emphatically expresses that 'unknown ideal life'.[13]

Understanding the relationship between the relative and the absolute, the Known and the Unknown, perplexed more than just Spencer – practically the whole of Victorian (and much nineteenth-century European) culture seemed to be consumed by it, not least Christian thinkers grappling with the spiritual implications of Darwin's highly materialistic evolutionary thought. According to James Moore many rejected Darwinism outright (Christian Anti-Darwinists); others were uncertain (Christian Darwinisticists) and some positively embraced it (Christian Darwinists).[14] Crucial to all their positions is the role of divine agency – what today we might call intelligent design. Darwin himself vacillated on the topic:

[8] Herbert Spencer, 'The Origin and Function of Music', *Fraser's Magazine* (Oct. 1857), in Herbert Spencer, *Literary Style and Music: Including Two Short Essays on Gracefulness and Beauty*, New York: Philosophical Library, 1951, 70.

[9] Herbert Spencer, *First Principles*, 4th ed., New York: D. Appleton and Co., 1896, 325.

[10] Herbert Spencer, in F. Howard Collins, *An Epitome of the Synthetic Philosophy*, London and Edinburgh: Williams and Norgate, 1894, 13.

[11] Rev. George T. Ladd, *Lectures on the Unknown God of Herbert Spencer: And the Promise and Potency of Prof. Tyndall*, Milwaukee: I. L. Hauser & Co., 1900, 24.

[12] Spencer, in Collins, *Epitome*, 63. [13] Spencer, 'The Origin and Function of Music', 74.

[14] James R. Moore, *The Post-Darwinian Controversies: A Study of the Protestant Struggle to Come to Terms with Darwin in Great Britain and America: 1870–1900*, Cambridge: Cambridge University Press, 1979, 218.

My theology is a simple muddle: I cannot look at the Universe as the result of blind chance, yet I can see no evidence of beneficent design. As for each variation that has ever occurred having been preordained for a special end, I can no more believe in it than that the spot on which each drop of rain falls have been specially ordained.[15]

Others were more certain. For Frederick Temple, Archbishop of Canterbury and Arch-Christian Darwinist, either God created nature as it will be in perpetuity, or more probably he imbued creation with the ability to create itself perpetually:

In the one case the Creator made the animals at once such as they now are; in the other case he impressed on certain particles of matter which, either at the beginning or at some point in the history of His creation he endowed with life, such inherent powers that in the ordinary course of time living creatures such as the present were developed. The creative power remains the same in either case; the design with which that creative power was exercised remains the same. He did not make the things, we may say; no, but He made them make themselves.[16]

The problem with Temple's position is one that affected many evolutionary 'designers', namely the impression that God took his foot off the pedal and after an initial burst of creative activity let nature simply coast along. American music theologian John Harrington Edwards certainly gives that impression when he talks about 'design in design', leaving music with a trace, ghost presence of the divine: 'it is more in accord with the natural workings of minds like ours to see purposed ends in results of such a character as those attained by music, than it is to imagine a pseudo-immanence of finality directing a system of sounds so complicated, perfectly adjusted to the rest of the universe'.[17] Although not uncommon Temple and Edward's brand of divine passivity eventually failed the teleological litmus test because Christian thinkers wanted more from God than just a great big bang. They wanted intimate divine involvement. Darwin's friend the American botanist and devout Protestant Christian Asa Gray suggests this need when he defends Darwin's evolutionary thought in a way Darwin himself would never recognize – as expressing

[15] Charles Darwin, 'Recollections of the Development of my Mind and Character' [Autobiography (1876–4.1882)] CUL-DAR26.1–121, transcribed Kees Rookmaaker, 62–63, Darwin Online http://darwin-online.org.uk/,accessed 26/11/15.

[16] Frederick Temple, 'Lecture IV: Apparent Conflict between Religion and the Doctrine of Evolution', in *The Relations between Religion and Science: Eight Lectures Preached Before the University of Oxford in the Year 1884*, London, 1903, Project Gutenberg, www.gutenberg.org/files/17194/17194.txt, accessed 26/11/15.

[17] John Harrington Edwards, *God and Music*, New York: The Baker and Taylor Co., 1903, 197.

a 'wider teleology': 'Darwin's great service to Natural Science [was] in bringing it back to Teleology: so that instead of Morphology *versus* Teleology, we shall have Morphology wedded to Teleology'.[18] Gray's wider teleology expresses what Alister McGrath calls 'directionality within the natural world' – a 'narrative of increasing complexification from the origins of the universe to the present day'.[19] But directionality as McGrath understands it needs amplification because like many of his contemporaries Gray uses directionality to illustrate the two interrelated forms of spiritual progress: *relative* directionality (material, natural, historical time), and *absolute* directionality (spiritual, divine, eschatological eternity). They unite in the Great Chain of Being to form spiritual progress:

> Darwinian teleology has had the special advantage of accounting for the imperfection and failures as well as for successes . . . Without the competing multitude, no struggle for life; and without this, no natural selection and survival of the fittest, no continuous adaptation to changing surroundings, no diversification and improvement, leading from lower up to higher nobler forms.[20]

Spiritual progress also forces Gray to revise his interpretation of the cornerstone to Darwin's evolutionary theory, natural selection:

> Natural selection is not the wind that propels the vessel, but the rudder which, by friction, now on this side and now on that, shapes the course. The rudder acts while the vessel is in motion, effects nothing when the vessel is at rest. Variation answers to the wind. 'Though hearest the sound thereof, but canst not tell when[ce] it cometh and wither it goeth'. [John 3:8] Its course is controlled by natural selection, the action of which at any given moment, is seemingly small or insensible; but the ultimate results are great.[21]

While most Darwinians were apt to agree that natural selection was a key mechanism of evolution its pious demotion to 'rudder' was widely shared by Christian evolutionists eager to assert a theory of special creation – what elsewhere Gray had described as the 'mystery of a beginning' laying hidden in each variation.[22] In Victorian musical culture the problem of creation

[18] Asa Gray, 'Scientific Worthies: Charles Robert Darwin', *Nature* 10/240 (1874), 81.

[19] Alister E. McGrath, *Darwinism and the Divine: Evolutionary Thought and Natural Theology*, The 2009 Hulsean Lectures, University of Cambridge, Malden, MA, and Oxford: Wiley Blackwell, 2011, 187.

[20] Asa Gray, *Darwiniana: Essays and Reviews Pertaining to Darwinism*, New York: D. Appleton and Co., 1889, 378.

[21] Gray, *Darwiniana*, 386.

[22] Asa Gray, *Natural Science and Religion: Two Lectures Delivered to the Theological School of Yale College*, New York: Charles Scribner's Sons, 1880, 72.

prompted two separate but inter-related debates representing different sides of the classic scientific and religious divide: one about the natural and temporal origin of music through impassioned speech (Spencer) or descended through sexual selection (Darwin); the other about the spiritual and cosmic origin of music in God. Neither side, however, relinquished default mysticism when speculating on the origin of music. For someone whose science many derided as rankly materialist Darwin's reflections on music evolution are surprisingly – or perhaps expectedly – evasive when it comes to sourcing its origin: 'As neither the enjoyment nor the capacity of producing musical notes are faculties of the least direct use to man in reference to his ordinary habits of life, they must be ranked amongst the most mysterious with which he is endowed'.[23] Spencer is no clearer though ostensibly equally materialistic. Although he believes that music begins in impassioned speech – effectively in emotion – he also believes that not only is music the 'unknown ideal life' but that the origin of the emotions (as the mind) is itself completely unknowable:

> That no idea or feeling arises, save as a result of some physical force expended in producing it, is fast becoming a commonplace of science. How this metamorphosis takes place . . . how it is possible for aerial vibrations to generate the sensation we call sound, or for the forces liberated by chemical changes in the brain to give rise to emotion – they are mysteries which it is impossible to fathom.[24]

For many Protestant evolutionists that mystery begins with a complex set of Christological propositions teleologically linking the relative material immanence of man to the absolute spiritual transcendence of God through the 'evolutionary' perfectibility embodied in the Incarnation of Christ. American Congregationalist Lyman Abbot espouses this view in *The Theology of an Evolutionist* (1897):

> The evolutionist . . . thinks of Christ, not as a strange, inexplicable God-man who was neither God nor man, not as a being who went through life doing some things as God and some things as man, but as God *in* man, God so perfectly possessing one unique human soul that in that soul we see reflected at once the image of God and the perfection of manhood. Christ is God manifest in the flesh; that is, such a manifestation of God is possible in human life.[25]

[23] Charles Darwin, *The Descent of Man, and Selection in Relation to Sex*, 2 vols. London: John Murray, Vol. 2, 1871, 333.

[24] Spencer, in Collins, *Epitome*, 27–8.

[25] Lyman Abbott, *The Theology of an Evolutionist*, Boston and New York: Houghton, Mifflin and Co., 1897, 189.

Abbott reflects a classic Schleiermachian interpretation of the Incarnation of Christ as a sempiternal embodiment of God within man and man within nature. For Schleiermacher

> the most rigorous view of the difference between Him [Christ] and all other men does not hinder us from saying that his appearing, even regarded as the incarnation of the Son of God, is a natural fact . . . as certainly as Christ was a man, there must reside in human nature the possibility of taking up the divine into itself, just as did happen in Christ. So that the idea that the divine revelation in Christ must in this respect be something absolutely supernatural will simply not stand the test.[26]

Schleiermacher applies this incarnational principle to a deeply embodied music theology based on human feeling. Although natural feelings must be attuned by reason, attuned feelings are developed into calmed passion and cleansed before becoming music. This creates a 'timeless oneness of life'[27] and the expression of feeling leading to musical generation. According to Jeremy Begbie for Schleiermacher the feeling (*Gefühl*) in music outstrips language, and it makes music the most inward art because it is related to our immediate self-consciousness. As with religious experience, hearing or listening to music is essentially a receptive matter yet it offers access to a universal plane of experience beneath or underlying particular things; music experience shares religious self-consciousness; and music permeates our being with the infinite.[28] While Schleiermacher resists transcendentalizing music or indeed the Incarnation of Christ he does, as this suggests, transform our understanding of time to account for Christ's ontologically human and divine natures: 'that Christ even as a human person was ever coming to be simultaneously with the world itself'.[29] Although Christ became incarnate and lived within time as man (relatively) he was nonetheless always in the process of incarnation (absolutely); in other words,

> incarnation did not take place on a particular date [when Jesus was born] but was a process that had been going on during a very long period of preparation [eternity]. The suggestion seems to be that incarnation is an aspect of world-history or an element in the cosmic process, so that even as the world was developing and

26 Friedrich Schleiermacher, *The Christian Faith* ed. H. R. Mackintosh and J. S. Stewart (*Der Christliche Glaube*, 1st ed., 2 vols. Berlin: G. Reimer, 1821–22; 2nd ed., 2 vols. Berlin: G. Reimer, 1830–31), 2 vols. in 1 vol., London and New York: T & T Clark, 1999, 64.

27 Friedrich Schleiermacher, *Ästhetik: Über den Begriff der Kunst*, ed. Thomas Lehnerer, Hamburg: Felix Meiner, 1831–32/1984, 64.

28 Jeremy Begbie, *Resounding Truth: Christian Wisdom in the World of Music*, London: Society for Promoting Christian Knowledge, 2007, 148–49.

29 Schleiermacher, *Christian Faith*, 64.

assuming its form through the ages in which it has been in existence, Jesus Christ too as a physical existence was in process of formation in these events.[30]

The Incarnation also blurs into an eschatological future. Abbot sets this out in sustained theological detail in *The Theology of an Evolutionist* (1897): 'The evolutionist . . . believes in Incarnation; this is, in the indwelling of God in his children, of which Incarnation the type and pattern is seen in Him who is at once the manifestation of God to man and the revelation to men of what humanity is to be when God's work in the world is done, – perfect God and perfect man, because God perfectly dwelling in a perfect man. He believes in atonement; that is, in a true reconciliation between God and man, making them at one through the Incarnation and Passion of Jesus Christ, who lived and suffered, not to relieve men from future torment, but to purify and perfect them in God's likeness by uniting them to God . . . He believes in immortality, not as a mere endless existence, but as an undying nature, which is superior to death, because it shares with God, its Redeemer, the power of an endless life'.[31]

'Undying nature' not only embodies Christological concepts of spiritual progress uniting relative nature and absolute spirit in a teleological projection up the Great Chain of Being, it also plays into the hands of recapitulationary interpretations of the Incarnation. Where Schleiermacher sees feeling as recapitulating the essence of embodied Christian incarnation music is the theological epitome of a longstanding recapitulatory principle. Used to describe Irenaeus's theory on the Incarnation recapitulation is a term adapted from Ephesians 1:10 where we read that all people will be gathered up in Christ the fullness of time. Irenaeus views Jesus as an 'antitype of Adam',[32] expiating sin through his death:

For by summing up in Himself the whole human race from the beginning to the end, he also summed up its death . . . Now he [Christ] died on the day in which he did eat [the Last Supper]. For God said, 'in that day on which ye shall eat of it, ye shall die by death'. The Lord, therefore, recapitulating in Himself this day, underwent His sufferings upon the day preceding the Sabbath, that is, the sixth day of the creation, on which man was created; thus granting him a second creation by means of His passion, which is that [creation] out of death.[33]

[30] Macquarrie, *Jesus Christ in Modern Thought*, 205.

[31] Abbott, *The Theology of an Evolutionist*, 190–91.

[32] Hans Schwarz, *Christology*, Grand Rapids, MI, and Cambridge: William B. Eerdmans Publishing Company, 1998, 254.

[33] Irenaeus, *Against Heresies*, in Rev. Alexander Roberts and James Donaldson (eds.), *The Writings of the Fathers down to A.D. 325*, 10 vols. vol. 1: Apostolic Fathers, New York: Charles Scribner's Sons, 1913, 551.

It is easy to see why Irenaeus would appeal to nineteenth-century spiritual progressivists, fed as they were on a wide range of implicitly recapitulationary evolutionary paradigms. Just as spiritual progressivist Schleiermacher considers music to recapitulate the feeling of divine oneness which is 'essential to a full humanity and a full Christianity'[34] likewise he considers the progression from Jesus to Christ to recapitulate the totality of natural life-history. In scientific terms the (relative) ontogeny of Jesus recapitulates the (absolute) phylogeny of Christ: 'What comes into existence through the being of God of Christ is all perfectly human, and in its totality constitutes a unity, the unity of a natural life-history, in which everything that emerges is purely human'.[35] Abbot broadens this classic recapitulationary line of thinking in *The Evolution of Christianity* (1892) to incorporate a definitional statement of spiritual progress:

> In the spiritual, as in the physical, God is the secret and source of life; phenomena, whether material or spiritual, are the manifestations of his presence; but he manifests himself in growth, not in stereotyped and stationary forms; and this growth is from lower to higher, from simpler to more complex forms, according to well defined and invariable laws, and by a force resident in the growing object itself. That unknown force is God – God in nature, God in the church, God in society, and God in the individual soul.[36]

Spiritual progress and Incarnational Christology cross-fertilize the theologised science of Victorian culture. This is perhaps no better articulated than in James Iverach's *Christianity and Evolution* (1894):

> The process of revelation was slow, evolutionary, progressive. Revelation was always related to the natural, proceeded on it, assumed, rectified it, and transformed it to a higher character and use. Even when, as Christians believe, revelation had been complete and redemption had been in essence realised in the work of Christ, then began again a slow course of evolution ... upwards. This new evolution was with a view to realise, in man and by man, a standard actually, historically, and concretely given in Jesus Christ ... It makes no difference to the character of the process of evolution, looked at as a process "ever climbing after some ideal end," whether we consider the end as one already given and known, or as one not yet realised, and so far unknown. The process of evolution of Christian life is a process towards Christ. He is the aim, the goal, the end towards which

[34] Steven Guthrie, 'Carmen Universitatis: A Theological Study of Music and Measure', PhD diss., University of St Andrews (2000), 65.

[35] Schleiermacher, *Christian Faith*, 409.

[36] Lyman Abbott, *The Evolution of Christianity*, Boston and New York: Houghton, Mifflin and Co.; and Cambridge: The Riverside Press, 1892, v.

Christian life in the individual and in society every tends, and He is also the means without which the end can never be reached.[37]

Iverach's synthesis expresses what many felt about the Incarnation – we can know transcendence through immanence; the spiritual through the material; the absolute through the relative and the eternal through history. Perhaps the greatest advocate of this belief – and its greatest influence on sacred musical culture – is England's most renowned Protestant convert to Catholicism John Henry Newman. Catholic Newman seems at first an unlikely candidate to support Protestant Iverach yet even at his most polemical Newman's theological message is clear. History (the relative) without revelation (the absolute) can never explain the fullness of religion:

no Religion has yet been a Religion of physics or of philosophy. It has ever been synonymous with Revelation. It has never been a deduction from what we know: it has ever been an assertion of what we are to believe. It has never lived in a conclusion; it has never been a message, or a history, or a vision.[38]

For Newman Christianity progressed from natural to revealed religion – from a general sense of moral obedience to an authority outside human consciousness to a religion revealed through dogma.[39] Frank Turner puts it aptly when he considers Newman's Christianity to resemble 'more nearly an Aristotelian form that must realize itself only through a material embodiment, in this case in the embodied material life of human beings in human society. Christianity was what Christianity had become and presumably was becoming'.[40] In *An Essay on the Development of Christian Doctrine* (1845) Newman takes 'the Incarnation as its [Christianity's] central aspect'[41] but without apparently according it the higher status of an idea encapsulating its totality.[42] Newman's concept of the Incarnation would have a profound impact on certain quarters of Victorian musical culture, especially those linked to the Gregorian chant revival. Christopher Dawson suggests Newman 'saw that it was only in history that the process

[37] James Iverach, *Christianity and Evolution*, London: Hodder and Stoughton, 1894, 226–27.

[38] John Henry Newman, *Discussions and Arguments on Various Subjects*, London: Basil Montegu Pickering, 1872, 296.

[39] Frank M. Turner, 'John Henry Newman and the Challenge of a Culture of Science', *The European Legacy* 1/5 (1996), 1697.

[40] Turner, 'Newman', in Fergusson (ed.), *Blackwell Companion to Nineteenth-Century Theology*, 128.

[41] J. H. Newman, *An Essay on the Development of Christian Doctrine*, London: Basil Montegu Pickering, 1878, 54.

[42] Drew Phillip Morgan, 'Hermeneutical Aspects of John Henry Newman's Essay on the Development of Christian Doctrine', *Horizons* 16/2 (1989), 228.

of progressive revelation and spiritual renovation could be fulfilled'.[43] Henry Formby, Newman's ideologically pungent friend, and musical thorn in his theological side, would eventually apply this principle by sacramentalizing chant – by elevating it to a position of sacredness akin to God himself. As their correspondence reveals later in this chapter Newman was not amused.

While for Newman progressive revelation and spiritual renovation unite in the doctrine of Incarnation[44] the idea the doctrine projects is entirely embodied in the image of Christ. Indeed the relative material and absolute spiritual culminate in Newman's image of Christ as the recapitulation of historical man within eternal God. Newman conceives the 'Thought or Image of Christ'[45] as the original – and therefore ongoing – instrument of conversion and principle of fellowship of the Christian community. But for him music is only part of that process of conversion. It is not the object of spiritual progress but a theological means to a spiritual end. The object is the 'central Image' we recognize as the 'vivifying idea both of the Christian body and of individuals in it'.[46] As Newman conceptualizes it the Thought, Image or Idea of Christ is inextricably bound to the history and physical body of the church's fraternity both in its individual members and the lifeblood of its collective communal structure, and it is Christ who encapsulates what 'is found, through His preachers, to have imprinted the Image or Idea of Himself in the minds of His subjects individually; and that Image, apprehended and worshipped in individual minds, becomes a principle of association, and a real bond of those subjects one with another, who are thus united to the body by being united to that Image'.[47] It is in many respects this vision of Christ – Newman's vision of a church *really* bound to Christ as Thought or Image or Idea – that informs and inspires the most aspirationally Christological genres of Victorian music, Gregorian chant. According to Formby and his comrade in chant the architect Augustus Welby Northmore Pugin it is the Christ of chant who has imprinted the Image, the Idea, the Thought of Himself within His music, and through it has united and reunited Himself to His church and His Church to his creation.

[43] Christopher Dawson, James Oliver and Christina Scott (eds.), *Religion and World History: A Selection from the Works of Christopher Dawson*, Garden City, NY: Doubleday, 1975, 292.

[44] Terrance Merrigan, 'Revelation', in Ian Ker and Terrance Merrigan (eds.), *The Cambridge Companion to John Henry Newman*, Cambridge: Cambridge University Press, 2009, 57.

[45] John Henry Newman, *An Essay in Aid of a Grammar of Assent*, London: Burns, Oates, & Co., 1874, 465.

[46] Newman, *Essay in Aid of a Grammar of Assent*, 464.

[47] Newman, *Essay in Aid of a Grammar of Assent*, 464–65.

Music Theology

Christian Musical Representation

There were many Victorians united to the Image of Christ, not least people involved in the activities of composing, performing, listening to and writing about music. But amongst musicians there were also religious thinkers who eschewed a pointedly Christian conception of the world. For them the spiritual world need not be specifically Christian – it could be transcendent, mysterious, ineffable, sublime or divine without subscribing to the entrenched dogmas of historical religion. For these people music expresses those qualities because it combines the relative and the absolute in perfect harmony, as it were. Inevitably perhaps Victorians reflected their German counterparts in philosophical idealism, especially in musical models which viewed the ontogeny of the relative as recapitulating the phylogeny of the absolute.[48] Romantic recapitulationist Daniel Schubart, for example, writes of the symphony that 'It is not the mere din of voices . . . it is a musical whole whose parts, like emanations of spirit, form a whole again'.[49] Schubart touches upon an unresolved tension in theories of musical absolutism, which is whether music hierarchizes divinity over nature by seeking to overcome its material content (the relative) through its spiritual form (the absolute). Schiller highlights this dilemma when he divulges that 'the real artistic secret of the master, [is] *that he eradicates the material through form*'.[50]

Admiration for form and the apparently concomitant demotion of material is pervasive across a certain brand of German idealism. Johann Gebhard Ehrenriech Maaß writes that 'the beauty of a thing resides in its form'.[51] The elevation of form over content – and by implication the absolute over the relative – is an aesthetic chauvinism with Christological parallel prompted by what theologians usually call something like 'Lessing's ditch'. Lessing's ditch is effectively the sometimes yawning gap which exists between the content of the historical Jesus and the form of the

[48] See Carl Dahlhaus, *The Idea of Absolute Music*, trans. Roger Lustig, Chicago and London: University of Chicago Press, 1978/1991, and Daniel Chua, *Absolute Music and the Construction of Meaning*, Cambridge: Cambridge University Press, 1999.

[49] Christian Friederich Daniel Schubart, *Leben und Gesinnungen*, Stuttgart: Gebrüdern Mantler, 1791–93, cited in Dahlhaus, *The Idea of Absolute Music*, 10.

[50] Friedrich Schiller, *Über die äesthetische Erzeihung des Menschen* (Letter 22), cited in Mark Evan Bonds, *Absolute Music: The History of an Idea*, Oxford: Oxford University Press, 2014, 99.

[51] Johann Gebhard Ehrenriech Maaß, 'Ueber die Instrumentalmusick', Neue Bibliothek der schönen Wissenschaften 48 (1792), 3–40, cited in Bonds, *Absolute Music*, 101.

Christian religion. Critics complain, for example, that David F. Strauss's putatively materialist *Life of Jesus* (1834) embraces the historical Jesus at the expense of organized religion, but in fact he finishes his book by resorting to 'a Hegelian interpretation of Jesus as the realisation of the Absolute in the realm of history'.[52] Hegel has a lot to answer for when it comes to depreciating the material value of the musical absolute. Indeed spiritual progress for Hegel amounts to nothing more than material regress, or recess; aesthetician Edward Lippman is perceptive on the matter:

> comprehensive view of the historical and logical unfolding of the world-spirit, the earthly form of the absolute Idea, music represents a particular phase in the series of the arts, preceded by architecture, sculpture, and painting, in that order, and followed by literature, while the arts as a whole are then succeeded by philosophy.
>
> The successive order of the arts represents the stages of self-realization of the world-spirit, with the corresponding stages of the reduction of the weight of its material embodiment. In music, spatial objectivity is annihilated, and we apprehend, through the ear rather than eye, the ideal activity of the soul rather than a material object.[53]

The spiritually driven suppression of embodied musical content is an elusive concept, however. Schleiermacher, for example, insists that the dematerialization of musical content is an integral component in the materialization of its spiritual function in man. The composer's greatest triumph, he opines, 'is when he bids adieu to language altogether and embodies, in this endlessly changing wealth of tonal sequences and harmonies, all the tremors of life that can pass through the soul'.[54] Schleiermacher makes an explicit evolutionary point when he refers to the tremors of life *passing* through the soul, because he believes that, like Christianity progressing from Jesus to Christ, music must progress from content to form, from elementary to organic perfection.[55] As Gunter Scholtz says, however, for Schleiermacher 'Content is that "inner form," the principle giving shape (*Gestaltung*) to the tones. Thus our customary separation of form and content becomes blurred'.[56]

[52] Colin J. D. Greene, *Christology in Cultural Perspective: Marking Out the Horizons*, Grand Rapids, MI, and Cambridge: William B. Eerdmans Publishing Company, 2003, 143.

[53] Edward Lippman, *A History of Western Musical Aesthetics*, Lincoln and London: University of Nebraska Press, 1992, 233.

[54] Schleiermacher, *Ästhetik*, 174.

[55] Gunter Scholtz, 'Schleiermacher', in Stefan Lorenz Sorgner and Oliver Fürbeth (eds.), *Music in German Philosophy: An Introduction*, trans. Susan H. Gillespie, Chicago and London: University of Chicago Press, 2010, 59.

[56] Scholtz, 'Schleiermacher', in Sorgner and Fürbeth, *Music in German Philosophy*, 61.

Blur is a common feature of Victorian musical absolutism which confuses the ontological boundaries of content and form – effectively the boundaries of relative material *creation* and absolute eternal *design*. Henry Wylde, Gresham Professor of Music at the Royal Academy of Music, adamantly refuses to ascribe compositional agency to a specifically Christian God but does concede the existence of a musical 'Creator' or 'Divine Teacher'. Yet Wylde also conceives of his Creator as an aloof musical originator – not as an intimate collaborator with man, and never as Creation's silently embodied partner continuingly recreating the musical work from within the material content of its spiritual form. Like founding father of modern natural theology William Paley and his conceptual successor Frederick Temple Wylde lets God remove his foot from the creationary pedal, leaving man to drive his worldy work forward: 'The recorded history of arts and sciences, no less than a philosophical study of man, assures us that the Creator has implanted in our very being the germ of all talents, and an ever-unfolding capacity for the acquisition of knowledge, as the progressive tendencies of every age imparts it'.[57] For Wylde the direction of creative evolutionary traffic flows uni-directionally from God the Creator to man yet even 'the great executants' can struggle to recreate the Creator's original template (what Wylde calls 'the primal endowment of the organization') without distortion.[58]

Wylde attributes this to a sometimes disappointing mismatch of content and form: 'If the knowledge of form be wanting ... his most sublime conceptions of the beautiful are but aerial shapes, lacking material embodiments ... great originality, and even beauty in conception, cannot wholly compensate for crudity in form, and poverty of design'.[59] Wylde is also unrelenting over the achievement of perfect balance between content and form, yet he stresses the significance of form to the detriment of content. Form, 'the graceful order and design in which Nature is compelled by creative wisdom to array herself', comes from the Creator through the comprehensible set of attributes apparent in 'the same mathematical laws and definite proportions that underlie the natural sciences'.[60] If Wylde blurs the evolutionary boundaries between human material creation and divine absolute design he also confuses the nature of spiritual progress in music which God Creates and Men – even great executants – reassemble. In the best human re-assemblages music simply retains the original

[57] Henry Wylde, *Music in Its Art-Mysteries,* London: L. Booth, 1867, 107.

[58] Wylde, *Music in Its Art-Mysteries*, 3; 4.

[59] Wylde, *Music in Its Art-Mysteries*, 3.

[60] Wylde, *Music in Its Art-Mysteries*, 3; 4.

imprint of its divine inscription, and it is left for man to pick up the broken jigsaw and piece it together in time. But this is not an explicitly theological message designed to direct man's spiritual attention progressively towards God through music, even if it does begin with the proposition of a Creator or Divine Teacher. It is effectively a natural theological statement about the limits of music's divine representationality.

Popular Victorian music theology tried to overcome the problem of representationality created by natural theology. It did this by uniting relative content and absolute form with didactic, catechistical theology in a variety of genres which aim to recapitulate the theology they represent. Within non-liturgical contexts oratorio is probably the most obvious genre to supply that medium, but Jeremy Begbie admits that even at its most indicative – in Elgar's, for example – it can only represent 'a theological conviction' despite going very deeply beyond contemporary cultural expectation.[61] Elgar's adaptation of Newman's *Dream of Gerontius* – the protracted story of a solemn Christian death – takes theology seriously as a form of self-expression and human empathy, and Begbie claims that Elgar focuses even more intensely than Newman on the character of Gerontius's physical, emotional and spiritual experience: 'he is Everyman, with who fallen and frail humanity we are invited to identify'.[62] According to Paul Harper Scott Elgar ingeniously captures Gerontius's halted progression towards heaven (he glimpses it from purgatory) by playing havoc with conventional teleological expectation; Elgar's structures are 'anti-heroic',[63] and in oratorios like *Gerontius* Elgar also uses diatonicism 'to represent states of spiritual peace or grace' and chromaticism, spiritual turbulence.[64]

According to Byron Adams all of these characteristics unite in producing a biographical composition portraying Elgar's supreme self-loathing or 'tortured personality';[65] for Begbie they conversely reflect 'a purely immanent and, in principle, eternal, irresolvable struggle'[66] in Elgar's

61 Jeremy Begbie, 'Confidence and Anxiety in Elgar's Dream of Gerontius', in Martin V. Clarke (ed.), *Music and Theology in Nineteenth-Century Britain*, Farnham: Ashgate, 2012, 198.

62 Begbie, 'Confidence and Anxiety in Elgar's Dream of Gerontius', 199.

63 J. P. Harper-Scott, *Edward Elgar, Modernist,* Cambridge: Cambridge University Press, 2006, 106.

64 Matthew Riley, 'Heroic Melancholy: Elgar's Inflected Diatonicism', in J. P. E. Harper-Scott and Julian Rushton (eds.), *Elgar Studies*, Cambridge: Cambridge University Press, 2007, 285, 300.

65 Begbie, 'Confidence and Anxiety in Elgar's Dream of Gerontius', 207; see Byron Adams, 'Elgar's Later Oratorios: Roman Catholicism, Decadence and the Wagnerian Dialectic of Shame and Grace', in Daniel M. Grimley and Julian Rushton (eds.), *The Cambridge Companion to Elgar*, Cambridge: Cambridge University Press, 2005, 81–105.

66 Begbie, 'Confidence and Anxiety in Elgar's Dream of Gerontius', 212.

fluctuation between hopefulness and anxiety. But no matter how closely and diversely its musical content is read as theological form no interpretation of Elgar's content accedes to elevate the theology of *Gerontius* to a higher level of form other than one of plain musical representation. Begbie only rises to consider 'irreducibly theological factors in play'.[67] In his equally incisive reading of *The Apostles* Charles McGuire offers similar findings; for him Elgar composes 'a series of studied theological layerings'.[68] Twentieth-century theologian Hans Urs von Balthasar might therefore describe Elgar's oratorios as 'a shimmer of the absolute, fascinating preciousness, *Kostbarkeit*, of God'.[69]

Intentionally sacred musical genres try to 'shimmer' even more brightly. Like oratorios, hymns radiate broadly catechistical ideas in their theological intention and aim to encapsulate through music what they consider to be the universal teaching of their church. Tractarian hymn writer John Mason Neale claims that 'Church hymns must be the life-expression of all hearts'.[70] John Richard Watson glosses Neale, contending that

> In saying this, Neale was pointing to a sense in which hymns should be concerned with the life of the heart, using the heart as a metonymy for the whole person, spiritual, emotional, physical, the feeling and thinking human being. He was concerned, too, with all hearts, with something that can be shared by all humanity . . . For hymns are not Christian Dogmatics, or Systematic Theology, but the expression of all the varieties of human religious experience.[71]

Many Victorian hymns aim to fulfil this universal promise by using textual content to recapitulate musical form, so that hymn music does not merely *represent* its words but through music *presénts* (displays) and *présents* (embodies) its Christian God.[72] This aspiration drove the Victorian hymn towards an almost Augustinian notion of beauty which understands music

[67] Begbie, 'Confidence and Anxiety in Elgar's Dream of Gerontius', 209.

[68] Charles Edward McGuire, 'Elgar, Judas, and the Theology of Betrayal', *19th-Century Music* 23/3 (Spring 2000), 237.

[69] Hans Urs von Balthasar, 'Christliche Kunst und Verkündigung' in J. Feiner and M. Löhrer (eds.), *Mysterium Salutis; Grundriss heilsgeschichtlicher Dogmaktik*, 4 vols. Einsiedeln, Zurich and Cologne, 1965–73, vol. 1 (1965), 711, cited in Aidan Nichols O.P., *Redeeming Beauty: Soundings in Sacral Aesthetics*, Aldershot: Ashgate, 2007, 56.

[70] John Mason Neale, 'Introduction', *Hymns of the Eastern Church*, London: J. T. Hayes, 1862, xvii.

[71] John Richard Watson, *The English Hymn: A Critical and Historical Study*, Oxford: Oxford University Press, 1999, 4.

[72] Bennett Zon, 'Music', in *The Oxford Handbook of Nineteenth-Century Christian Thought*, Oxford: Oxford University Press, 2016, 459–70.

not merely in the outward appearance of its species but 'that which constitutes the existence of a particular thing in the fullest sense of the word: it is *forma* which gives it its life and being'.[73] In the best examples Victorian hymns (Protestant and Catholic) succeed in achieving Augustinian *forma*, especially when hymns combine Trinitarian theology, a good dose of musical word painting and the poetic 'gimmick' of using three verses, like 'Bromley' and Neale's translation of the Ambrosian hymn *O Lux beata Trinitas*, 'O Trinity of Blessed Light' (1852). Catherine Winkworth and William H. Havergal's Lutheran creedal hymn 'We Believe in One True God' (1861/3) is another good example amongst many.

Despite their uncontested quality as musical compositions hymns like these and oratorios like Elgar's raise broader questions over the musical limits of theological representation: can any musical genre express theology without resorting to representationality; is it truly possible to depict (presént) *and* embody (présent); can any musical genre can express Jesus and embody Christ without resorting to empty representationality? Can music go even further than merely representing theology – further than even Maeve Heaney suggests in the title of her book *Music as Theology: What Music Has to Say about the Word* (2012)? Can music actually *be* God? According to Joseph Harper, Tractarian theologian and friend of Newman John Keble believed it possible, as a function of the proto-Eucharistic incarnational theology he calls church 'type'. This suggests 'that Christ himself was synonymous with a figuration or metaphor of "the Church"'.[74] Like his friend Newman Keble enters scientific territory when discussing type, and their incarnational message is not dissimilar. For Keble, like the church itself man is suffused in the embodiment of the incarnated Christ. Thus Christ is not merely representative of the external historical edifice and structure of the church but everything the church witnesses from within itself through time and eternity, including every aspect of its expression and spiritual progress. The Church is representative of Christ and the spiritual progression of its time through history 'from David to Christ and in the Revelation from Christ to the Church'. Christ is therefore 'the type of the Church',[75] Keble opines; He is 'the end of the

[73] Carol Harrison, *Beauty and Revelation in the Thought of St Augustine*, Oxford: Oxford University Press, 1992, 38.

[74] Joseph Harper, 'Towards an Understanding of Tractarian Hymnody: A Critical Appraisal of the Interaction between Theology, Poetry and Music in Anglican Hymnody between 1840 and 1900', PhD diss., Durham University, 2010, 13.

[75] John Keble, 'On the Mysticism Attributed to the Early Fathers of the Church', *Tracts for the Times*, Vol. 5, London: J. G. F. and J. Rivington, 1840, 129.

ancient types and the beginning of a new series. In Him all that has happened before was as it were, brought to a point; and all again that should come after, was but so many developments of what He said, did, and suffered among us'.[76] As Georgiana Battiscombe suggests, 'Christ is the "type" of the Church in the sense that his mystery is effectually present, and actually realising itself, in the Church'.[77] This belief is implicit in Keble's aspirationally incarnational attitude towards its music. Contemporary John Campbell Shairp records Keble as having defined music as 'the audible *embodiment* [current author's emphasis] of some unknown reality'.[78] And of course for Keble's friend Newman theology is more than mere words – it expresses the inexpressible 'divinity in the theology of the Church'; and music is more than mere notes, and expresses the sublimity and beauty of God. Both are incomprehensible, and neither perishes; neither the theology of the church nor music

> comes and goes, and begins and ends in itself. It is not so; it cannot be. No; they [musical notes] have escaped from some higher sphere; they are the outpourings of eternal harmony in the medium of created sound; they are echoes of our Home; they are the voices of Angels, or the *Magnificat* of Saints, or the living laws of Divine Governance, or the Divine Attributes.[79]

Plainchant

If Victorian oratorios and hymns struggle to do more than *represent* Christ they are in good musical company generically speaking, because there is only one genre which Victorians truly believed to *be* Christ, one church in which they said so, and possibly only one person in that one church who claimed it to be true. Heard surreptitiously in churches from the period of English Catholic exile in the sixteenth century, and legalized in public worship after the first repeal of the penal code in 1778 Gregorian chant resurfaced at the beginning of the nineteenth century not only as a genre battered and bruised by its clandestine experience but as an unexpected survivor – a fittest musical survivor. By middle part of the nineteenth century the English plainchant revival was in full swing in both Roman

[76] Keble, 'On the Mysticism', 130.

[77] Georgiana Battiscombe, *John Keble: A Study in Limitations*, Constable: London, 1963, 292; 292.

[78] John Campbell Shairp, *John Keble: An Essay on the Christian Year*, Edinburgh: Edmonston & Douglas, 1866, 18

[79] John Henry Newman, *Fifteen Sermons Preached Before the University of Oxford*, 3rd ed., London, Oxford and Cambridge; Rivingtons, 1872, 347.

Catholic and High Anglican churches, with a wide range of denominationally specific publications, from practical guides on performing chant to polemical, controversial writings on its history, aesthetics and theology. Whether pro- or anti-Gregorian, chant was a hot topic and everyone had something to say about it, yea or nay.[80] The apologetics which drove the nineteenth-century revival in Anglican and Catholic churches were very similar. Amongst pro-Gregorians there was a general appreciation of its historical authenticity as the earliest music of the church, despite the historical corruptions which had crept in over time. Because of its authenticity there was a sense in which chant expressed a mystical otherness not found in contemporary music through the ages. There was also the undeniable simplicity in its marriage of words and tones – in itself a source of loathing and loving in turns – and its consequently unusual rhythmic structure, half melody/half recitation, liminally placed between speaking and singing. A writer for the Anglican magazine *The Parish Choir* may as well be writing for Catholics as for High Anglicans when he recommends

> the Gregorian chants as the best psalm-chants, not only because they breathe of the fervent piety of the early Christians, but also because they were adopted and sanctioned by the English Reformers, and because, from their antique simplicity, their manly and dignified tone, and their entire difference from all common and profane music they are the most congenial to the character of our ancient and Apostolic Liturgy. We believe, too, they can be sung with great ease by uneducated persons, through the simplicity and boldness of their melodies.[81]

The combination of its historicity and musical difference heightened its allure by intensifying its progressive spiritual liminality. A correspondent for the Anglican journal *The Ecclesiastic* of 1846 captures this quality by charting its spiritual progression almost prodigally, as chant gradually enwraps (recapitulates) the individual within the totality of the church – in essence evolving from the relative materiality of music towards the absolute expression of Augustinian *forma*: Gregorian chant

> grows upon the devout soul, that it fills and possesses it, so that it would choose no other sounds to express thoughts of holy meditation . . . what was at first strange and wild, harsh and uncouth, becomes in the highest degree expressive. We get to lose each several mood, because each brings its own associations with it, and has

[80] For the most comprehensive exploration of the topic, see Bennett Zon, *The English Plainchant Revival*, Oxford: Oxford University Press, 1999.

[81] No author, 'On the Gregorian Tones, and their Adaptation to English Psalmody', *The Parish Choir* 1/9 (Oct. 1846), 87.

the feeling which it expresses, discriminated. We gain a reverent familiarity with these melodies. They become the songs of our home, the heavenly Jerusalem, the city of our God.[82]

The transformative nature of spiritual progression is an expectedly common feature in chant apologetics, Anglican or Catholic, and in the highest Anglican commentaries it comes into greater theological focus because of their preponderant Catholic sympathies. High Church periodical *The Ecclesiologist* registers its pro-Gregorian membership with an increasingly incarnational message for its readers, topping and tailing an article from the broad generality of music to the specificity of chant:

Music expresses every feeling of which our minds are capable; and the elder Church by command of God, as well as the Christian Church by the example of God incarnate, and acting under the special guidance of the Holy Spirit, adopted it as the mode of conveying her prayers and praises to His throne of grace . . . there is a severity, and frequently a penitential tone, about Gregorian music, which especially marks it as being well adapted for the worship of the Catholic Church [Anglican or Roman Catholic], which is a religion of self-denial, and which inculcates on her children that each in his own sphere must be ready to bear his Cross, after his Divine Lord.[83]

Although as this suggests chant occasionally brings Anglican perspectives on the incarnational attributes of music into High relief (if it may be put that way) Anglican pro-Gregorianism was generally less theological than it was broadly practical. The work of celebrated Anglican plainchant spokesman the Rev. Thomas Helmore provides one of the best Victorian examples. Like many popular handbooks of the time *Plain-Song* (1850) eschews theological controversy by framing potentially contentious apologetics in denominationally emollient terms. On the use of Gregorian psalm tones, for example, he appeals to the congregational ethos of the Anglican Reformation:

besides Choir Music, there is People Music . . . As the restoration of the proper Psalm Melodies of the Church had been found, wherever well done, conducive in a high degree to the end in view, it was believed that the translation and noting of the ancient hymns with their ancient melodies, would prove an additional boon to

[82] No author, 'The Ecclesiastical Chant', *The Ecclesiastic* 1 (Jan.–June 1846), 335, cited in Zon, *English Plainchant Revival*, 324.

[83] No author, 'Ecclesiastical Music', *The Ecclesiologist*, new series vol. 10 (Aug. 1846–April 1850), 209; 216–17.

the Church at large. And one which the best and most far-sighted of her Bishops have earnestly desired from the Reformation to the present day.[84]

The same ideological reserve is only partially true of Roman Catholic comparators. On first glance Roman Catholic and High Anglican pro-Gregorianism do look almost identical. The generality of apologetics echoed the same concerns for authenticity in scores and performance, appreciation of its historical significance within the development of the church, its transcendence, simplicity and solemnity.

Roman Catholics had one thing Anglicans did not, however: a protean ecclesiastical voice steeped in German philosophy and trained in both Anglican and Roman Catholic theology: the Reverend Henry Formby. Henry Formby (1816–84) came from an affluent Catholic family which kept their faith until the eighteenth century, when (apart from one branch of the family) they closed the chapel in their ancestral home in Lancashire and embraced Anglicanism. Formby was educated at Clitheroe Royal Grammar School, Charterhouse and Brasenose College, Oxford, and subsequently became vicar of Ruardean, Gloucestershire. There in 1843 he began a long literary and theological career, publishing amongst other things a pamphlet reprinted from *The English Churchman* called 'Parochial Psalmody Considered' (1845). He left Ruardean in 1845, and subsequently travelled to Belgium. *The Catholic Encyclopedia* stresses how profoundly influenced he was by the Oxford Movement, and soon after Newman became Catholic he followed suit in 1846, taking holy orders at Oscott seminary in Birmingham in 1847 followed by a brief stint as an Oratorian in 1848. In Birmingham Formby become attached to St Chad's Cathedral, where he wrote his three seminal – if controversial – treatises on Gregorian chant: *The Catholic Christian's Guide to the Right Use of Christian Psalmody of the Psalter* (1846), *The Plain Chant the Image and Symbol of the Humanity of the Divine Redeemer and the Blessed Virgin Mary* (1848), and *The Roman Ritual and Its Canto Fermo, Compared with the Works of Modern Music, in Point of Efficiency and General Fitness for the Purpose of the Catholic Church* (1849). Later writing continued his musical interest with *Congregational Singing: Two Instructions on the Duty and Privileges of Congregational Singing, Preached in the Convent Church of St. Mary's, Handsworth, on Christmas Day, and the Following Sunday* (1849), *The Young Singer's Book of Songs* (1852), *School Songs and Poetry to Which Music Is Adapted* (1852) and *First Series of Hymns and Songs for the Use of*

[84] No author, *The Ecclesiologist*, new series vol. 9 (1851), 175.

Catholic Schools and Families (1853). Other works include an extensive array of religious, social, educational, devotional and polemical topics.[85]

That Formby was a cantophile extremist is obvious from Newman's correspondence. One letter recognizes him as 'mad on the subject of Gregorians';[86] another reveals his interest in translating chant books – books he presumably encountered while studying in Louvain,[87] and many more refer to his dealings with the plainchant publisher John Lambert.[88] That Formby was also not uncontroversial – that he was a strong (overbearing), independently minded (pig-headed) and frequently uncontrollable force (a loose cannon) – is equally evident. Despite their friendship Newman clearly saw his friend as something of a maverick. Responding to friend and Provost of the London Oratory Frederick William Faber Newman offers advice on managing Formby's disobedient lack of fraternity: 'he [Formby] will have a great trial [as an Oratorian], if he remains with us. He has to overcome *self*'.[89] Ian Ker portrays Formby in similar terms, as pushing Newman's patience to grant him permission to publish on chant: Formby 'considered that Plain Chant was 'pre-destinated', and was anxious to burst into print on the subject. But Newman was ruthless about such 'literary vanity'.[90] More importantly Newman was deeply and increasingly critical of Formby's theological approach, especially after receiving the advertisement or preface to *The Roman Ritual* Formby had sent him for approval:

> My dear F. Henry
>
> On consulting with F[ather] Bernard, I think I must say that it is satisfactory to neither of us, to read in your proposed Preface that 'words and music are contemplated as one whole in the Divine Mind'. It is as much as saying that the Gregorian Chant was predestinated to be used in the Church, which we do not think a safe proposition. And we cannot let it pass.

[85] Edwin Burton, 'Henry Formby', in Charles G. Herbermann et al. (eds.), *Catholic Encyclopedia: An International Work of Reference on the Constitution, Doctrine, Discipline, and History of the Catholic Church*, 15 vols. and index, New York: The Encyclopedia Press, Inc., 1907–12/1913, vol. 6, 139.

[86] John Henry Newman, letter to J. D. Dalgairns (22 March 1847), in John Henry Newman, *The Letters and Diaries of John Henry Newman, July 1850 to December 1851*, vol. 12: Rome to Birmingham (January 1847–December 1848), ed. Charles Stephen Dessain, London: Thomas Nelson and Song Ltd, 1963, 65.

[87] Newman, *Letters and Diaries of John Henry Newman*, vol. 12, fn. 1, 65; fn. 3, 181.

[88] Newman, *Letters and Diaries of John Henry Newman*, vol. 12, 434 (index).

[89] John Henry Newman, letter to F. W. Faber (17 Oct. 1848), cited in Peter C. Wilcox, *John Henry Newman, Spiritual Director 1845–1890*, Eugene, OR: Pickwick Publications, 2013, 245.

[90] Ian Ker, *John Henry Newman: A Biography*, Oxford: Oxford University Press, 1988/2010, 340.

Also, you have not quoted the passage from the last page of your work, in which you seem to say that God suffered for the plain chant, and explained that you only mean that His death has been the indirect means of introducing the ritual, and so inclusively the music of the ritual, into the world.[91]

Newman continues defensively:

The more you bring out your meaning the less I like to commit the Oratory to it, and especially your hint that it is connected with the new German theological School alarms me. If there is any province which does not directly belong to the Oratory it is that of dogmatics . . . if there is any thing which we must avoid . . . it is new views in Theology. This consideration has so weighed with myself, that, many as the misrepresentations which have been published on my Essay on Development, I have written nothing on the subject since I was a Catholic.[92]

Formby defended his stand on chant 'in the Divine Mind' by comparing the Spirit and understanding – 'The Idea on which the comparison is based, he suggests, 'is I believe one that rather belongs to the recently school of German Theology',[93] but neither Newman nor his brothers in the Oratory were convinced. Faber considered him truculent and delusional:

He thinks God called or calls him to be the apostle of Gregorians, he suspected this before he joined us . . . you . . . told him there was nothing to hinder his pursuit of his favourite study in the Oratory – he joins – what you said to him the other day utterly floors the apostolate aforesaid – now what is he to do? He thinks of leaving.[94]

Newman also took umbrage over Formby's tactless presumption in going to press without waiting for approbation:

My dear F. Henry

We had an anxious conversation yesterday on the subject of your work, anxious from the lively feelings which we entertain of attachment to you. It will be well for us indeed if, as the Oratory proceeds, we gain from members men of the intellectual powers, the unaffected goodness and kindness of heart, and the liking for us personally which we recognize in you – but this

[91] John Henry Newman, letter to Henry Formby (17 Oct. 1848), in Newman, *Letters and Diaries of John Henry Newman*, vol. 12, 300.

[92] John Henry Newman, letter to Henry Formby (21 Oct. 1848), in Newman, *Letters and Diaries of John Henry Newman*, vol. 12, 302.

[93] Henry Formby, letter to John Henry Newman (date unspecified), Newman, *Letters and Diaries of John Henry Newman*, vol. 12, 302.

[94] Frederick William Faber, letter to Newman (15 Oct. 1848), in Newman, *Letters and Diaries of John Henry Newman*, vol. 12, fn. 1, 298–99.

> conviction must not be allowed to decide the matter before us, and that for your sake quite as much as ourselves, as you will perfectly understand.
>
> I must say then at once that our judgment about your publication is what it was – we cannot sanction it; and as to the Advertisement it should not have been sent to the [Catholic magazine] Tablet without our knowledge; the Preface to the work was not out of our hands; you had sent me one, which we did not like, and we asked for something further – we have given you no permission to go to Press or advertise.
>
> We do not object to your publishing grammars of plain chant, or writing in defence of it; but the line of defence which is not historical but doctrinal, involving doctrine as we think of a disputable character.[95]

Recalcitrant to the bitter end Formby appears to end the matter with a letter to Newman on 21 November, effectively resigning his novitiate at the Oratory: 'the thought of abandoning it presents no other notion to my mind than that of spiritual extinction'.[96]

As Newman's correspondence suggests, from a theological standpoint Formby's Gregorian odyssey begins rather badly, without official approval of the Oratorians. But why did Newman and the Oratorians react the way they did? The answer can be found in Newman's previous communication: 'the line of defence which is not historical but doctrinal'. Formby sets out his theological stall in a sustained argument across three increasingly incarnational works, *The Catholic Christian's Guide to the Right Use of Christian Psalmody of the Psalter* (1846), *The Plain Chant: The Image and Symbol of the Humanity of our Divine Redeemer and the Blessed Mary* (1848) and *The Roman Ritual and Its Canto Fermo Compared with the Works of Modern Music in Point of Efficiency and General Fitness for the purposes of the Catholic Church* (1849). *The Catholic Christian's Guide* begins with an explication of seven motives behind God's creation of music. Much of *The Catholic Christian's Guide* concerns the potential use of music to overcome the evil purposes of the Devil so these motives illustrate how God saves Christians from evil by imbuing music with increasingly redemptive forces graduating (progressing spiritually) from ordinary religious diversion to communion with Jesus Christ crucified:

1. music is a religious diversion from idleness, and therefore a pious delight.
2. music is providentially medicinal

[95] John Henry Newman, letter to Henry Formby (24 Oct. 1848), in Newman, *Letters and Diaries of John Henry Newman*, vol. 12, 307.

[96] Henry Formby, letter to Newman in Newman (21 Nov. 1848), *Letters and Diaries of John Henry Newman*, vol. 12, 344.

3. music is a chief element in social and Christian life
4. music possesses charm which can lead to virtue (or when misused, vice)
5. music is the voice of praise and occasionally prayer
6. music can help obtain the prayers of the Christian saints
7. the Christian singer is raised to the company of spiritual heights, including The Virgin, St Paul and the Prophets, and Jesus Christ (who sang on the night of his crucifixion)[97]

The type of music best suited to encapsulate the redemptive qualities of music is Gregorian Chant, or what Formby calls variously Plain Song or Plain Chant. He prosecutes his argument in a further set of seven points:

1. vocal music in harmony is 'artificial' (uncongregational)
2. chant is 'a people's song' which unlike vocal harmony is not affected or compromised when singers of independent parts are absent
3. vocal harmony requires women
4. harmony becomes wearisome
5. chant expresses the words directly, and speaks 'to the understanding of the people'
6. chant uses scales from time immemorial
7. chant gives fallen Christians – the church – a means of approaching the Divine Presence which celebrates the mercy of their redemption[98]

Like the first set of seven this set progresses spiritually from the congregational nature of chant to its redemptive properties in the context of the church at large. This time, however, Formby expands upon the role of the individual in relation to the church by claiming that something which has greater potential for incompleteness (vocal harmony) cannot be redemptive. Vocal harmony is simply not theological because it does not reflect the narrative unity of time, people and place that the congregational singing of plain song achieves. Singers of plain song not only sing but theologically harmonize with the Saints and the church in its totality through history. Although it would be reasonable to expect that vocal music teaches 'Diversity in unity, "the plurality of members, and the unity of the body"', according to Formby it is the unassailable unity of content and form which gives plain song its uniquely theological meaning: 'what is more precious practical truth', he muses, 'than Unity, the "Communio Sanctorum"? and

[97] Rev. Henry Formby, *The Catholic Christian's Guide to the Right Use of Christian Psalmody and of the Psalter*, London, Dublin and Derby: Thomas Richardson and Son, 1846, 12–16.

[98] Formby, *Catholic Christian's Guide*, 25–28.

how can there be on earth a more perfect typical expression of this, than the vast unisonous song of a multitude, who, with one mouth and one heart, glorify their God and their Redeemer – where the voices of all, young men and maidens, old men and children, are, as it were, the voice of one person?'[99] Plain song, moreover, is 'a foretaste of the unity and happiness of Heaven',[100] anticipated in the psalms through which Christ sings in the anticipatory 'before' of the Old Testament (through David); in the incarnationally fulfilled 'present' of the New Testament (through Jesus); and in the glorified 'after' of an eschatological future (through Christ). To put Formby's words in Schleiermacher's mouth:

> the most rigorous view of the difference between Him [plain song] and all other men [other music] does not hinder us from saying that his [plain song's] appearing, even regarded as the incarnation of the Son of God, is a natural fact . . . as certainly as Christ [plain song] was a man [music], there must reside in human nature the possibility of taking up the divine into itself, just as did happen in Christ [plain song]. So that the idea that the divine revelation in Christ [plain song] must in this respect be something absolutely supernatural will simply not stand the test.[101]

Although widely positive, press reviews of *The Catholic Christian's Guide* suggest that the incarnational message was nonetheless too theologically extreme. Formby was characteristically unrelenting, and responded with a series of seven increasingly defensive essays in *The Tablet* under the title 'Ecclesiastical Song'. The first article provides the conceptual framework in seven points:

I. The Existence of a positive Authority, Ecclesiastical and Moral, for the Ancient Song.
II. The complete and effective System of the Ecclesiastical Song Books.
III. The Moral Fitness of the Ecclesiastical Song as a Sacrificial Song.
IV. The Fitness of the Ecclesiastical Song for the Offices of the Church, and for the People as a Congregational Song.
V. The Fitness of the Ecclesiastical Song to be a medium or vehicle of Divine Truth among the People.
VI. The remarkable Popularity of many of its Compositions.
VII. The Catholicity or Companionship of the Catholic Doctrines over the Globe.[102]

99 Formby, *Catholic Christian's Guide*, 34; 34.
100 Formby, *Catholic Christian's Guide*, 40.
101 Schleiermacher, *Christian Faith*, 64.
102 Rev. Henry, Formby, 'The Ecclesiastical Song', *The Tablet* (3 April 1847), 3.

These points are developed in *The Plain Chant*, especially chant's underlying authority as *the* congregational song of the universal church. Not only does plain song recapitulate the totality of Christ from the relativity of history to the absoluteness of revelation, significantly it also expresses Eucharistic properties. Plain song has become elevated to a sacramental level; it no longer merely embodies Christ's humanity: it *is* the very essence of His sacrifice and offering. *The Plain Chant*'s 'advertisement' expresses this in no uncertain terms:

The idea contained in the following pages, of the Song of the Divine Office being a symbol of our Lord's Incarnation, is but part of an idea capable of being exemplified in every means that the Catholic Church has taken to manifest the Godhead Incarnate, who kingdom she is to men. In architecture, in ceremonial, in the vestments and ornaments of public worship, in literature, &c., the same great fact, that God has become Man, is displayed in the character now stamped upon them in the Catholic Church. The doctrine of the Incarnation of God suffices to explain all that strikes the senses as strange and peculiar in them. And if persons who thoughtlessly complain that the Plain Chant is unpleasantly remarkable for a strange unearthly and unmusical character, would but reflect a moment, they would perceive it to be much stranger, that it should profess to be the Song that commemorates God becoming Flesh, without possessing that solemn, unearthly character which, if it be judged, (though this can never be a Christian judgment,) apart from the Incarnation which it symbolises, is perhaps not of a nature to be immediately relished.[103]

On first glance Formby's incarnational proposition seems to be redolently Newmanian: it utilizes the Incarnation to locate plain song within musical history; it uses what Christopher Dawson calls Newman's 'progressive revelation and spiritual renovation'[104] to illustrate how plain song helps people progress from natural to the revealed and the historical to the doctrinal; and it paraphrases Newman's theological vocabulary: is plain song not an example of what 'is found, through His preachers, to have imprinted the Image or Idea of Himself in the minds of His subjects individually; and that Image, apprehended and worshipped in individual minds, becomes a principle of association, and a real bond of those subjects one with another, who are thus united to the body by being united to that Image'.[105] Again,

[103] Rev. Henry Formby, *The Plain Chant: The Image and Symbol of the Humanity of Our Divine Redeemer and the Blessed Mary*, London: James Burns, 1848, 4.

[104] Christopher Dawson, *Religion and World History: A Selection from the Works of Christopher Dawson*, ed. James Oliver and Christina Scott, Garden City, NY: Doubleday, 1975, 292.

[105] Newman, *Grammar of Assent*, 464–65.

seven progressively incarnational points reinforce that sacrificial Image from the humanity of Christ to his role as Life-Giver in the Eucharist:

1. song has the inherent capability of being a manifestation of the Lord's humanity
2. because Christ bequeathed the church a Divine Office which should be sung, that song is a manifestation of it
3. there is in the existence of plain song 'some very perceptible, and, I may add, beatific manifestation of Christ'
4. the devil hates plainchant because it is 'the most engaging manifestation of our blessed Redeemer'
5. plain song 'is a *sacrifice* and *offering* . . . But the blessed Eucharist is also a sacrifice and offering. If, then, the sacrifice of the blessed Eucharist be the true real humanity of Christ, the verum corpus natum ex Maria Virgine, why may we not also believe the sacrifice of the Song of the Divine Offices to be the representative of His humanity'.
6. people who sing in the Divine Office 'perceive it to be a real manifestation of Christ to their hearts'.
7. there is a striking correspondence between plain song and the characteristics of Christ's humanity including his royalty (grandeur and solemnity); priesthood (sanctity and dignity); prophetic Office (purity of expression); Life-Giver in the Eucharist ('instinct with a life that never fails'); virgin beauty and sweetness (characteristics of his mother); sympathy (tenderness); and warriorship (against evil).[106]

Regardless of their metaphoric properties Formby considered sacramental, Eucharistic, incarnational correspondences between plain song and Christ to serve a manifestly devotional purpose which provides the basis for spiritual progression from the relative to the absolute, from content to form, from the individual to the corporate unity of the church, and ultimately from man to God: 'It seems to point out to us', he reflects, 'that what we have to do is, first to regard the Song itself as gifted with the mysterious power of symbolising the Man-God, and manifesting Him in a sacramental but intelligible manner to all who hear, and in an especial degree, to those who sing; and then to meditate continually upon all the characteristics of our blessed Lord and His virgin Mother in their life and actions upon earth'.[107] In *The Roman Ritual* spiritual progression is taken to a higher level as metaphoric correspondences tip precariously into even

[106] Formby, *Plain Chant*, 10–11. [107] Formby, *Plain Chant*, 13.

more profoundly tendentious doctrinal pronouncements. Formby begins uncontentiously enough, structuring his apologetics with a summary of previously rehearsed ideas:

I. Authority: 1, ecclesiastical; 2, moral.
II. Claim to the completeness and order of a System.
III. Moral fitness: 1, as a Sacrificial Song; 2, as a Song for the Offices of the Church.
IV. Fitness for passing among the people as a Congregational Song.
V. Moral influence in the formation of character.
VI. The medium or vehicle for Divine truth passing among the people.
VII. Medicinal virtue.
VIII. Capacities for durable popularity.
IX. Security against abuse.
X. Catholicity, or companionship of the Catholic doctrines over the globe.[108]

Soon, however, his previous scholastic approach begins to change theological direction and Formby returns to first principles, beginning with a classic definition of divine simplicity. The Godhead subsists in the Trinity in a perfect and undivided cooperative whole which cannot change – in infinite, divine simplicity:

To God the Father theologians assign the Eternal Idea, or the conception from all eternity of the Idea or Form of Creation.

To God the Son, the realisation of the Idea of the Father, or the act of bringing created things into being out of nothing, in conformity with the Idea of the Father.

To God the Holy Ghost, the bringing creation to its perfection through the period of its development or growth.[109]

Like Formby himself the doctrine of divine simplicity has its own deeply contentious theological and philosophical background. Essentially a study of the nature of God's unity the doctrine of divine simplicity espoused an understanding of God in which God can be nothing other than himself by virtue of his self-sufficiency and perfection; according to James Dolezal 'God is "without parts" and is "most absolute"'.[110] According to

[108] Rev. Henry Formby, *The Roman Ritual and Its Canto Fermo Compared with the Works of Modern Music in Point of Efficiency and General Fitness for the purposes of the Catholic Church*, London: James Burns, 1849, 10–11.

[109] Formby, *Roman Ritual*, viii.

[110] James E. Dolezal, *God Without Parts; Divine Simplicity and the Metaphysics of God's Absoluteness*, Eugene, OR: Pickwick Publications, 2011, Kindle loc. 118.

seventeenth-century divine Steven Charnock 'God being infinitely simple, hath nothing in himself which is not himself, and therefore cannot will any change in himself, by being his own essence and existence'.[111] Today many theologians criticise divine simplicity for being conceptually incoherent but in the nineteenth century it received a boost from German Idealism postulating the interrelatedness of relative material content and absolute spiritual form. One of the greatest proponents of this type of approach is Catholic Tübingen theological Idealist Franz Anton Staudenmaier. It is obvious why Staudenmaier would appeal to Formby. According to Staudenmaier the world revolved around history and that history constituted a revelation: 'Christendom ... is the *system of God's activities.* God's activities cannot but be conceived historically, and that is when the historical conception of the world (*die historische Weltanschauung*) of Christendom generates itself'.[112] In the context of the Tübingen School God is a free agent yet 'in the changeless infinity of his own mind'[113] conceives his plan through historical human development. That notion of development is in itself a theological reinterpretation of a Schiller-like progress in which Absolute form invariably progresses in order to dematerialize Relative content. In contrast the realization of God's continuous *creation* rather than *succession* within the world forms a primary revelation of history, an Idea. For Staudenmaier and Formby in order to create that historical development God the Father must first have an Idea of the form of creation so when creation is realized it corresponds to His Idea. Formby explains this through Staudenmaier's own words: 'The world is God's idea of the world brought into being, and the perfection of the original world consisted in the fact, that it absolutely corresponded to the Divine idea'.[114]

For Staudenmaier the history of creation itself reveals the Divine idea through its spiritual progression towards the Absolute. At the same time God subsists in his creation through history for the purposes of manifesting

[111] Stephen Charnock, *The Existence and Attributes of God*, 2 vols. 1682; reprint, Grand Rapids, MI: Baker, 1979, vol. 1, 333.

[112] Franz Anton Staudenmaier, *Geist der göttlichen Offenbarung, oder Wissenschaft der Geschichtsprincipien des Christenthums*, Giessen: B. G. Ferber, 1837, cited in Henning Graf Reventlow and William Farmer (eds.), *Biblical Studies and the Shifting of Paradigms, 1850–1914*, Sheffield: Sheffield Academic Press, 1995, 156.

[113] Gerald A. McCool, S.J., *Nineteenth-Century Scholasticism: The Search for a Unitary Method*, New York: Fordham University Press, 1999, 71.

[114] Franz Anton Staudenmaier, *Die Lehre con der Idee. In Verbindung mit einer Entwicklungsgeschichte der Ideenlehre und der Lehre vom göttlichen Logos*, Giessen: B. G. Ferber, 1840, cited in Formby, *Roman Ritual*, xi.

himself within the Relative, within man. According to Aidan Nichols, for Staudenmaier 'the historical process is itself a revelation of the Absolute: the meaning of history is to be identified as the history of God . . . [as] a vision of history where, as in a covenant dialogue, divine hypostases are disclosed *for the sake of personal union* with their human images'.[115] Thus, according to Formby the 'Idea of the Creator is to man . . . the exemplar, or pattern, of his perfection'.[116] And 'What is the likeness, or Ideal, that a Christian, as far as the power is given to him, should seek to aim at bringing himself to resemble? It is answered by the fact of the incarnation of the Song of God. He is the Incarnate Exemplar, or Pattern'.[117] If, therefore, Christ is the Idea of God the Absolute, and the Ideal of man the Relative, it is in His Church that the truest form of music can be found, and the truest music in His Church is not modern music but inescapably Plain Song. Plain song is the only music which has the theologically 'harmonic' properties conducive to the realization – and ultimately recapitulation – of the Divine Idea: it is the only music which unites the relative content of words to the absolute form of music in the same way that the Trinity is hypostatically united in divine simplicity; and it is the only music which effectively recapitulates the Divine Idea. Plain song is therefore synonymous with the Divine Idea as Song (music) and as Truth (the Word or Logos, not merely words), and although discreet Song and Truth are 'but practically only one, for in practice they are indissolubly linked together, and constitute one moral whole, as body and soul together make up but one living being'.[118] As Divine Simplicity itself plain song, like Christ himself, also recapitulates the Divine Idea by expressing itself through the Church in the concourse of time and eternity. Now Formby's spiritual progression – his plain song pilgrimage – is incarnationally complete, taking man on spiritual journey into elliptical orbit around God, from the Divine Idea of God (The Trinity) to the Divine Ideal of Christian Man (Christ), the Divine Idea of Plain Song as Song and Truth (Jesus in his humanity), the Church on the ground in which Plain Song is the music of Jesus (through the psalms), the music of Christ (through his Church) and the music of God the Divine Idea in the simplicity of the Trinity:

115 Aidan Nichols, 'Catholic Theology of the Trinity in the Nineteenth Century', in Gilles Emery, O.P., and Matthew Levering (eds.), *The Oxford Handbook of the Trinity*, Oxford: Oxford University Press, 2010, 285.

116 Formby, *Roman Ritual*, xiii.

117 Formby, *Roman Ritual*, xvi.

118 Formby, *Roman Ritual*, 7.

the advocate of Gregorian Song will be found among the first to submit to the spirit of the time, and will be content to labour for at least some Christian use of so noble a gift, which a merciful Creator gave, a loving Redeemer has recovered, and a sanctifying Spirit has directed and approved – the Holy Trinity, Three Persons and One God, blessed for ever.[119]

Conclusion

At its most basic Formby's apologetical use of divine simplicity is part of a wider Victorian effort to understand the relation between the relative and absolute, content and form, representation and reality, and to place that understanding within an interpretable framework of spiritual progression which makes sense of future pasts, or eschatological history. Through Gregorian chant Formby showed that developmental models of history transform into eternity when theologically infused – when Christologized through the doctrine of divine simplicity. If Formby's interpretation of chant is as unique as it was extreme it was, however, not alone in using historicism to situate idealized chant at the centre of Victorian spiritual progression. Indeed Victorian plainchant apologetics are full of religiously motivated theological indignance designed to rectify and re-cultivate the spiritually developmental function of music. The plainchant apologetic closest to Formby's in content, if not form, comes from someone with a remarkably similar personality and background: architect, designer and icon of British neo-Gothicism, Augustus Welby Northmore Pugin. Also a convert, and renowned for his equally extreme views on the degradation of Catholic ritual, Pugin published *An Earnest Appeal for the Revival of the Ancient Plain Song* in 1850. It is evident from Pugin's letters that he and Formby were in contact many times over the years but the extent and content of their communications over Gregorian chant is unknown in any theological detail.[120] Nevertheless, Pugin's treatise bears some of the hallmarks of Formby, even if it lacks the theological knowledge or motivation of his contemporary. Without Formby's profoundly doctrinal superstructure Pugin prosecutes his argument on the basis of the purity of English Catholic antiquity. Decrying the current state of religious services he, like Formby, rails against the inappropriateness of church architecture,

[119] Formby, *Roman Ritual*, 78.

[120] See for example Augustus Welby Northmore Pugin, *The Collected Letters of A. W. N. Pugin 1851–1852*, ed. Margaret Belcher, 5 vols. Oxford: Oxford University Press, 2015, vol. 5, 710.

vestments, furnishings and music – everything experienced within the liturgical context which debilitates the congregation's ability to participate fully. Music, he says, suffers from 'a want of reality'.[121]

It suffers because its development has been thwarted by historical accretions which distort and transgress the unity which it originally provided for the nascent church. 'Want of reality' is another name for want of unity because for Formby and Pugin unity – like simplicity – is the essence of spiritual progress. Unity must be preserved and enhanced to give the historical church its future and the future church its past. Unity, like simplicity, humanizes as it spiritualizes. Without it worshippers become listeners, 'precluded from taking any real part in the service of Almighty God'.[122] And under these conditions 'want of reality' becomes a self-fulfilling and dehumanizing state of vicarious musical experience. But the effect goes even further than merely degrading the aesthetic quality of religious services, or the Divine Office. It erodes the very essence of spiritual progression by abnegating the authority of the church itself:

> What can be more perfect, what more edifying and consoling than that Divine Office, the compilation of so many saints and glorious men, and which is so wonderful in the perfection of its system and composition, that the more it is studied, the more it gains on our reverence and love! What appropriate fitness in all the antiphons – what noble simplicity in the hymns! while the Chaunt of the Psalter has an almost sacramental power in calming a troubled spirit and leading the soul to God; these were the divine Chaunts that penetrated the heart of St Augustine, and though many centuries have elapsed, they have not lost one fraction of their influence.[123]

The enduring, incorruptible nature of Gregorian chant (and the authority which has engendered it) radiates powerful symbolic value, especially when spiritual progress aims towards the historic past and eschatological future simultaneously within the present. 'May he grant us to see a restoration not only of the external glory of His temple but of the reverent service which is alone suited to its ancient symbolism'.[124]

Like Formby, Pugin made enemies by letting his passions run riot with statements like these. Newman was inevitably hostile because *An Earnest Appeal* directly attacked the aesthetics of the Oratory and their

[121] Augustus Welby Northmore Pugin, *An Earnest Appeal for the Revival of the Ancient Plain Song* (1850), London: R. & T. Washbourne; and New York, Cincinnati and Chicago: Benziger Bros., 1905, 7.

[122] Pugin, *An Earnest Appeal*, 8.

[123] Pugin, *An Earnest Appeal*, 16–17.

[124] Pugin, *An Earnest Appeal*, 20.

modern Italianate, Ultramontane (Rome-centred) services. Writing to Faber jubilantly in 1850 on the restoration of the Roman Catholic Hierarchy (it had been dismantled during the Reformation) Newman retorts: 'Pugin, not to be wanting to himself, has paid us the delicate compliment of turning a plea for plain chant into a plain chant for our virtues. A *profound silence*, I suppose by your Very-reverence will agree with me, is the only way of bearing such blushing honors'.[125] Implicitly Newman's hostility also belies an underlying theological antagonism over virtue and the dispensation of morality within music. Although the ontogeny of music can recapitulate the phylogeny of Catholicism, music is emphatically not revelatory; it is 'the expression of ideas greater and more profound than any in the visible world, ideas which centre, indeed, in Him whom Catholicism manifests, who is the seat of all beauty, order, and perfection whatever, [but] still ideas after all which are not those on which Revelation directly and principally fixes our gaze'.[126] For all his musical sympathy and experience, Newman considered music more broadly unable to promote revelation, and if unable to promote revelation music was only partially redemptive, no matter what type was under consideration. And according to J. Oswald Dykes Newman was also fundamentally sceptical about Gregorian music: 'In Gregorian music Newman could see no beauty whatever – none at any rate in the usual antiphons and "tones". An exception must be made in favour of those familiar chants occurring in the mass for the *Sursum Corda*, the *Prefaces*, and the *Paternoster*'.[127] Dykes suggests that the combination of his Protestant musical upbringing and Roman musical experience (where plainchant was uncommon) prejudiced Newman's attitude towards chant; in fact Newman was a key figure in the introduction of vernacular hymns into Catholic liturgical repertoire.[128] Another friend of Newman, Edward Bellassis, portrays Newman with more generosity of feeling towards chant, from a light-hearted suggestion

[125] John Henry Newman to William Frederick Faber, *The Letters and Diaries of John Henry Newman, July 1850 to December 1851*, vol. 14: Papal Aggression (July 1850–December 1851), ed. Charles Stephen Dessain and Vincent Ferrer Blehl, S.J., London: Thomas Nelson and Song Ltd, 1963, 103.

[126] John Henry Newman, *The Idea of the University Defined and Illustrated*, London, New York and Bombay: Longmans, Green, and Co., 1852/1905, 80.

[127] J. Oswald Dykes, 'Personal Reminiscences of Newman', in Arthur Wollaston Hutton (ed.), *Personal Reminiscences of Newman*, London: Hodder and Stoughton 1890, 233–40, https://archive.org/stream/a622619500huttuoft#page/n0/mode/2up, cited in Edward Bellassis, 'Newman as a Musician', London: Kegan Paul, Trench, Trübner, and Co., 1892, https://ia601003.us.archive.org/21/items/cardinalnewmanas26427gut/26427–8.txt.

[128] Dykes, 'Personal Reminiscences of Newman', 234.

to intersperse the festival performance of a Cherubini mass with chant (to give it a more devotional flavour) to a more serious demand for some Gregorian-only services.[129] What neither Dykes nor Bellassis recognize, however, was that Newman's limited appreciation of chant stemmed not just from his Protestant and Catholic musical experiences but his view of development which accepts the changing – evolutionary, if you will – nature of authority within the church. Authority can change over time, and so, therefore, can musical allegiances. This is where he disagrees with Pugin and Formby. Despite Formby's incarnational Christology and despite Pugin's devoutly aestheticised antiquarianism neither recognized or accepted the paradigm of developmental authority Newman espoused within the history of the church, and although they postulate a pattern for spiritual progression their music's authority has been superseded by changes in style and appreciation: according to Newman, in Victorian Britain there is effectively no reason to believe that plain chant is any more authoritative than Palestrina.

Perhaps this is where the Victorian obsession with the relative and the absolute, content and form, the material and the spiritual come together. To reiterate, Newman claims that

> we are ever measuring and sounding, framing cross classes and cross divisions, and thereby rising from particulars to generals, that is from images to notions. In processes of this kind we regard things, not as they are in themselves, but mainly as they stand in relation to each other . . . He [man] is attenuated into an aspect, or relegated to his place in a classification. Thus his appellation is made to suggest, not the real being which he is in this or that specimen of himself, but a definition . . . he is made the logarithm of his true self.[130]

Clearly both Formby and Pugin use the particularity of Gregorian chant to rise 'from images to notions', but for Newman something is lost in any process of generalization – the genuinely human, 'the true self'. Something is lost in a progression which seeks the Absolute at the expense of the Relative, form at the expense of content, even the spiritual at the expense of the material. For Newman, generalization must never be confused with development, and development in the church must never be confused with evolution in nature. Newman never considered the realms of theology and science to be contradictory; they simply answer different

129 Bellassis, *Newman as a Musician.*

130 John Henry Newman, *An Essay in Aid of A Grammar of Assent*, London: Burns, Oates, & Co., 1870, 28–29.

questions.[131] Nor did he appear troubled with Darwin and evolutionary science: 'It is strange', he mused, 'that monkeys should be so like men, with no historical connection between them, as that there should be no course of facts by which fossil bones got into rocks . . . I will either go whole hog with Darwin, or, dispensing with time and history altogether, hold not only the theory of distinct species but that also of the creation of fossil bearing rocks'.[132] But like many theological evolutionists and evolutionary theologians such as Iverach, Abbot, Gray and Temple Newman was troubled by any evolutionary principle which seemed to disempower the incarnational process inherent within the Godhead – a developmental process operating in the very opposite direction of generalization, towards a mysteriously relativized manifestation of God in history. For Newman as for Pugin, Formby and much Victorian musical culture if, in reverse expectation the Absolute could become Relative, spiritual progression would always have a place in the theology, history and doctrine of the church.

[131] Don O'Leary, *Roman Catholicism and Modern Science: A History*, New York and London: Continuum, 2006, 17–18.

[132] John Henry Newman, unpublished papers at the Birmingham Oratory, note of 9 December 1863, quoted in J. M. Cameron, 'Newman and the Empiricist Tradition', in John Coulson and A. M. Allchin (eds.), *The Rediscovery of Newman: An Oxford Symposium*, London: Sheed & Ward, 1967, 90.

Epilogue

Victorian Musical Culture and the Great Chain of Becoming

'Not only had the originally complete and immutable Chain of Being been converted into a Becoming, in which all genuine possibles are, indeed, destined to realization grade after grade, yet only through a vast, slow unfolding in time; but now God himself is placed in, or identified with, this Becoming'.[1] When Lovejoy imagined the Great Chain of Being metamorphosing into a Great Chain of Becoming he expressed an idea which would have pleased Victorians eager to reconcile rather than repudiate the relationship between religion and science. Subscribers to a complementarity theory were by necessity both Darwinian and non-Darwinian and often believed in a Great Chain of Being culminating in God, for Christians – as we have seen – very probably formulated in Christological terms. Much has been said about the history of a supposed war between religion and science, and today similarly minded historians (complementarians) challenge the warfare narrative, to use Alister McGrath's phrase.[2] For McGrath there are failings in the warfare narrative, however; indeed the relationship both today and, by implication, historically is 'more porous and provisional' than we think. It is even possible to conceive of the relationship as a possibly enriching co-existence.[3] As McGrath notes the warfare narrative is even contested amongst historians of science, some claiming that the conflict was not between religion and science but over different understandings of science.[4] Irrespective, the warfare narrative makes for an easy, convenient – if intractable – polemic, and like the Land Without Music its purpose is largely political. In the cultural politics of Victorian Britain the magisteria of religion and science were definitely not non-overlapping (if that is not a double negative), and musical culture clearly influenced that relationship. One way it did that was to reveal and reflect music's centrality within current evolutionary debates informed by the Great

1 Arthur O. Lovejoy, *The Great Chain of Being: A Study of The History of An Idea*, New York: Harper Torchbooks, 1936/1960, 325–26.

2 Alister McGrath, *Inventing the Universe: Why We Can't Stop Talking About Science, Faith and God*, London, Hodder & Stoughton Ltd, 2015, Kindle loc. 596.

3 McGrath, *Inventing the Universe*, Kindle loc. 658.

4 See Matthew Stanley, *Huxley's Church and Maxwell's Demon: From Theistic Science to Naturalistic Science*, Chicago and London: University of Chicago Press, 2015, 242–63.

Chain of Being. It did that through a variety of musical interdisciplines, each with its own non-Darwinian evolutionary imagination. The Great Chain of Being gave the interdisciplinary framework of Victorian musical culture a spinal conceptual structure from Zoomusicology to Music Theology. The top of the Great Chain of Being (Music Theology) admittedly plays into the hands of the warfare narrative, pushing the relationship between religion and science to its utmost extreme with an unrepentantly Christological evolutionary narrative. The Christological narrative would seem to be the very upper limit of the Great Chain of Musical Being, but Lovejoy (and my reading of Victorian musical culture) suggests otherwise. For Lovejoy there is a Great Chain of Becoming, and for Victorian culture a Great Chain of Musical Becoming.

The Great Chain of Becoming is not only bound strictly to Victorian religion but surfaces in secular culture as well. In secular culture it finds its form in a transcendental future through mechanisms of utopianism linked to socio-cultural evolutionary progression. According to Kenneth Roemer in literary contexts utopianism is

> a fiction that invites readers to experience vicariously an alternative reality that critiques theirs by opening intellectual and emotional spaces that encourage readers to perceive the realities and potentialities of their cultures in new ways. If the author and/or readers perceive the imaginary culture as being significantly better than their 'present' reality, then the work is a literary eutopia (or more commonly, a utopia).[5]

That fictional world could easily apply to music, especially in the nineteenth century more broadly. Ruth Levitas cites as emblematic Schubert's lied 'An die Musik', for example, a setting of Franz von Schober's poetry which 'celebrates the capacity of music to transport the listener or performer into a better world'.[6] Levitas would probably disagree with Roemer's analogue definition of utopianism; hers is a broader, more forgiving definition: 'utopia is the expression of the desire for a better way of being or living. It is not a classificatory definition that says this is utopian, that is not'.[7] Levitas aligns her capacious utopianism with a range of thinkers, but in thinking about music locates Ernst Bloch at her conceptual core: for Bloch 'the power of music as a utopian form rests in its nonverbal, nonrepresentational, abstract character and its consequent capacity to

[5] Kenneth M. Roemer, 'Paradise Transformed: Varieties of Nineteenth-Century Utopias', in Gregory Claeys (ed.), *The Cambridge Companion to Utopian Literature*, Cambridge: Cambridge University Press, 2010, 79.

[6] Ruth Levitas, 'In eine bess're Welt entrückt: Reflections on Music and Utopia', *Utopian Studies* 21/2 (2010), 215.

[7] Levitas, 'In eine bess're Welt entrückt', 217.

transcend the utterable. It does not merely prefigure a better world but invokes it'.[8] Bloch himself claims that

> because the expressive power of music lies beyond all known names, in the end expression in music is no longer under discussion at all but *music itself as expression*. This means the *entirety of its intending, signifying, depicting and of that which it depicts in such a clouded but, in both senses of the word, moving way* . . . Musical expression as a whole is thus ultimately a *vice-regent for an articulation which goes much further than anything so far known* . . . Thus music is that art of pre-appearance which relates most intensively to the welling core of existence (moment) of That-Which-Is and relates most expansively to its horizon;– *cantus essentiam fontis vocat* [very broadly translated, song voices the source of being].[9]

Given their similarities it can come as no surprise that Ernst Bloch and Arthur C. Lovejoy are roughly contemporary, even though Bloch's *Principle of Hope*, written in the 1930s and 1940s, was not published until the 1950s. There is a strong strand of utopianism subsisting within Lovejoy's *Great Chain of Being*, especially in his idea of plenitude. Bianca Michi identifies this connection when suggesting that a good definition of utopian plenitude hinges on a sense for the superabundance – or spilling over – of potentiality in the world. For Michi a genuine potentiality 'must include all the necessary conditions for its actualization'.[10] Lovejoy puts it this way: the principle of plenitude is a

> strange and pregnant theorem of the 'fullness' of the realization of conceptual possibility in actuality . . . the universe is a *plenum formarum* in which the range of conceivable diversity of kinds of living things is exhaustively exemplified . . . no genuine potentiality of being can remain unfulfilled . . . the extent and abundance of the creation must be as great as the possibility of existence and commensurate with the productive capacity of a 'perfect' and inexhaustible Source, and that the world is the better, the more things it contains.[11]

Note how easily Lovejoy's better world – a 'becoming' – compares with Bloch's: 'thus music as a whole stands at the frontiers of mankind, but at those where mankind with new language and the *call-aura around captured intensity, attained We-World*, is still only forming. And precisely the order in musical expression intends a house, indeed a crystal, but from future freedom, a star, but as a new earth'.[12]

[8] Levitas, 'In eine bess're Welt entrückt', 220–21.

[9] Ernst Bloch, *The Principle of Hope*, trans. Neville Plaice, Stephen Plaice and Paul Knight, 3 vols., Oxford: Basil Blackwell, 1986, Vol. 3, 1069–70.

[10] Bianca Beatric Michi, 'The Principle of Plenitude, Utopia and Heterotopia', in Ligia Tomoiaga, Minodora Barbul and Ramona Demarcsek (eds.), *From Francis Bacon to William Golding; Utopias and Dystopias of Today and of Yore*, Newcastle: Cambridge Scholars Publishing, 2012, 96.

[11] Lovejoy, *Great Chain of Being*, 52.

[12] Bloch, *Principle of Hope*, 1103.

As the principle of plenitude is built into the very fabric of the Great Chain of Being – and the Great Chain of Being is itself built into the very fabric of Victorian culture – evidence for utopian plenitude must exist plentifully, as it were, across a wide range of Victorian cultural contexts. One of the easiest places to find it is in theological concepts of eschatology. The theology of eschatology is by its very nature paradoxically plenitudinous, expressing the liberation – or 'becoming' – that comes with an end to ends, when potentiality and actuality collapse. Jerry Walls sums it up when he defines eschatology as 'the study of the final end of things, the ultimate resolution of the entire creation'.[13] Christian eschatology comes in two forms, personal (the final state of the individual at the end of time) and cosmic (the final state of creation as a whole), and it is both a temporal (concerning the end of time) and teleological concept (concerning the fulfilment of God's creation). Like all eschatology Victorian eschatology is inevitably inflected by denominational priorities and structures, but certain general concerns apply across the theological board, amongst them death, individual judgment, heaven, hell, the Second Coming, the resurrection of the dead, the end of the world and last judgment. Understandably given the omnipresence of death Victorians dwelt seriously on these matters, constructing theologies to help rationalize the hardship of their experience, and conversely constructing music to help rationalize the consolation of their theologies. A crucial element in this theo-musical experience is hope ultimately expressed in the concept of heaven, a place of reward for the blessed, 'the goal, the fulfilment, the consummation of the Christian's "lively hope"'.[14] As Michael Wheeler points out, however, heaven does not fit easily into a narrative of 'becoming' because it was characteristically portrayed as static in both Evangelical and Tractarian theological traditions; heaven to Victorians was pie-in-the-sky, a place where friends, family (and subversively lovers) were blissfully reunited.[15] Victorian heaven was not merely a place; it was a plenitudinous relationship – a becoming – between the two worlds of earth and heaven. Quaker F. D. Maurice, for example, conflates them in a characteristically early Victorian fashion in *The Kingdom of Christ* (1837, rev. 1842): 'For that men are not to gain a kingdom hereafter, but are put in possession of it now, and that through their chastisements and the oppositions of their evil nature they are to learn its character and enter into its privilege, is surely taught in every verse'.[16] The eschatological expression of

[13] Jerry L. Walls, *The Oxford Handbook of Eschatology*, Oxford: Oxford University Press, 2008, 4.

[14] Michael Wheeler, *Heaven, Hell, and the Victorians*, Cambridge: Cambridge University Press, 1994, 119.

[15] Wheeler, *Heaven, Hell, and the Victorians*, 20.

[16] Frederick Denison Maurice, *The Kingdom of Christ; or, Hints to a Quaker Respecting the Principles, Constitution and Ordinances of the Catholic Church*, 2 vols., 3rd ed., London: Macmillan, 1883, vol. 1, 299.

contiguity between earth and heaven is frequently realized in hymns, for example, and their often dominant use of the present tense. Wheeler identifies three types of application: an 'eternal' present of heaven as an affirmed transcendental reality, a 'locutionary' present in which the hymn itself is sung and an 'existential' present of the mortal lifespan as against its future post-mortem existence. The characteristic transition from eternal to locutionary and existential present is found in the character of the angels, in the transition in verses 2 and 3 in J. M. Neale's 'Around the Throne of God a Band' (1842). Angels signal eternity by serving a dual purpose, to stand by in heaven and to guard us day and night (through life and death); they are entreated in verse 3 and express the present of time future in verse 4.[17]

Around the Throne of God a band
Of glorious Angels ever stand;
Bright things they see, sweet harps they hold,
And on their heads are crowns of gold.

Some wait around Him, ready still
To sing His praise and do His will;
And some, when He commands them, go
To guard His servants here below.

Lord, give thy Angels every day
Command to guide us on our way,
And bid them every evening keep
Their watch around us while we sleep.

So shall no wicked thing draw near,
To do us harm or cause us fear;
And we shall dwell, when life is past.
With Angels round The throne at last.[18]

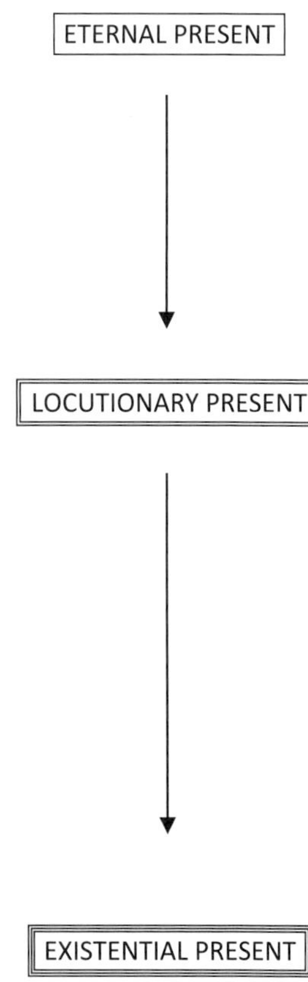

[17] Wheeler, *Heaven, Hell, and the Victorians*, 136.

[18] J. M., Neale, 'Around the Throne of God a Band', originally in *Hymns for Children*, 1842, and subsequently revised and reduced in later publications to four stanzas, www.hymnary.org/text/around_the_throne_of_god_a_band, accessed 15/4/16, cited in Wheeler, *Heaven, Hell, and the Victorians*, 136–37.

Contiguity or modulation between the immanence and transcendence of life is also a fundamental part of music's fundamentally utopian ontology – its plenitudinous 'becoming'. Ruth Levitas argues, as do many, that music is problematical precisely because of its unique combination of performative contingency and non-verbal abstraction.[19] In more music theological contexts Borthwick, Hard and Monti claim that

> the object of Christian hope is ... a *transcendent* one, looking both upward and forward, beyond the imaginable range of this world's possibilities, to an action of God so drastic and radical that biblical writers are finally compelled to picture it as a 'new creation', something of which God alone is capable, and which will leave the world as we know it difficult to recognize.[20]

Looking to Mahler, they suggest that 'It is this very possibility of "temporality with an open future" ... that is suggested by the model of closure we have seen in Mahler. Whether by fading into an ethereal dominant, as in the Fourth [Symphony], or by dissolving into a palpable silence, as in the Ninth, Mahler's gestures of closure suggest ... that completion need not imply an ending. Rather, completion may suggest an opening out onto that which is without end or limit – that is, onto infinity or, better perhaps, the transcendent future of God's promise'.[21] As this suggests, God's promise is inherent within time. Jeremy Begbie likens it to 'time-out' from our lives – music can 'suggest or approximate to a time*less* condition'.[22] Reflecting upon Augustine he asserts that musical time is also deeply embodied, and through a matrix of rhythm and metre mediated by the mind it can offer a 'time-bridging glimpse of the unified order of eternity'.[23] Begbie goes further than any other current music theologian in reflecting a plenitudinous approach to Christian becoming. For him '*different pieces of music are temporal in different ways*,' and within 'a piece of music there is usually *a multiplicity of temporal continua, operating concurrently*'; moreover, '*those who make music and those who hear or listen to it are time-involved creatures*' and as 'physical creatures we inhabit the *multiple temporal continua of the physical world*'.[24]

Victorians might just have recognized Begbie's multi-layered, multi-valenced interpretation of time. For late and post-Victorians time was in

[19] Levitas, 'In eine bess're Welt entrückt', 215.

[20] Alastair Borthwick, Trevor Hard and Anthony Monti, 'Musical Time and Eschatology', in Jeremy Begbie and Steven Guthrie (eds.), *Resonant Witness: Conversations Between Music and Theology*, Grand Rapids, MI, and Cambridge: William B. Eerdmans Publishing Company, 2011, 278.

[21] Borthwick, Hard and Monti, 'Musical Time and Eschatology', 283.

[22] Jeremy S. Begbie, *Theology, Music and Time*, Cambridge: Cambridge University Press, 2000, 34.

[23] Begbie, *Theology, Music and Time*, 84.

[24] Begbie, *Theology, Music and Time*, 35.

a heightened state of transition from organic to mechanical modes of existence. Lewis Mumford, Lovejoy and Bloch's contemporary, blames the invention of the clock, which made time master of its keepers.[25] For Mumford the progression from organic to mechanical time was accompanied by plenitude but not becoming; it emphasized the quantitative rather than qualitative experience exemplified in the mathematicization of what Kuhn would call the scientific revolution.[26] Moreover, it gave rise to the wrong kind of plenitude – the increasingly self-serving and hedonistic kind: 'the goal of eotechnic civilization as a whole until it reached the decadence of the eighteenth century was not more power alone but a greater intensification of life: color, perfume, images, music, sexual ecstasy, as well as daring exploits in arms and thought and exploration'.[27] As Jarlath Killeen remarks it was, however, the clock which Paley used to reinstantiate and reinvigorate organicism within mechanism[28] – inverting the metaphor of mechanisation to redeem man's increasing technological alienation from nature.

In Victorian fiction the dialectic is embodied in Trollope's use of music in *The Warden* (1855). Septimus Harding is the Warden of Hiram's Hospital and (musical) precentor of Barchester Cathedral. He is being threatened with legal action for allegedly receiving disproportionately generous remuneration from the hospital charity, and travels to London to tender his resignation with Sir Abraham Haphazard, a London barrister, Attorney General and legal advisor to the Church. Perceptively, Killeen describes his moment of resignation as one of triumph. The fidgety and contemplative eccentric Harding was always given to playing an imaginary cello (like playing the air guitar today), but under the strain of resignation his habit erupts into an almost trance-like mechanism of self-defence:

> His meeting with Sir Abraham Haphazard is a confrontation between organic and mechanical time, or even between sacred and secular time . . . the Warden escapes triumphantly into another discourse, moving (briefly) back into that organic [musical] time . . . From the point of view of the realist stuck in mechanical time,

[25] Lewis Mumford, *Technics and Civilisation*, London: Routledge and Kegan Paul Ltd, 1934, 14.

[26] See Richard S. Westfall, 'The Scientific Revolution of the Seventeenth Century: A New World View', in John Torrence (ed.), *The Concept of Nature*, Oxford: Oxford University Press, 1992, 63–93.

[27] Mumford, *Technics and Civilisation*, 149–50.

[28] Jarlath Killeen, 'Emptying Time in Anthony Trollope's *The Warden*', in Trish Ferguson (ed.), *Victorian Time: Technologies, Standardizations, Catastrophes*, London: Palgrave Macmillan, 2013, 47–48.

Harding's wild gesticulations, his playing on the invisible cello looks like a kind of psychosis; but Harding is playing upon the deeper harmony.[29]

Harding not only plays, he gesticulates wildly in an act (and caricature) of plenitudinous becoming, encapsulating Wheeler's eschatological progression from eternal to locutionary to existential presents:

as he finished what he had to say, he played up such a tune as never before had graced the chambers of any attorney-general. He was standing up, gallantly fronting Sir Abraham, and his right arm passed with bold and rapid sweeps before him, as though he were embracing some huge instrument, which allowed him to stand thus erect; and with the fingers of his left hand he stopped, with preternatural velocity, a multitude of strings, which ranged from the top of his collar to the bottom of his coat. Sir Abraham listened and looked in wonder. As he had never before seen Mr Harding, the meaning of these wild gesticulations was lost upon him; but he perceived that the gentleman who had a few minutes since been so subdued as to be unable to speak without hesitation, was now impassioned, – nay, almost violent.[30]

As a study in time Trollope's Harding reveals many truths about music, but set against a backdrop of evolutionism his character says even more. Music helps Harding survive, and not just the ordeal of his own resignation. In its silent scream music not only empowers him to redeem his life, it also sends Haphazard a message about its redemptive, eschatological properties. It becomes 'a signal of transcendence'.[31] In some respects it is also an act of general forgiveness, a self-recognition of what Philip David describes as the fallen, occluded and deficient motivations of Trollope's characters.[32] Richard Gibson substantiates this view when he suggests that Trollope explores 'how forgiveness can be achieved – what forms it might take – when humans fall short, as they inevitably will'.[33] In *The Warden*'s case the form forgiveness takes is musical because music is uniquely positioned to express a condition beyond itself – if it is not tautological to say so, a condition of becoming.[34]

[29] Killeen, 'Emptying Time in Anthony Trollope's *The Warden*', 51–52.

[30] Anthony Trollope, *The Warden* (1855), Kindle loc. 2377.

[31] Peter L. Berger, *A Rumor of Angel: Modern Society and the Rediscovery of the Supernatural*, Garden City, NY: Doubleday, 1969, 53.

[32] Philip David, *The Victorians*, The Oxford English Literary History, 13 vols., Oxford: Oxford University Press, 2002, vol. 8, 382.

[33] Richard Hughes Gibson, *Forgiveness in Victorian Literature: Grammar, Narrative, and Community*, London and New York: Bloomsbury, 2015, 82.

[34] A not-dissimilar interpretation of Harding's use of the cello is found Elaine Hadley, *Living Liberalism: Practical Citizenship in Mid-Victorian Britain*, Chicago and London: University of Chicago Press, 2010, 81–120. Hadley, however, develops a theory of secular liberal aesthetics

Music breaks the Great Chain(s) of Human Bondage because unlike other arts music can become its own Becoming. Walter Pater puts it succinctly if elusively: 'All art constantly aspires to the condition of music'.[35] Decoding Pater's riddle-like axiom is an academic industry; everyone has their opinion. In fact, although music and sound permeate his writing Pater seldom focuses his attention exclusively on the matter of music. 'The School of Giorgione' is a brief exception, but then he passes comment on music only while critiquing Giorgione's art:

> *All art constantly aspires towards the condition of music.* For while in all other kinds of art it is possible to distinguish the matter from the form, and the understanding can always make this distinction, yet it is the constant effort of art to obliterate it. That the mere matter of a poem, for instance, its subject, namely, its given incidents or situation – that the mere matter of a picture, the actual circumstances of an event, the actual topography of a landscape – should be nothing without the form, the spirit, of the handling, that this form, this mode of handling, should become an end in itself, should penetrate every part of the matter: this is what all art constantly strives after, and achieves in different degrees.[36]

He continues – only music can perfectly collapse (or obliterate) form and content:

> It is the art of music which most completely realizes this artistic ideal, this perfect identification of matter and form. In its consummate moments, the end is not distinct from the means, the subject from the expression; they inhere in and completely saturate each other; and to it, therefore, to the condition of its perfect moments, all the arts may be supposed constantly to aspire.[37]

Understandably, musicologists and aestheticians have been greatly exercised by Pater's remarks, often identifying them with concepts of absolute music (purely instrumental music like symphonies, chamber music and sonatas). Even when Pater says elsewhere that 'Modern thought is distinguished from ancient by its cultivation of the "relative" spirit in place of the "absolute",'[38] Patricia Herzog, for example, can believe that absolute music

without considering the influence of Trollope's eschatological contexts. She describes Harding as 'feeling abstracted to the register of the altruistic' (94); his manliness is 'not imbricated in class, status, or other conventional instantiations of manhood ... but rather in liberal eccentricity' (95).

35 Walter Pater, 'The School of Giorgione', in Adam Phillips (ed.), *The Renaissance: Studies in Art and Poetry*, Oxford and New York: Oxford University Press, 1986, 86.

36 Pater, 'School of Giorgione', 86. 37 Pater, 'School of Giorgione', 88.

38 Walter Pater, 'Coleridge's Writings', *Westminster Review*, New Series 29 (January 1866), 107.

is what Pater had in mind because it 'contains nothing extraneous to the medium of music itself, a medium consisting solely of tonally moving forms arranged melodically, harmonically and rhythmically. The form and the content of absolute music would thus appear to be identical'.[39] Perhaps more helpfully, others have problematized Pater's axiom by teasing out two interrelated concepts, ekphrasis and Anders-streben. Ekphrasis 'produces images for the mind's eye by means of words'.[40] Pater uses an ekphrastic imagination to conceptualize painting as music; in describing Titian's *Concert* (1510–12), for example, he praises 'the skill with which he caught the waves of wandering sound, and fixed them forever on the lips and hands of the performers'.[41] At the same time Pater also believes in 'what German critics term as *Anders-streben* – a partial alienation from its own limitations, through which the arts are able, not indeed to supply the place of each other, but reciprocally to lend each other new forces'.[42] Anders-streben – literally other-seeking – lies at the heart of Pater's musical becoming, not because all arts aspire to achieve a hierarchical apogee in music (because it collapses form and content) but because music is by its very nature the most artistically plenitudinous of the arts:

> Thus, again, sculpture aspires out of the hard limitation of pure form towards colour, or its equivalent; poetry also, in many ways, finding guidance from the other arts, the analogy between a Greek tragedy and a work of Greek sculpture, between a sonnet and a relief, of French poetry generally with the art of engraving, being more than mere figures of speech; and all the arts in common aspiring towards the principle of music; music being the typical, or ideally consummate art, the object of the great Anders-streben of all art, of all that is artistic, or partakes of artistic qualities.[43]

Just what exactly Anders-streben *is* continues to befuddle theorists. I would like to weigh in and ask an alternative question about it: what does Anders-streben *do* for Pater, and why? Anders-streben defines a kind of plenitudinous force (a becoming) inherent within the arts which by their very nature seek the condition of music (whatever that might be), and one thing that music might be is not what it is, but what it was. Pater constructs a complex, opportunistic understanding of historiography –

[39] Patricia Herzog, 'The Condition to Which All Art Aspires: Reflections on Pater on Music', *British Journal of Aesthetics* 36/2 (April 1996), 125.

[40] Lydia Goehr, 'How to Do More with Words: Two Views of (Musical) Ekphrasis', *British Journal of Aesthetics* 50/4 (October 2010), 389.

[41] Pater, 'School of Giorgione', 71. [42] Pater, 'School of Giorgione', 85.

[43] Pater, 'School of Giorgione', 85–86.

part Hegelian, part Heracleatian, part evolutionary. In evolutionary terms he is a paradigmatic consistently inconsistent non-Darwinian, questioning Darwinian and non-Darwinian thought in his estimation of evolutionary change:

> The entire modern thought of 'development' . . . what is it but old Heracliteanism awake once more in a new world . . . It is the burden of Hegel on the one hand, to whom nature, and art, and polity, and philosophy, eye, and religion too, each in its long historic series, are but so many conscious movements in the secular process of the eternal mind; and on the other hand of Darwin and Darwinism, for which 'type' itself properly is not but *is* only always *becoming.*[44]

Much has been written about Pater's Darwinism, but seldom does it appear to be aligned with music (generally Pater's interest in Darwin is grouped in with his attitudes towards history). Kanarakis Yannis presents a potentially useful thesis according to which the choices made by Pater's critic reflect the plenitudinous becoming of the art itself: 'from the fact that it [natural selection] is presented as a life-giving principle, since the critic or the artist, in the manner of Botticelli "plays fast and loose with. . .data, rejecting some and isolating others, and always combining them anew"'.[45] In his effort to Darwinize Pater's methodology, however, Yannis misses a crucial element in his 'combining anew', which is to say, music – the art to whose condition all other art aspires. Yannis misses the importance of the power of Pater's ekphratic imagination to 'combine anew', and the way it not only selects but re-présents music as the art of becoming. There are larger concerns with implications for music as becoming. While, according to Iser, Pater is fundamentally antagonistic to any kind of metahistorical narrative, such as Hegel's eschatology (the 'self-revelation' of the spirit) and even Darwinian evolution (the struggle for survival), his concept of 'becoming' supersedes these. For Pater 'evolution is neither futuristic nor chiliastic; it is historical. The beginning and end of becoming are of no significance beside the actual process of change,'[46] and it is the historical within the present which is important: we come into the world, each of us,

> 'not in nakedness', but by the natural course of organic development clothed . . . in a vesture of the past, nay fatally shrouded, it might seem, in those laws or tricks or

[44] Walter Pater, *Plato and Platonism* (1893), cited in Wolfgang Iser, *Walter Pater: The Aesthetic Moment*, trans. David Henry Wilson, Cambridge: Cambridge University Press, 1960/1987, 76.

[45] Kanarakis Yannis, 'The Aesthete as a Scientist: Walter Pater and Nineteenth-Century Science', *Victorian Network* 2/1 (Summer 2010), 99.

[46] Iser, *Walter Pater*, 77.

> heredity which we mistake for our volitions: in the language which is more than one half of our thoughts; in the moral and mental habits, the customs, the literature, the very houses, [and I would interpolate here, also music] which we did not make for ourselves; in the vesture of a past, which is (so science would assure us) not ours, but of the race, the species.[47]

Pater's concern over evolutionary thought is palpable, and although evolutionisms tended to blur in his mind – as vision or spectre, as resistance or reinterpretation, to use Kate Hext's words[48] – above all else Pater is deeply concerned to preserve the individual within this potentially alienating process. According to Hext Pater suggests 'an organic vision of temporal continuity in which history is defined not by teleology, or Victorian Progress of any other kind, but by a directionless modification of its present conditions'.[49] In this evolution the individual is reduced to an instant of speciation, moulded by uncontrollable and irresistible evolutionary forces, but those forces are forces of becoming, and becoming is musical:

> Man's physical organism if *played upon* not only by the physical conditions about it, but by remote laws of inheritance, *the vibrations of long-past acts* in the midst of the new order of things in which he lives. When we have estimated these conditions he is not yet simple and isolated; for the mind of the race, the character of age *sway in him this way or that* through the medium of language and current ideas . . . he is so receptive, all the influences of nature and society, *ceaselessly playing upon him*, so that every hour in his life is unique, changed altogether by a stray word, or glance, or touch.[50]

Hext makes an excellent point, which is that evolution plays the individual like an instrument. Angela Leighton defines the effect as a loss of identity: 'the difference between self and not self is lost'.[51] But even as just a metaphor Hext gives the impression, perhaps inadvertently, that for Pater music is controlling instead of liberating, and that evolution is a performer playing on, rather than releasing, the individual by its plenitudinous becoming. In fact I would argue the exact opposite. All art aspires to the

[47] Pater, *Plato and Platonism* (1893), in Iser, *Walter Pater*, 77–78.

[48] Kate Hext, *Walter Pater: Individualism and Aesthetic Philosophy*, Edinburgh: Edinburgh University Press, 2013, 132; 133.

[49] Hext, *Walter Pater*, 134.

[50] Walter Pater, 'Coleridge', in Walter Pater (ed.), *Appreciations, With an Essay on Style*, London and New York: Macmillan & Co., 1895, 67.

[51] Angela Leighton, 'Aesthetic Conditions: Returning to Pater', in Laurel Brake, Lesley Higgins and Carolyn Williams (eds.), *Walter Pater: Transparencies of Desire*, 1880–1920 British authors series 16, Greensboro, NC: ELT Press, 2002, 12–23.

condition of music because the beauty of music helps individuality survive the ordeal of evolution; music, for Pater, is a kind of survival mechanism. As Hext says, 'Though subject to external forces and threatened by evolutionary history, the individual confirms his own significance because he can reconceive time passing and find within it eternity'.[52] Man, woman – every man, every woman, every human individual – is, in a sense, the art that aspires to the condition of music; and like music each individual has evolved to become his or her own Anders-streben – a partial alienation from one's own limitations. As the highest form of aesthetic contemplation music liberates man from time through time. As Hext so beautifully expresses it: for Pater

> As aesthetic time expands, the imagination, set free, flits between the unity of time; our experience is enhanced and intensified. Eternity is in each moment if only we can perceive it . . . Thus does art [and I would argue for Pater, supremely music] save the individual from the apparent meaninglessness of a single life under the conditions of modernity.[53]

***

For Pater and so many Victorians – Trollope and Neale, not to mention Lovejoy and even the very young Mumford at a pinch – there was an evolutionary Becoming which exceeded even the Christological limitations of the non-Darwinian Great Chain of Being – music itself. That Becoming overflowed with plenitude, and in Victorian culture that plenitude overflowed into all the disciplines which intersected with music to form interdisciplines with their own literatures in zoomusicology, ethnomusicology, folk musicology, music pedagogy, music biography, music history and music theology. If Anders-streben can serve as an analogy for modern interdisciplinarity then these Victorian disciplines are the Anders-streben of musical culture. But did all Victorian disciplines aspire to the condition of musicology therefore? – it is an admittedly rhetorical question. The intense value disciplines accorded music is indicative of something meaningful, however – and something indispensible to the evolution of Victorian culture. Victorian musical culture valued music because just like today music helped Victorians survive, and it helped them survive because it signalled Becoming in all the vernacular plenitude of their unreconstructedly non-Darwinian Great Chain of Being – from their Unknown to God, Christ, Jesus, Great Men, children, peasants, savages, animals and the most

[52] Hext, *Walter Pater*, 144. [53] Hext, *Walter Pater*, 145.

rudimentary forms of nature. Today music may be 'both meaningful and meaningless';[54] musical worlds may 'seem to constitute their own, independent spheres of reality';[55] and music's 'lack of representation' may well have become 'the ineffable intimations of the noumenal self',[56] but music is also something people would simply struggle to live without, no matter how seemingly non-adaptive. Herbert Spencer speaks for the Victorians as much as he does for us today when he claims that 'Those vague feelings of unexperienced felicity which music arouses – those indefinite impressions of an unknown ideal life which it calls up, may be considered as a prophecy, to the fulfilment of which music is itself aids'.[57] Music may have aroused vague feelings of unexperienced felicity for Victorians, but whether it was deemed genuinely essential to survival is another matter. Certainly Spencer believed it was essential, and he was certainly not alone; for him music's ultimate evolutionary purpose was the development of altruism and the creation of human sympathy:

> In its bearings upon human happiness, we believe that this emotional language which musical culture develops and refines is only second in importance to the language of the intellect; perhaps not even second to it. For these modifications of voice produced by feelings are the means of exciting like feelings in others. Joined with gestures and expressions of face, they give life to the otherwise dead words in which the intellect utters its ideas; and so enable the hearer not only to *understand* the state of mind they accompany, but to *partake* of that state. In short, they are the chief media of sympathy.[58]

Be that as it may; that music was an integral feature in the evolutionary landscape is undeniable. No Victorian worth his or her evolutionary salt would have called music cheesecake, even at its most decadent. Whether its purpose was sympathy or sexual selection, and whether it emerged from language or language emerged from it, music was too indispensible a part of the Victorian evolutionary toolbox to be summarily dismissed. On the contrary, music gave a profoundly non-Darwinian Victorian culture answers to some of its greatest questions on the origin of man and man's

[54] Vladimir Jankélévitch, *Music and the Ineffable*, trans. Carolyn Abbate, Princeton, NJ: Princeton University Press, 2003, 11.

[55] James Jirtle, 'Understanding Music's Theological Significance: A Kantian Approach', PhD diss., Durham University, 2010, 140.

[56] Daniel Chua, 'Music as the Mouthpiece of Theology', in Begbie and Guthrie (eds), *Resonant Witness*, 154.

[57] Herbert Spencer, 'The Origin and Function of Music' (*Fraser's Magazine*, October 1857), in Herbert Spencer (ed.), *Literary Style and Music*, New York: Philosophical Library, 1951, 74.

[58] Spencer, 'Origin and Function of Music', 74–75.

place in nature – Victorian answers which continue to feed the cheesecake wars of today. There may have been *the* Land Without Music, but in Victorian musical culture there never really was *a* Land Without Music, no matter how ostensibly primitive. Music, like Victorian culture more generally, was in a constant state of Becoming, and evolutionary thought proved it, after a fashion. In the originally complete and immutable Chain of Being music became a Becoming, and in the evolution of Victorian musical culture music helped evolution evolve.

Bibliography

Abbott, Lyman, *The Evolution of Christianity*, Boston and New York: Houghton, Mifflin and Co., and Cambridge: The Riverside Press, 1892.

The Theology of an Evolutionist, Boston and New York: Houghton, Mifflin and Co., 1897.

Abrams, M. H., *The Mirror and the Lamp: Romantic Theory and the Critical Tradition*, Oxford: Oxford University Press, 1953.

Adams, Byron, 'Elgar's Later Oratorios: Roman Catholicism, Decadence and the Wagnerian Dialectic of Shame and Grace', in Daniel M. Grimley and Julian Rushton (eds.), *The Cambridge Companion to Elgar*, Cambridge: Cambridge University Press, 81–105.

Adams, H. G., *Favourite Song Birds*, London: Groombridge and Sons, 1887.

Adams, John, *The Evolution of Education Theory* (review), *Philosophical Review* 21/5 (1912), 608–9.

Adler, Guido, 'Umfang, Methode und Ziel der Musikwissenschaft', *Vierteljahrsschrift für Musikwissenschaft* 1 (1885), 5–20.

Airy, George Biddell, *On Sound and Atmospheric Vibrations, with the Mathematical Elements of Music*, London and Cambridge: Macmillan & Co., 1868.

Aldrich, Richard, *School and Society in Victorian Britain: Joseph Payne and the New World of Education*, Abingdon: Routledge, 2012.

Altick, Richard D., *Lives and Letters: A History of Literary Biography in England and America*, New York: Alfred A. Knopf, 1965.

Amigoni, David (ed.), *Life Writing and Victorian Culture*, Aldershot: Ashgate, 2006.

'Animal Love of Music', *The Leisure Hour: A Family Journal of Instruction and Recreation* 272 (12 March 1857), 165–67.

'Animals at Work and Play: Their Activities and Emotions', *The Quarterly Review* 187/374 (April 1898), 471–89.

Archibald, J. David, *Aristotle's Ladder, Darwin's Tree: The Evolution of Visual Metaphors for Biological Order*, New York: Columbia University Press, 2014.

'Art. IV. – *The Psalter, or Psalms of David, pointed for Chanting. 24mo.* London: Burns. 1840' (review), British Critic (Oct. 1840), 374–90.

Ashcroft, Bill, *Post-Colonial Transformation*, London and New York: Routledge, 2001, 3.

Ashton, Rosemary, *The German Idea: Four English Writers and the Reception of German Thought, 1800–1860*, Cambridge: Cambridge University Press, 1980.

Asma, Stephen T., *Stuffed Animals and Pickled Heads: The Culture of Evolution of Natural History Museums*, Oxford: Oxford University Press, 2003.

Atkinson, Juliette, *Victorian Biography Reconsidered: A Study of Nineteenth-Century 'Hidden' Lives*, Oxford: Oxford University Press, 2010.

Auerbach, Jeffrey A., and Peter H. Hoffenberg (eds.), *Britain, the Empire, and the World at the Great Exhibition of 1851*, Aldershot: Ashgate, 2008.

Augstein, Hanna Franziska (ed.), *Race: The Origins of the Idea, 1760–1850*, Bristol: Thoemmes Press, 1996.

Axon, William E. A., *Folk Song and Folk-Speech in Lancashire*: On the Ballads and Songs of the County Palatine; with notes on the Dialect in which Many of them are Written, and an Appendix on Lancashire Folk-Lore, Manchester: Tubbs & Brook, 1870.

Backsheider, Paula R., *Reflections on Biography*, Oxford: Oxford University Press, 1999.

Baer, Karl Ernst von, *Uber die Entwickelungsgeschichte der Thiere: Beobachtung und Reflexion*, Königsberg: Bornträger, 1828.

Bagchi, Jasodhara, 'Order and Progress: The Reception of Comteian Positivism in Nineteenth-Century Bengal', in G. R. Taneja and Vinod Sena (eds.), *Literature: East and West*, Bombay: Allied Publishers Ltd, 1995, 150–62.

Baillie-Hamilton, J., 'On the Application of Wind to String Instruments', *Proceedings of the Musical Association* 1 (1874–75), 42–46.

Bain, Alexander, *Education as a Science*, The International Scientific Series 25, London: Kegan Paul, Trench, & Co., 1879/1886.

The Emotions and the Will, 2nd ed., London: Longman's Green and Co., 1865.

On the Study of Character including an Estimate of Phrenology, London: Parker, Son & Bourn, 1861.

Baldwin, James Mark, *Mental Development in the Child and the Race: Methods and Processes*, New York and London: Macmillan and Co., 1895.

Balthasar, Hans Urs von, 'Christliche Kunst und Verkündigung', in J. Feiner and M. Löhrer (eds.), *Mysterium Salutis; Grundriss heilsgeschichtlicher Dogmatik*, 4 vols., Einsiedeln, Zurich and Cologne: Benziger, 1965–73, vol. 1, 708–26.

Banfield, Stephen, 'What Do You Think of Stainer's Crucifixion? Current Victorian Musicology', *Journal of Victorian Culture* 15/1 (2010) 119–29.

Bannan, Nicholas (ed.), *Music, Language and Human Evolution*, Oxford: Oxford University Press, 2012.

Bantock, Geoffrey Herman, *Studies in the History of Educational Theory*, 2 vols., London and New York: Routledge, 1984 (vol. 1, Artifice and Nature, 1350–1765; vol. 2, The Mind and the Masses, 1760–1980).

Barrett, William Alexander, *English Church Composers*, London: Sampson, Low, Marston, Searle, & Rivington, and New York: Scribner and Welford, 1882.

Barrington, Daines, 'Experiments and Observations on the Singing of Birds, by the Hon., Vice Pres. R. S. In a Letter to Mathew Maty, M. D. Sec. R. S.', *Philosophical Transactions* 63 (1773–74 [1773]), 249–91.

Barrow, Mark V., *A Passion for Birds: American Ornithology After Audubon*, Princeton, NJ: Princeton University Press, 1998.

Bashford, Christina, and Leanne Langley (eds.), *Music and British Culture, 1785–1914: Essays in Honour of Cyril Ehrlich*, Oxford: Oxford University Press, 2000.

Basu, Prem Sundar (ed.), *Life and Works of Brahmananda Keshav*, Calcutta: Navavidhan Publication Committee, [1877]/1940.

Basu (Bose), Rajnarayan, *Sekal ar Ekal* (*Then and Now*), Calcutta: Valmiki Press, 1874.

Batchelor, John (ed.), *The Art of Literary Biography*, Oxford: Clarendon Press, 1995.

Bate, Jonathan, 'Shakespeare and Original Genius', in Penelope Murray (ed.), *Genius: The History of an Idea*, Oxford: Basil Blackwell, 1989, 76–97.

Battiscombe, Georgiana, *John Keble: A Study in Limitations*, London: Constable, 1963.

Bayard, Samuel, 'Prolegomena to a Study of the Principal Melodic Families of British-American Folk Song', *The Journal of American Folklore* 63/247 (Jan–March 1950), 1–44.

Bearman, C. J., 'Who Were the Folk?: The Demography of Cecil Sharp's Somerset Folk Singers', *The Historical Journal* 43/3 (Sept. 2000), 751–55.

Beddow, Michael, 'Goethe on Genius', in Penelope Murray (ed.), *Genius: The History of an Idea*, Oxford: Basil Blackwell, 1989, 98–112.

Beer, Gillian, *Darwin's Plots: Evolutionary Narratives in Darwin, George Eliot and Nineteenth-Century Fiction*, 2nd ed., Cambridge: Cambridge University Press, 1983/2000.

Beer, Gillian, Malcolm Bowie and Beate Perrey (eds.), *In(ter)discipline: New Languages for Criticism*, London: Legenda, 2007.

Beetham, Margaret, 'Towards a Theory of the Periodical as a Publishing Genre', in Laurel Brake, Aled Jones and Lionel Madded (eds.), *Investigating Victorian Journalism*, Basingstoke: Macmillan, 1990, 19–32.

Beeton's Book of Birds: Showing How to Rear and Manage Them in Sickness and in Health, London: Ward, Lock, & Co., [n.d.].

Begbie, Jeremy S., 'Confidence and Anxiety in Elgar's Dream of Gerontius', in Martin V. Clarke (ed.), *Music and Theology in Nineteenth-Century Britain*, Farnham: Ashgate, 2012, 197–213.

Resounding Truth: Christian Wisdom in the World of Music, London: Society for Promoting Christian Knowledge, 2007.

Theology, Music and Time, Cambridge: Cambridge University Press, 2000.

Bellassis, Edward, 'Newman as a Musician', London: Kegan Paul, Trench, Trübner, and Co., 1892, https://ia601003.us.archive.org/21/items/cardinalnewmanas26427gut/26427-8.txt.

Benedict, Ruth, 'Folk-lore', in Edwin R. A. Seligman and Alvin Johnson (eds.), *Encyclopaedia of Social Sciences*,Vol. 5, London: Macmillan Publishers, 1931.

Bennett, Gillian, 'Geologists and Folklorists: Cultural Evolution and "The Science of Folklore"', *Folklore* 105 (1994), 25–37.

Bentley, George, 'Sincerity in Biography', *Temple Bar* 11/63 (1881), 329–36.

Bentley, Michael, *Modern Historiography: An Introduction*, London and New York: Routledge, 1999.

Modernizing England's Past: English Historiography in the Age of Modernism, 1870–1970, Cambridge: Cambridge University Press, 2005.

Berger, Peter L., *A Rumor of Angel: Modern Society and the Rediscovery of the Supernatural*, Garden City, NY: Doubleday, 1969.

Berkhofer, Robert F. Jr, *Beyond the Great Story: History as Text and Discourse*, Cambridge, MA, and London: Harvard University Press, 1995.

Berlin, Isaiah, *The Roots of Romanticism: The A. W. Mellon Lectures in the Fine Arts, 1965, The National Gallery of Art*, Washington, DC, ed. Henry Hardy, London: Pimlico, 2000, 16–18.

Berneval, Jue de, *Music Simplified: A New Method to Propagate the Study of Music* (*La Musique Apprise sans Maître*, 1824), trans. E. W. W. G. Honseal, London: Printed and published by the author [c. 1840].

Berry, Laura C., *The Child, the State, and the Victorian Novel*, Charlottesville and London: University of Virginia Press, 1999.

Bevis, Matthew (ed.), *The Oxford Handbook of Victorian Poetry*, Oxford: Oxford University Press, 2013.

Bhabha, Homi K., *The Location of Culture*, London: Routledge, 1994.

Bhatt, Chetan, *Hindu Nationalism: Origins, Ideologies and Modern Myths*, Oxford: Berg, 2001.

Biber, Edward, *Henry Pestalozzi and His Plan of Education, Being an Account of His Life and Writings, With Copious Extracts from His Works, and Extensive Details Illustrative of the Practical Part of His Method*, London: John Souter, 1831.

Bingley, William, *Musical Biography: or, Memoirs of the Lives and Writings of the Most Eminent Musical Composers and Writers Who Have Flourished in the Different Countries of Europe During the Last Three Centuries*, 2 vols., London: Henry Colburn, 1814/1834.

Bircham, Peter, *A History of Ornithology*, London: HarperCollins Publishers Ltd, 2007.

Birkhead, Tim R., *Bird Sense: What It's Like to Be a Bird*, New York: Walker and Co., 2012.

The Wisdom of Birds: An Illustrated History of Ornithology, London: Bloomsbury, 2008.

'Alfred Newtown's Contribution to Ornithology: A Conservative Quest for the Facts Rather Than Grand Theories', *Ibis* 154 (2012), 887–905.

Birkhead, Tim, Jo Wimpenny and Bob Montgomerie, *Ten Thousand Birds*, Princeton and Oxford: Princeton University Press, 2013.

Blacking, John, *How Musical Is Man?*, Seattle and London: University of Washington Press, 1973/2000.

Blackmore, Elizabeth, 'The Angelic Choir: Rethinking Female Voices in Anglican Sacred Music, c. 1889', MMus diss., Durham University, 2015.

Blaserna, Pietro, *The Theory of Sound in its Relation to Music*, London: H. S. King & Co., 1876.

Bloch, Ernst, *The Principle of Hope*, trans. Neville Plaice, Stephen Plaice and Paul Knight, 3 vols., Oxford: Basil Blackwell, 1986.

Blum, B. B. E., 'Solmization in Nineteenth-Century American Sight-Singing Instruction', PhD diss., University of Michigan, 1968.

Blum, Stephen, 'European Musical Terminology and the Music of Africa', in Bruno Nettl and Philip V. Bohlman (eds.), *Comparative Musicology and Anthropology of Music*, Chicago and London: University of Chicago Press, 1991, 3–36.

Blunt, Wilfred, *The Ark in the Park: The Zoo in the Nineteenth Century*, London: Hamish Hamilton, 1976.

Bohlman, Philip V., *The Study of Folk Music in the Modern World*, Bloomington and Indianapolis: Indiana University Press, 1988.

World Music: A Very Short Introduction, Oxford: Oxford University Press, 2002.

Bonds, Mark Evan, *Absolute Music: The History of an Idea*, Oxford: Oxford University Press, 2014.

Bonnet, Charles, *Contemplation de la nature*, 1764, reproduced in vol. 4 (1781) of Charles Bonnet, *Œuvres d'histoire naturelle et de philosophie*, 8 vols., Amsterdam: Chez Marc-Michel Rey M., 1779–83.

Œuvres d'histoire naturelle et de philosophie, 8 vols., Amsterdam: Chez Marc-Michel Rey M., 1779–83.

Traité d'insectologie, 1745, reproduced in vol. 1 (1779) of Charles Bonnet, *Œuvres d'histoire naturelle et de philosophie*, 8 vols., Amsterdam: Chez Marc-Michel Rey M., 1779–83.

Borthwick, Alastair, Trevor Hard and Anthony Monti, 'Musical Time and Eschatology', in Jeremy Begbie and Steven Guthrie (eds.), *Resonant Witness: Conversations Between Music and Theology*, Grand Rapids, MI, and Cambridge: William B. Eerdmans Publishing Company, 2011, 271–94.

Bosanquet, R. H. M., 'Temperament; Or, the Division of the Octave' (Part I), *Proceedings of the Musical Association* 1 (1874–75), 4–17.

'Temperament; Or, the Division of the Octave (Part II)', *Proceedings of the Musical Association* 1 (1874–5), 112–58.

Bossche, Chris R. *Vanden, Thomas Carlyle, Histories*, Berkeley and London: University of California Press, 2002.

Bovberjerg, Dana, and Jeremy Iggers, *The Joy of Cheesecake*, Hauppauge, NY: Barron, 1980.

Bowler, Peter J., *Darwin Deleted: Imagining a World Without Darwin*, Chicago and London: Chicago University Press, 2013.

Evolution: The History of an Idea, 3rd ed., Berkeley, Los Angeles and London: University of California Press, 2003.

'Herbert Spencer and Lamarckism', in Mark Francis and Matthew W. Taylor (eds.), *Herbert Spencer: Legacies*, Abingdon and New York: Routledge, 2015, 203–21.

The Invention of Progress: The Victorians and the Past, Oxford: Basil Blackwell, 1989.

The Non-Darwinian Revolution: Reinterpreting a Historical Myth, Baltimore, MD, and London: Johns Hopkins University Press, 1988.

Boyes, Georgina, *The Imagined Village: Culture, Ideology and the English Folk Revival*, Manchester and New York: Manchester University Press, 1993.

Brake, Laurel, Aled Jones and Lionel Madded (eds.), *Investigating Victorian Journalism*, Basingstoke: Macmillan, 1990.

Brake, Laurel, Lesley Higgins and Carolyn Williams (eds.), *Walter Pater: Transparencies of Desire*, 1880–1920 British authors series 16, Greensboro NC: ELT Press, 2002.

Breakspeare, Eustace J., *Mozart*, London: J. M. Dent & Sons, Ltd, and New York: E. P. Dutton & Co., 1902/1922.

Brehony, Kevin J., 'The Fröbel Movement in England, 1850–1911: Texts, Readings and Readers', in Helmut Heiland, Michael Gebel and Karl Neumann (eds.), *Perspektiven der Fröbelforschung*, Würzburg: Königshausen and Neumann Verlag, 2006, 49–64.

Brewin, Robert, *The Life of Haydn, in a Series of Letters Written at Vienna, Followed by The Life of Mozart, with Observations on Metastasio, and on the Present State of Music in France and Italy*, London: John Murray, 1817, a translation of Bombet Louis-Alexandre-César, *Lettres écrites de Vienne en Autriche, sur le célèbre compositeur Jh. Haydn: suivies d'une vie de Mozart, et de considérations sur métastase et l'état présent de la musique en France et en Italie*, Paris: P. Didot L'Ainé, 1814.

Broadwood, Rev. John, *Old English Songs*, London: Balls, 1847.

Brotman, Charles M., 'Alfred Wallace and the Anthropology of Sound in Victorian Culture', *Endeavour* 25/4 (2001), 144–47.

'The Power of Sound: Evolutionary Naturalism and Music in Anglo-American Victorian Culture', PhD diss., University of Rochester, 2005.

Brown, C. Mackenzie, *Hindu Perspectives on Evolution: Darwin, Dharma, and Design*, London and New York: Routledge, 2012.

Brown, Richard Henry, 'Cultural Representation and Ideological Domination', *Social Forces* 71/3 (March 1993), 657–76.

Bruce, Tina, Anne Findlay, Jane Read and Mary Scarborough, *Recurring Themes in Education*, London: Paul Chapman Publishing Ltd, 1995.

Buckle, Henry Thomas, *History of Civilization in England*, 2 vols., London: J. W. Parker and Son, 1857 (vol. 1) and 1861 (vol. 2).
Buckley, Arabella B., *The Fairy-Land of Science*, London: Edward Stanford, 1880.
Life and Her Children: Glimpses of Animal Life, from the Amoeba to the Insects, London: Stanford, 1880.
Winners in Life's Race or the Great Backboned Family, London: Edward Stanford, 1883.
Burke, John G., 'The Wild Man's Pedigree: Scientific Method and Racial Anthropology', in Edward J. Dudley and Maximillian E. Novak (eds.), *The Wild Man Within: An Image in Western Thought from the Renaissance to Romanticism*, Pittsburgh: University of Pittsburgh Press, 1973, 259–80.
Burney, Charles, *A General History of Music from the Earliest Ages to the Present Period. To which is Prefixed, a Dissertation on the Music of the Ancients*, 4 vols., London: Printed for the Author, 1776–89.
Burrow, John W., *A Liberal Descent: Victorian Historians and the English Past*, Cambridge: Cambridge University Press, 1981.
'Historicism and Social Evolutionism', in Benedikt Stuchtey and Peter Wende (eds.), *British and German Historiography 1750–1950: Traditions, Perceptions, and Transfers*, Oxford: Oxford University Press, 2000, 251–64.
Burrowes, J. F., *The Piano-Forte Primer; Containing the Rudiments of Music: Calculated Either for Private Tuition, or Laurence Gomme Teaching in Classes*, London: published by the author, 1849.
Burton, Edwin, 'Henry Formby', in *Catholic Encyclopedia*, ed. C. G. Herbermann et al., Vol. 6, New York: Robert Appleton Co., 1913, 139.
Busby, Thomas, *A General History of Music From the Earliest Times To the Present; Comprising the Lives of Eminent Composers and Musical Writers*, 2 vols., London: G. And W. B. Whittaker and Simpkin and Marshall, 1819.
Butler, Samuel, *Life and Habit*, London: Trübner and Co., 1878, 104.
Butterfield, Herbert, *The Whig Interpretation of History*, London: G. Bell and Sons, 1931.
Cameron, J. M., 'Newman and the Empiricist Tradition', in John Coulson and A. M. Allchin (eds.), *The Rediscovery of Newman: An Oxford Symposium*, London: Sheed & Ward, 1967, 76–96.
Campbell, James, 'Stubbs, Maitland, and Constitutional History', in Benedict Stuchtey and Peter Wende (eds.), *British and German Historiography 1750–1950: Traditions, Perceptions, and Transfers*, Oxford: Oxford University Press, 2000, 99–122.
Campbell, Olive, and Cecil Sharp, *English Folk Songs from the Southern Appalachians*, London: G. P. Putnam's Sons, and New York: The Knickerbocker Press, 1917.
Cantor, Geoffrey, Gowan Dawson, Graeme Gooday, Richard Noakes, Sally Shuttleworth and Jonathan R. Ropham (eds.), *Science in the*

Nineteenth-Century Periodical: Reading the Magazine of Nature, Cambridge: Cambridge University Press, 2004.
Capwell, Charles, 'Marginality and Musicology in Nineteenth-Century Calcutta: The Case of Sourindro Mohun Tagore', in Bruno Nettl and Philip V. Bohlman (eds.), *Comparative Musicology and Anthropology of Music: Essay on the History of Ethnomusicology*, Chicago and London: University of Chicago Press, 1991, 228–43.
Carlyle, Thomas, 'Biography', *Fraser's Magazine* 5/27 (April 1832), 253–60.
A Carlyle Reader, ed. G. B. Tennyson, Cambridge: Cambridge University Press, 1984.
'Characteristics', review of Thomas Hope, *An Essay on the Origin and Prospects of Man*, 3 vols., London: John Murray, 1831, *The Edinburgh Review* 54 (Dec. 1831), 351–83.
Critical and Miscellaneous Works, Philadelphia: Carey & Hart, 1845.
The French Revolution: A History, London: James Fraser, 1837.
'The Hero as Divinity' (1840), in Thomas Carlyle (ed.), *On Heroes, Hero-Worship and the Hero in History*, London: J. M. Dent & Sons, Ltd, 1840/1911, 239–77.
'The Hero as a Man of Letters' (1840), in Thomas Carlyle (ed.), *On Heroes, Hero-Worship and the Hero in History*, London: J. M. Dent & Sons, Ltd, 1840/1911, 383–421.
'The Hero as Priest' (1840), in Thomas Carlyle (ed.), *On Heroes, Hero-Worship and the Hero in History*, London: J. M. Dent & Sons, Ltd, 1840/1911, 346–82.
History of Friedrich II of Prussia, called Frederick the Great, London: Chapman and Hall, 1858.
Life of Friedrich Schiller, London: Taylor and Hessey, 1825.
The Life of John Sterling, Boston: Phillips, Sampson and Co., 1851.
Oliver Cromwell's Letters and Speeches, with Elucidations, London: Chapman and Hall, 1845.
On Heroes, Hero-Worship and the Hero in History, London: J. M. Dent & Sons, Ltd, 1840/1911.
'On History', *Fraser's Magazine* 2/10 (Nov. 1830), 413–18.
Two Note Books of Thomas Carlyle, from 23rd March, 1822, to 16th May, 1832, ed. Charles Eliot Norton, New York: The Grolier Club, 1898.
Carpenter, Lant, *Principles of Education, Intellectual, Moral and Physical*, London: Longman, Hurst, Rees, Orme, and Brown, 1820.
Carroll, Joseph, 'Steven Pinker's Cheesecake for the Mind', *Philosophy and Literature* 22/2 (1998), 478–85, https://muse-jhu-edu.ezphost.dur.ac.uk/journals/philosophy_and_literature/v022/22.2carroll.html.
Catholic Encyclopedia: An International Work of Reference on the Constitution, Doctrine, Discipline, and History of the Catholic Church, ed. Charles G. Herbermann et al., 15 vols. and index, New York: The Encyclopedia Press, Inc., 1907–12/1913.

Chamberlain, Alexander Francis, *The Child: A Study in the Evolution of Man*, London: The Walter Scott Publishing Co., Ltd, and New York: Charles Scribner's Sons, 1902.
Chambers, Robert, *Vestiges of the Natural History of Creation*, John Churchill: London, 1844.
Chappell, William, *A Collection of National English Airs* (1838), 2 vols., London: Chappell and Simpkin, Marshall & Co., 1838–40.
The History of Music, London: Chappell and Co., 1874.
Popular Music of the Olden Time: A Collection of Ancient Songs, Ballads and Dance Tunes, Illustrative of the National Music of England, London: Chappell & Co., 1855–59.
Charnock, Stephen, *The Existence and Attributes of God*, 2 vols., 1682; reprint, Grand Rapids, MI: Baker, 1979.
Chatterjee, Partha, *Nationalist Thought and the Colonial World: A Derivative Discourse?*, Minneapolis: University of Minnesota Press, 1993.
Chattopadhaya, Bankimchandra, 'Anukaraṇ', in Jogesh Chandra Bagal (ed.), *Baṅkim Racanābalī*, vol. 2, Calcutta: Sahitya Samsad, 1965.
Baṅkim Racanābalī, ed. Jogesh Chandra Bagal, vol. 2, Calcutta: Sahitya Samsad, 1965.
'Mill, Darwin and Hindu Religion' (1875), in *Sociological Essays*, trans. and ed. S. N. Mukherjee and Marian Maddern, Calcutta: Rddhi-India, 1986, 60–70.
Christensen, Dieter, Hans-Peter Renecke and Klaus P. Wachsmann (eds.), *Hornbostel Opera Omnia* 1, The Hague: Nijhoff, 1975.
Chua, Daniel, *Absolute Music and the Construction of Meaning*, Cambridge: Cambridge University Press, 1999.
'Music as the Mouthpiece of Theology', in Jeremy Begbie and Steven Guthrie (eds.), *Resonant Witness: Conversations Between Music and Theology*, 2011, 271–94.
Claeys, Gregory (ed.), *The Cambridge Companion to Utopian Literature*, Cambridge: Cambridge University Press, 2010.
Clarke, Martin V., *Music and Theology in Nineteenth-Century Britain*, Farnham: Ashgate, 2012.
Clayton, Martin, 'Musical Renaissance and Its Margins in England and India', 1874–1914, in Martin Clayton and Bennett Zon (eds.), *Music and Orientalism in the British Empire, 1780s–1940s: Portrayal of the East*, Aldershot: Ashgate Publishing Ltd, 2007, 71–93.
Clayton, Martin, and Bennett Zon (eds.), *Music and Orientalism in the British Empire, 1780s–1940s: Portrayal of the East*, Aldershot: Ashgate Publishing Ltd, 2007.
Clayton, Martin, Trevor Herbert and Richard Middleton (eds.), *The Cultural Study of Music: A Critical Introduction*, New York and London: Routledge, 2003.

Clifford, James L. (ed.), *Biography as an Art: Selected Criticism, 1560–1960*, London, New York and Toronto: Oxford University Press, 1962.
Clifford, William Kingdon, *Lectures and Essays by the Late William Kingdon Clifford*, ed. Leslie Stephen and Frederick Pollick, London: Macmillan and Co., 1879.
Cockshut, A. O. J., *Truth to Life: The Art of Biography in the Nineteenth Century*, London: Collins, 1974.
Cohen, Francis 'Folk-Song Survivals in Jewish Worship-Music', *Journal of the Folk-Song Society* 1/2 (1 Jan. 1900), 32–59.
Coleridge, Samuel Taylor, *Biographia Literaria; or Biographical Sketches of My Literary Life and Opinions*, ed. James Engell and Walter Jackson Bate, 2 vols., London: Rest Fenner, 1817.
The Complete Works of Samuel Taylor Coleridge, 7 vols., New York: Harper & Brothers, 1853.
'On the Constitution of Church and States' (1830), in Samuel Taylor Coleridge, *The Complete Works of Samuel Taylor Coleridge with an Introductory Essay upon his Philosophical and Theological Opinions*, ed. W. G. T. Shedd, 7 vols., New York: Harper & Brothers, vol. 6, 1853.
'A Prefatory Observation on Modern Biography', *The Friend* 21 (25 Jan. 1810), 337–40.
Collins, F. Howard, *An Epitome of the Synthetic Philosophy*, London and Edinburgh: Williams and Norgate, 1894.
Conlin, Jonathan, *Evolution and the Victorians: Science, Culture and Politics in Darwin's Britain*, London and New York: Bloomsbury, 2014.
Cook, John, *Sound and the Physical Basis of Music*, London and Edinburgh: W. & R. Chambers, 1877.
Cooke, Ebenezer, 'Editor's Preface', in Johann Heinrich Pestalozzi (ed.), *How Gertrude Teaches Her Children: An Attempt to Help Mothers Teach their Children and An Account of the Method by Johann Heinrich Pestalozzi*, trans. Lucy E. Holland and Francis C. Turner, and ed. with introduction and notes by Ebenezer Cooke, London: Swann Sonnenschein, and Syracuse: C. W. Bardeen, 1894.
Cooper, Colin, *Individual Differences*, 2nd ed., London: Arnold, and New York: Oxford University Press, 2002.
Cornish, C. J., *Life at the Zoo: Notes and Traditions of the Regent's Park Gardens*, New Ed., London: Seeley & Co. Ltd, 1897.
Coulson, John and A. M. Allchin (eds.), *The Rediscovery of Newman: An Oxford Symposium*, London: Sheed & Ward, 1967.
Cowie, Helen, *Exhibiting Animals in Nineteenth-Century Britain: Empathy, Education, Entertainment*, Houndmills: Palgrave Macmillan, 2014.
Crook, David, *Darwinism, War and History*, Cambridge: Cambridge University Press, 1994.
Crowest, Frederick James, *Beethoven*, London: J. M. Dent & Sons, Ltd, and New York: E. P. Dutton & Co., 1899/1913.

Dictionary of British Musicians, London: Jarrold and Sons, 1895.
Phases of Musical England, London: Remington, 1881.
The Story of British Music (From the Earliest Times to the Tudor Period), London: Richard Bentley and Son, 1896.
Culler, A. Dwight, *The Victorian Mirror of History*, New Haven and London: Yale University Press, 1985.
Cummings, William H., *Henry Purcell, 1658–1695*, London: Sampson Low, Marston & Co., Ltd, n.d. [1880s?].
Curwen, Annie Jessy (Mrs Curwen), *Mrs Curwen's Pianoforte Method* (The Child Pianist)
Pianoforte Method, London: J. Curwen and Sons, 1886/c. 1913.
Psychology Applied to Music Teaching, London: Curwen, 1913.
The Teacher's Guide to Mrs. Curwen's Pianoforte Method. (The Child Pianist.) Being a Practical Course of the Elements of Music, London: J. Curwen & Sons, Ltd, 1886.
Curwen, John, 'Lessons on Singing', *Independent Magazine* (1842), 23–24.
The Standard Course of Lessons and Exercises in the Tonic Sol-Fa Method of Teaching Music, London: Tonic Sol-Fa Agency, 1876.
A Tract on Musical Statics: An Attempt to Show the Bearing of the Most Recent Discoveries in Acoustics on Chords, Discords, Transitions, Modulations, and Tuning, as Used by Modern Musicians, London: Tonic Sol-Fa Agency, 1876.
Curwen, John Spencer, *Memorials of John Curwen*, London: J. Curwen & Sons, 1882.
Da Sousa Correa, Delia, *George Eliot, Music and Victorian Culture*, Houndmills and New York: Palgrave Macmillan, 2003.
Dagg, Joachim, *Abteilung für Entomologie*, Institut für Phytopathologie und Pflanzenschutz, Göttingen, www.victorianweb.org/science/science_texts/spencer_dev_hypothesis.html, accessed 14/5/13.
Dahlhaus, Carl, *The Idea of Absolute Music*, trans. Roger Lustig, Chicago and London: University of Chicago Press, 1978/1991.
Dale, Catherine, *Music Analysis in Britain in the Nineteenth and Early Twentieth Centuries*, Aldershot: Ashgate, 2003.
Dannreuther, Edward, 'Beethoven and His Works', *Macmillan's Magazine* (July 1876), 193–209.
Darwin, Charles, *Charles Darwin's Natural Selection: Being the Second Part of His Big Species Book written from 1856 to 1858*, ed. R. C. Stauffer, Cambridge: Cambridge University Press, 1975.
Darwin's Notebooks on Transmutation of Species, ed. Sir Gavin de Beer, Part II, Second Notebook [C], February to July 1938, Bulletin of the British Museum (Natural History), Historical Series 2/3 (May 1960), 75–118, Darwin Online, http://darwin-online.org.uk/content/frameset?keywords=the

%20differ%20of%20men%20chiefly%20races&pageseq=33&itemID=F1574b&viewtype=text.

The Descent of Man, and Selection in Relation to Sex, 2 vols., London: John Murray, 1871; 2nd ed. rev. and augmented, 1874.

'Essay of 1844', in Francis Darwin (ed.), *The Foundations of the Origin of Species: Two Essays Written in 1842 and 1844 by Charles Darwin*, Cambridge: Cambridge University Press, 1909.

The Expression of the Emotions in Man and Animals, London: John Murray, 1872.

On the Origin of Species by Means of Natural Selection, or the Preservation of Favoured Races in the Struggle for Life, London: John Murray, 1859.

'Recollections of the Development of My Mind & Character' [Autobiography (1876–4.1882)] CUL-DAR26.1–121, transcribed Kees Rookmaaker, 62–63, *Darwin Online*, http://darwin-online.org.uk/.

Darwin, Erasmus, *The Botanic Garden*, Part II. The Loves of the Plants, New York: T. & J. Swords, 1790–91/1798.

The Temple of Nature: Or the Origin of Society: A Poem, with Philosophical Notes, London: J. Johnson, 1803.

Zoonomia; or the Laws of Organic Life, Philadelphia: Edward Earle, 1794–96/1818.

Dauney, William, 'Observations with a View to an Inquiry into the Music of the East', *Journal of the Royal Asiatic Society*, old series 6 (1841), 1–10.

Davey, Henry, *History of English Music*, London: J. Curwen and Sons, 1895.

History of English Music, revised and rewritten with appendix, London: J. Curwen and Sons, 1895/1921.

David, Philip, *The Victorians*, The Oxford English Literary History, 13 vols., volume 8, Oxford: Oxford University Press, 2002.

Davies, James Q., and Ellen Lockhart (eds.), *Sound Knowledge: Music and Science in London, 1789–1851*, Chicago and London: Chicago University Press, 2017.

Dawson, Christopher, *Religion and World History: A Selection from the Works of Christopher Dawson*, ed. James Oliver and Christina Scott, Garden City, NY: Doubleday, 1975.

Dawson, Gowan, and Bernard Lightman, 'Introduction', in Gowan Dawson and Bernard Lightman (eds.), *Victorian Scientific Naturalism: Community, Identity, Continuity*, Chicago and London: University of Chicago Press, 2014.

De Caro, Francis A., 'G. L. Gomme: The Victorian Folklorist as Ethnohistorian', *Journal of the Folklore Institute* 19/2–3 (May–Dec. 1982), 107–117.

De Quincey, Thomas, *Recollections of the Lake Poets* (1834–40), London: Adam & Charles Black, 1862.

De Val, Dorothy, *In Search of Song: The Life and Times of Lucy Broadwood*, Farnham: Ashgate, 2011.

Dennett, Daniel, *Darwin's Dangerous Idea: Evolution and the Meaning of Life*, New York: Touchstone, 1995.

Desmond, Adrian, and James Moore, *Darwin's Sacred Cause: Race, Slavery and the Quest for Human Origins*, London: Allen Lane, 2009.

Did-Huberman, Georges, 'The Surviving Image: Aby Warburg and Tylorian Anthropology', *Oxford Art Journal* 25/1 (2002), 61–69.

Dolezal, James E., *God Without Parts; Divine Simplicity and the Metaphysics of God's Absoluteness*, Eugene, OR: Pickwick Publications, 2011.

Donovan, J., *From Lyre to Muse: A History of the Aboriginal Union of Music and Poetry*, London: Kegan Paul, Trench, Trübner, & Co., Ltd, 1890.

Dorson, Richard, *The British Folklorists: A History*, Chicago: University of Chicago Press, and London: Routledge & Kegan Paul Ltd, 1968.

'The Great Team of English Folklorists', *The Journal of American Folklore* 64/251 (Jan–March 1951), 1–10.

Draper, John William, *History of the Conflict between Religion and Science*, London: Henry S. King, 1875.

Drummond, Henry, *Natural Law in the Spiritual World*, London: Hodder and Stoughton, 1896.

Dudley, Edward J., and Maximillian E. Novak (eds.), *The Wild Man Within: An Image in Western Thought from the Renaissance to Romanticism*, Pittsburgh: University of Pittsburgh Press, 1973.

Duff, William, *An Essay on Original Genius; and its Various Modes of Exertion in Philosophy and the Fine Arts, Particularly in Poetry*, London: Edward and Charles Dilly, 1767.

Dundes, Alan, 'The Devolutionary Premise in Folklore Theory', *Journal of the Folklore Institute* 6/1 (June 1969), 5–19.

Dykes, J. Oswald, 'Personal Reminiscences of Newman', in Arthur Wollaston Hutton (ed.), *Personal Reminiscences of Newman*, London: Hodder and Stoughton, 1890, 233–40, https://archive.org/stream/a622619500huttuoft#page/n0/mode/2up.

Eastham, Andrew, 'Walter Pater's Acoustic Space: "The School of Giorgione", Dionysian "Anders-streben", and the Politics of Soundscape', *The Yearbook of English Studies* 40/1 of 2, 'The Arts in Victorian Literature' (2010), 196–216.

'The Ecclesiastical Chant', *The Ecclesiastic* 1 (Jan.–June 1846).

'Ecclesiastical Music', *The Ecclesiologist*, new series vol. 10, Aug. 1846–April 1850), 208–17.

Edwards, John Harrington, *God and Music*, New York: The Baker and Taylor Co., 1903.

'The Effect of Music on Animals', *The Musical Standard* 9/231 (4 June 1898), 361.

Eichwald, Carl Edward von (ed.), *Zoologia Specialis quam expositis animalibus tum vivis, tum fossilibus potissimum Rossiae in universum, et Poloniae in specie, in usum lectionum publicarum in Universitate Caesarea Vilnensi habendarum*, Vilnae Josephi Zawadzki, 1829.

Eidensaul, Scott, *Of a Feather: A Brief History of American Birding*, New York: Harcourt, Inc., 2007.

Eliot, George, *Impressions of Theophrastus Such*, Edinburgh: William Balckwood and Sons, 1879.

'Value in Originality' (mid-1870s), in George Eliot, *The Works of George Eliot: Essays and Leaves from a Note-Book*, Edinburgh and London: William Blackwood and Sons, 1883, 303.

Ellingson, Ter, *The Myth of the Noble Savage*, Berkeley, Los Angeles and London: University of California Press, 2001.

Elliot, J. H, 'Music – and Morals', *Musical News and Herald* (19 June 1926).

Ellis, Alexander John, 'Illustrations of Just and Tempered Intonation', *Proceedings of the Musical Association* 1 (1874–75), 159–65.

Ellsworth, Therese, and Susan Wollenberg (eds.), *The Piano in Nineteenth-Century British Musical Culture: Instruments, Performers and Repertoire*, Aldershot: Ashgate, 2007.

Emery, Gilles, and Matthew Levering (eds.), *The Oxford Handbook of the Trinity*, Oxford: Oxford University Press, 2010.

Encyclopaedia of Social Sciences, ed. Edwin R. A. Seligman and Alvin Johnson, 15 vols., London: Macmillan Publishers, 1931–67.

Engel, Carl, *Descriptive Catalogue of the Musical Instruments in the South Kensington Museum*, London: HMSO, 1870.

An Introduction to the Study of National Music; Comprising Researches into Popular Songs, Tradition, and Customs, London: Longmans, Green, Reader, and Dyer, 1866.

The Literature of National Music, London: Novello, Ewer & Co., 1879.

'Music' in *Notes and Queries on Anthropology, For the Use of Travellers and Residents in Uncivilized Lands. [Drawn up by a Committee appointed by the British Association for the Advancement of Science.]*, London: Edward Stanford, 1874, 110–14.

The Music of the Most Ancient Nations, particularly of the Assyrians, Egyptians, and Hebrews; with Special Reference to Recent Discoveries in Western Asia and in Egypt, London: John Murray, 1864.

'English Literature in 1880', *Journal of the Statistical Society of London* 44 (March 1881), 95–97.

English Singing Birds: Their Songs, Homes, and Habits, London: W. B. Horner, no date.

Evans, Rev. W. E., *The Songs of the Birds; Or, Analogies of Animal and Spiritual Life*, London: Francis & John Rivington, 1845.

Fabian, Johannes, *Time and the Other: How Anthropology Makes Its Objects*, New York: Columbia University Press, 1983.

Fara, Patricia, *Erasmus Darwin: Sex, Science, and Serendipity*, Oxford: Oxford University Press, 2012.

Farber, Paul Lawrence, 'The Type-Concept in Zoology during the First Half of the Nineteenth Century', *Journal of the History of Biology* 9/1 (Spring 1976), 93–119.

Feiner, J., and M. Löhrer (eds.), *Mysterium Salutis; Grundriss heilsgeschichtlicher Dogmatik*, 2 vols., Einsiedeln: Benziger Verlag, 1965.

Fennell, James H., 'On the Effects of Music on Man and Animals', *The Mirror of Literature, Amusement, and Instruction* 1081 (16 Oct. 1841), 245–48.

Ferguson, Trish (ed.), *Victorian Time: Technologies, Standardizations, Catastrophes*, London: Palgrave Macmillan, 2013.

Fergusson, David A. (ed.), *The Blackwell Companion to Nineteenth-Century Theology*, Malden, MA, and Oxford: Wiley Blackwell, 2010.

Fétis, François-Joseph, 'Letters on the State of Music in London', *The Harmonicon* 7/11 (Nov. 1829), 275–81.

Findon, Benjamin William, *Sir Arthur Sullivan: His Life and Music*, London: J. Nisbet, 1904.

The Folk-Lore Society, First Annual Report of the Council, *The Folk-Lore Record* 2 (1879), 1–13.

Folk-Lore Society of Great Britain, *Handbook of Folklore*, ed. George Laurence Gomme, London: Published for the Folklore Society by D. Nutt, 1890.

'Folk-song', *Oxford English Dictionary*, Oxford University Press, 2016, www.oed.com.ezphost.dur.ac.uk/view/Entry/72553?redirectedFrom=folksong#eid.

Forbush, William Byron, *The Boy Problem: A Study in Social Pedagogy*, Boston, MA: The Pilgrim Press, 1901.

Forkel, Johann Nikolaus, *Life of Bach: With a Critical View of his Compositions*, trans. Stephenson, London: T. Boosey & Co., 1820, a translation of Johann Nikolaus Forkel, *Über Johann Sebastian Bachs Leben, Kunst und Kunstwerke*, Leipzig: Hoffmeister, 1802.

Formby, Rev. Henry, *The Catholic Christian's Guide to the Right Use of Christian Psalmody and of the Psalter*, London, Dublin and Derby: Thomas Richardson and Son, 1846.

Congregational Singing: Two Instructions on the Duty and Privileges of Congregational Singing, Preached in the Convent Church of St Mary's, Handsworth, on Christmas Day, and the Following Sunday, 1849.

'The Ecclesiastical Song', *The Tablet* (3 April 1847), 3.

The English Churchman called 'Parochial Psalmody Considered' (1845).

First Series of Hymns and Songs for the Use of Catholic Schools and Families, London, Dublin and Derby: Burns and Lambert and J. Duffy, 1853/1860.

'Parochial Psalmody Considered', reproduced from *The English Churchman*, London, 1845.

The Plain Chant: The Image and Symbol of the Humanity of our Divine Redeemer and the Blessed Mary, London: James Burns, 1848.

The Roman Ritual and Its Canto Fermo Compared with the Works of Modern Music in Point of Efficiency and General Fitness for the purposes of the Catholic Church, London: James Burns, 1849.

School Songs and Poetry to Which Music Is Adapted, London: Longmans and Co., 1852.
The Young Singer's Book of Songs, 4th ed., London: Longman and Co., 1852/1870.
Forster, John, *The Life of Charles Dickens*, London: Chapman and Hall, 1872–74.
Fowler, W. Warde, 'The Notes of Birds. *The Evolution of Bird-song*. By Charles A. *Witchell*', *Nature* (30 July 1896), 290–91.
France, Peter, and William St Clair (eds.), *Mapping Lives: The Uses of Biography*, Oxford: Oxford University Press, 2002.
Francis, Mark, and Matthew W. Taylor (eds.), *Herbert Spencer: Legacies*, Abingdon and New York: Routledge, 2015.
Freeman, Edward A., *The Chief Periods of European History: Six Lectures read in the University of Oxford in Trinity Term 1885*, London: Macmillan, 1886.
History of the Norman Conquest of England: Its Causes and Results, Oxford: Clarendon Press, 1867–79.
Frodeman, Robert, Julie Thompson Klein and Carl Mitcham (eds.), *The Oxford Handbook of Interdisciplinarity*, Oxford: Oxford University Press, 2012.
Froebel, Friedrich, *The Education of Man*, trans. Josephine Jarvis, New York: A. Lovell & Co., 1826/1885.
Friedrich Froebel's Education by Development: The Second Part of the Pedagogics of the Kindergarten, trans. Josephine Jarvis, London: Edward Arnold, 1899.
Froebel's Chief Writings on Education Rendered into English, trans. S. S. F. Fletcher and J. Welton, London: Edward Arnold, 1912.
Fröbel's Letters on the Kindergarten, ed., trans. and annotated by Emilie Michaelis and Henry Keatley Moore, London: W. Swan Sonnenschein & Co., 1891.
Journal for Friedrich Froebel's Educational Aims 1, 14 (1851–52).
Mother and Nursery Songs (*Mutter- und Kose-Lieder*, 1844), trans. Fannie E. Dwight and Josephine Jarvis and ed. Elizabeth P. Peabody, Boston: Lee and Shepard, 1898.
Froude, James Anthony, *The Divorce of Catherine of Aragon*, London: Longmans, 1891.
History of England from the Fall of Wolsey to the Defeat of the Spanish Armada, London: Parker, 1856–70.
'Inaugural Lecture', *Longman's Magazine* 21 (Dec. 1892), 140–62.
Thomas Carlyle: A History of the First Forty Years of His Life, 1795–1835, 2 vols., London: Longmans, Green, and Co., 1882 and 1884.
Fubini, Enrico, *The History of Music Aesthetics*, trans. Michael Hatwell, Houndsmill: Macmillan Press Ltd, 1964/1991.
Fuller, Sophie, and Nicky Losseff (eds.), *The Idea of Music in Victorian Fiction*, Aldershot: Ashgate, 2004.
Galton, Francis, 'Eugenics: Its Definition, Scope and Aims', *Nature* 70 (1904), 82.

Hereditary Genius: An Inquiry into its Laws and Consequences, London: Macmillan and Co., Ltd, 1869/1914.

Inquiries into Human Faculty and its Development, London: J. M. Dent and Co., and New York: E. P. Dutton & Co., 1883.

Geertz, Clifford, 'Blurred Genres', *American Scholar* 49 (1980), 165–79.

George, Rev. T. Ladd, *Lectures on the Unknown God of Herbert Spencer: And the Promise and Potency of Prof. Tyndall*, Milwaukee: I. L. Hauser & Co., 1900.

Gerard, Alexander, *An Essay on Genius*, London: W. Strahan and T. Cadell, and Edinburgh: W. Creech, 1774.

Gibson, Richard Hughes, *Forgiveness in Victorian Literature: Grammar, Narrative, and Community*, London and New York: Bloomsbury, 2015.

Glover, Sarah, *The History of the Norwich Solfa System*, 1844.

Scheme for Rendering Psalmody Congregational, Norwich: Jarrold and Sons, and London: Hamilton and Co., 1835.

Scheme to Render Psalmody Congregational, [n.p. Norwich?]: Hamilton and Jarrold, 1835.

Scheme for Rendering Psalmody Congregational, 1835; Together with the Sol-fa Tune Book, 1839, Kilkenny, Ireland: Boethius Press, 1982.

Godman, Stanley, 'John Broadwood, the Earliest English Folksong Collector', *West Sussex Gazette* (30 Jan. 1964).

Godwin, William, *Enquiry Concerning Political Justice and Its Influence on Morals and Happiness: 1793*, Oxford: Woodstock Books, [1793]/1992.

Goehr, Lydia, 'How to Do More with Words: Two Views of (Musical) Ekphrasis', *British Journal of Aesthetics* 50/4 (Oct. 2010), 389–410.

Golby, David, *Instrumental Teaching in Nineteenth-Century Britain*, Aldershot: Ashgate, 2004.

Golding, Rosemary, *Music and Academia in Victorian Britain*, Farnham: Ashgate, 2013.

Gomme, Allan, 'The Folk-Lore Society: Whence and Whither', *Folklore* 63/1 (March 1952), 1–18.

Gomme, George Laurence, *Folklore Relics of Early Village Life*, London: Stock, 1883.

The Folk-Lore Society, Seventh Annual Report of the Council, *The Folk-Lore Journal* 3/4 (1885), 385–96.

Primitive Folk-Moots; Or, Open-Air Assemblies in Britain, London: Sampson Low, Marston, Searle & Rivington, 1880.

The Village Community: With Special Reference to the Origin and Form of its Survivals in Britain, London: Scott, 1890.

Ethnology in Folkore, New York: D. Appleton and Co., 1892.

Folklore as an Historical Science, London: Methuen & Co., 1908.

Goodbrand, Robert, 'A Suggestion for a New Kind of Biography', *The Contemporary Review* 14 (April 1870), 20–28.

Gordon, Leonard A., *Bengal: The Nationalist Movement 1876–1940*, New York and London: Columbia University Press, 1974.

Gordon, W. J., *Popular Natural History for Boys and Girls*, London: The Religious Tract Society, 1894.

Gore, Charles (ed.), *Lux Mundi: A Series of Studies in the Religion of the Incarnation*, London: John Murray, 1890.

Gosse, Philip Henry, *The Aquarium: An Unveiling of the Wonders of the Deep Sea*, London, 1854.

Gould, Rev. Sabine Baring, and H. Fleetwood Sheppard, *Songs of the West: A Collection Made from the Mouths of the People*, London: Methuen & Co., 1891.

Gould, Stephen Jay, 'Introduction', in Carl Zimmer, *Evolution: The Triumph of an Idea*, London: William Heinemann, 2002, ix–xiv.

'Nonoverlapping Magisteria', *Natural History* 106 (March 1997), 16–22, in Stephen Jay Gould, *Leonardo's Mountain of Clams and the Diet of Worms: Essays on Natural History*, Cambridge MA: Harvard University Press, 1998/2011, 274.

'Nonoverlapping Magisteria', *Natural History* 106 (March 1997), 16–22, www.stephenjaygould.org/library/gould_noma.html

Ontogeny and Phylogeny, Cambridge, MA, and London: Belknap Press of Harvard University Press, 1977.

Gramit, David, *Cultivating Music: The Aspirations, Interests, and Limits of German Musical Culture 1770–1848*, Berkeley and Los Angeles: University of California Press, 2002.

Gray, Asa, *Darwiniana: Essays and Reviews Pertaining to Darwinism*, New York: D. Appleton and Co., 1889.

Natural Science and Religion: Two Lectures Delivered to the Theological School of Yale College, New York: Charles Scribner's Sons, 188.

'Scientific Worthies: Charles Robert Darwin', *Nature* 10/240 (1874), 79–81.

Green, John Richard, *Letters of John Richard Green*, ed. Leslie Stephen, London and New York: Macmillan, 1901.

A Short History of the English People, London: Macmillan and Co., Ltd, 1874/1899.

Greene, Colin, *Cultural Perspective: Marking Out the Horizons*, Grand Rapids, MI, and Cambridge, UK: William B. Eerdmans Publishing Company, 2003.

Greenslade, William, *Degeneration, Culture, and the Novel, 1880–1940*, Cambridge and New York: Cambridge University Press, 1994.

Great Exhibition of the Works of Industry of all Nations, 1851. Official Descriptive and Illustrated Catalogue, 3 vols., London: Spicer Brothers, 1851.

Gregory, E. David, *The Late Victorian Folksong Revival: The Persistence of English Melody, 1878–1903*, Lanham, MD, Toronto and Plymouth: The Scarecrow Press, Inc., 2010.

Victorian Songhunters: The Recovery and Editing of English Vernacular Ballads and Folk Lyrics, 1820–1883, Lanham, MD, Toronto and Oxford: The Scarecrow Press, Inc., 2006.

Griffin, David Ray, *Religion and Scientific Naturalism: Overcoming the Conflicts*, Albany: State University of New York Press, 2000.
Grimley, Daniel M., and Julian Rushton (eds.), *The Cambridge Companion to Elgar*, Cambridge: Cambridge University Press, 2005.
Grove, George, 'Beethoven', in George Grove (ed.), *A Dictionary of Music and Musicians (A.D. 1450–1880) By Eminent Writers, English and Foreign*, 4 vols. plus index, London: Macmillan and Co., 1879–90, vol. 1 (1879), 162–209.
(ed.), *A Dictionary of Music and Musicians (A.D. 1450–1880) By Eminent Writers, English and Foreign*, 4 vols. plus index, London: Macmillan and Co., 1879–1890.
Grove Music Online, *Oxford Music Online*, www.oxfordmusiconline.com, accessed 15/04/16.
Religion and Scientific Naturalism: Overcoming the Conflicts, Albany: State University of New York Press, 2000.
Gurney, Edmund, 'On Some Disputed Points in Music', *Fortnightly Review* (July 1876), 106–30.
The Power of Sound, London: Smith, Elder & Co., 1880.
Guthrie, Steven, 'Carmen Universitatis: A Theological Study of Music and Measure', PhD diss., University of St Andrews (2000).
Haddon, Alfred C. (ed.), *Reports of the Cambridge Anthropological Expedition to Torres Straits*, 6 vols. in 7, vol. 2: Physiology and Psychology (1903), Cambridge: Cambridge University Press, 1903, vol. 4: Arts and Crafts (1912), Cambridge: Cambridge University Press, 1901–35.
Hadley, Elaine, *Living Liberalism: Practical Citizenship in Mid-Victorian Britain*, Chicago and London: University of Chicago Press, 2010.
Hadow, W. H., *Studies in Modern Music: Hector Berlioz, Robert Schumann, Richard Wagner*, London: Seeley and Co., Ltd, 1911.
Haeckel, Ernst, *Generelle Morphologie des Organismen*, 2 vols., Berlin: Georg Reimer, 1866.
Riddle of the Universe at the Close of the Nineteenth Century, trans. Joseph McCabe, New York and London: Harper and Brother Publishers, 1900.
Hale, J. R., *The Evolution of British Historiography: From Bacon to Namier*, London and Melbourne: Macmillan, 1967.
Hamilton, Nigel, *Biography: A Brief History*, Cambridge, MA, and London: Harvard University Press, 2007.
Harker, David, *Fakesong: The Manufacture of British 'Folksong', 1700 to the Present Day*, Milton Keynes and Philadelphia: Open University Press, 1985.
Harper, Joseph, 'Towards an Understanding of Tractarian Hymnody: A Critical Appraisal of the Interaction between Theology, Poetry and Music in Anglican Hymnody between 1840 and 1900', PhD diss., Durham University, 2010.
Harper-Scott, J. P. E., *Edward Elgar, Modernist*, Cambridge: Cambridge University Press, 2006.
Harper-Scott, J. P. E., and Julian Rushton (eds.), *Elgar Studies*, Cambridge: Cambridge University Press, 2007.

Harrison, Carol, *Beauty and Revelation in the Thought of St Augustine*, Oxford: Oxford University Press, 1992.

Harrison, Frank, *Time, Place and Music: An Anthology of Ethnomusicological Observation c. 1550 to c. 1800*, Amsterdam: Frits Knuf, 1973.

Harrison, Frederick, *The Meaning of History and Other Historical Pieces*, London: Macmillan and Co., Ltd, 1916.

'The Subjective Synthesis', *The Fortnightly Review* 8 New Series (1 July – 1 Dec. 1870), 184–97.

Harrold, C. F., *Carlyle and German Thought, 1819–1834*, 1st AMS ed., New York: AMS, 1934/1978.

Hartland, Edwin Sidney, 'The Science of Folk-Lore', *Folk-Lore Journal* 3 (1885), 97–121.

Hartland, Edwin Sidney, C. Staniland Wake, Henry B. Wheatley and G. L. Gomme, 'Folk-Lore Terminology', *The Folk-Lore Journal* 2/11 (Nov. 1884), 340–48.

Hatfield, Gary, *The Natural and the Normative: Theories of Spatial Perception from Kant to Helmholtz*, Cambridge, MA, and London: MIT Press, 1990.

Haweis, Rev. H. R., *Music and Morals*, London: Strahan and Co., 1871.

Hawkins, John, *A General History of the Science and Practice of Music*, London: T. Payne, 5 vols., 1776.

A General History of the Science and Practice of Music, new (2nd) edition, London: Novello, Ewer & Co., and New York: J. L. Peters, 2 vols., 1776/1875.

Hawkins, Mike, *Social Darwinism in European and American Thought 1860–1945*, Cambridge: Cambridge University Press, 1997.

Hazlitt, William, *The Complete Works of William Hazlitt*, ed. P. P. Howe, 21 vols., J. M. Dent and Sons, 1930–34.

Head, Matthew, 'Birdsong and the Origins of Music', *Journal of the Royal Musical Association* 122/1 (1997), 1–23.

Heaney, Maeve Louise VDMF, *Music as Theology: What Music Has to Say about the Word*, Eugene, OR: Pickwick Publications.

Hearnshaw, L. S., *A Short History of British Psychology 1840–1940*, London: Methuen, 1964.

Heiland, Helmut, Michael Gebel and Karl Neumann (eds.), *Perspektiven der Fröbelforschung*, Würzburg: Königshausen and Neumann Verlag, 2006.

Henderson, William James, *The Story of Music*, New York: Longmans, Green & Co., 1889.

Heraud, John Abraham, 'On Poetic Genius, Considered as a Creative Power, *Fraser's Magazine* (Feb. 1830), 56–63.

Herbert, Christopher, *Victorian Relativity: Radical Thought and Scientific Discourse*, Chicago and London: University of Chicago Press, 2001.

Herle, Anita, and Sandra Rouse (eds.), *Cambridge and the Torres Strait: Centenary Essays on the 1898 Anthropological Expedition*, Cambridge: Cambridge University Press, 1998.

Herzog, Patricia, 'The Condition to Which All Art Aspires: Reflections on Pater on Music', *British Journal of Aesthetics* 36/2 (April 1996), 122–34.

Hett, Charles Louis, *A Dictionary of Bird Notes, To Which is Appended a Glossary of Popular Local and Old-Fashioned Synonyms of British Birds*, Brigg: Jacksons', 1989.

Hewitt, Martin, 'Victorian Studies: Problems and Prospects', *Journal of Victorian Culture* 6/1 (Spring 2001), 137–61.

Hext, Kate, *Walter Pater: Individualism and Aesthetic Philosophy*, Edinburgh: Edinburgh University Press, 2013.

Hickson, William Edward, *The Singing Master*, 4th ed., London: Taylor & Walton, 1840.

Higgins, David, *Romantic Genius and the Literary Magazine: Biography, Celebrity, Politics*, London and New York: Routledge, 2005.

Himmelfarb, Gertrude, *Darwin and the Darwinian Revolution*, Garden City, NY: Doubleday, 1959.

Hinsley, Curtis M., *Savages and Scientists: The Smithsonian Institution and the Development of American Anthropology 1846–1910*, Washington, DC: Smithsonian Institution Press, 1981.

Hodgen, Margaret T., 'The Doctrine of Survivals: The History of An Idea', *American Anthropologist* 33 (1931), 307–24, 1, http://aaanet.org/sections/gad/history/042hodgen.pdf.

Hogarth, William, *Musical History, Biography, and Criticism: Being A General Survey of Music, from the Earliest Period to the Present Time*, London: John W. Parker, 1835.

Holmes, Edward, *The Life of Mozart*, London: J. M. Dent & Sons, and New York: E. P. Dutton & Co., 1845.

Holmes, John, 'The Challenge of Evolution in Victorian Poetry', in Bernard Lightman, and Bennett Zon (eds.), *Evolution and Victorian Culture*, Cambridge: Cambridge University Press, 2014, 39–63.

Darwin's Bards: British and American Poetry in the Age of Evolution, Edinburgh: Edinburgh University Press, 2009.

Holmes, Richard, 'Biography: Inventing the Truth', in John Batchelor (ed.), *The Art of Literary Biography*, Oxford: Clarendon Press, 1995, 15–25.

Honeyman, Susan, *Elusive Childhood: Impossible Representations in Modern Fiction*, Columbus: Ohio State University Press, 2005.

Hornbostel, Erich Moritz von, 'The Problems of Comparative Musicology', trans. Richard Campbell (1905), in Dieter Christensen, Hans-Peter Renecke and Klaus P. Wachsmann (eds.), *Hornbostel Opera Omnia* 1, The Hague: Nijhoff, 1975, 247–70.

Hornby, Emma, and David Maw (eds.), *Essays on the History of English Music: Style, Performance, Historiography*, Woodbridge: Boydell & Brewer, 2010.

Howes, Frank, 'Recent Work in Folk-Music', *Proceedings of the Musical Association*, 64th session (1937–38), 39–69.

'Sharp, Cecil (James)', *Grove Music Online, Oxford Music Online*, www.oxfordmusiconline.com/subscriber/article/grove/music/25594?q=cecil+sharp&search=quick&pos=1&_start=1#firsthit.

Hueffer, Francis, *Richard Wagner and the Music of the Future: History and Aesthetics*, London: Chapman and Hall, 1874.

Hullah, John, *The History of Modern Music: A Course of Lectures Delivered at the Royal Institution of Great Britain*, London: Parker, Son, and Bourn, 1862.

The Third or Transition Period of Music History: A Course of Lectures delivered at the Royal Institution of Great Britain, London: Longmans, Green, and Co., 1876.

Time and Tune in the Elementary School, London: Longman, Green, Reader, & Dyer, 1875.

Huneker, John, *Mezzotints in Modern Music: Brahms, Tschaïkovsky, Chopin, Richard Strauss, Liszt and Wagner*, New York: Charles Scribner's Sons, 1899.

Hunt, Henry George. Bonavia, *A Concise History of Music from the Commencement of the Christian Era to the Present Time*, Cambridge: Deighton Bell, and Co., and London: George Bell and Sons, 1878.

Hurley, Kelly, *The Gothic Body: Sexuality, Materialism and Degeneration at the Fin de Siècle*, Cambridge: Cambridge University Press, 1996.

Hutton, Arthur Wollaston, *Personal Reminiscences of Newman*, London: Hodder and Stoughton, 1890.

Huxley, Thomas Henry, Prologue, in *Essays Upon Some Controverted Questions*, London: Macmillan, 1892.

'The Influence of Music on the Lower Animals', *All the Year Round* 30/735 (30 Dec. 1882), 538–41.

Irenaeus, *Against Heresies*, in The Rev. Alexander Roberts, Sir James Donaldson and Arthur Cleveland Coxe (eds.), *The Ante-Nicene Fathers: The Writings of the Fathers down to A.D. 325*, 10 vols., vol. 1: Apostolic Fathers, New York: Charles Scribner's Sons, 1913, 309–567.

Iser, Wolfgang, *Walter Pater: The Aesthetic Moment*, trans. David Henry Wilson Cambridge: Cambridge University Press, 1960/1987.

Ito, Takahashi, *London Zoo and the Victorians, 1828–1859*, Woodbridge: The Boydell Press for the Royal Historical Society, 2014.

Iverach, James, *Christianity and Evolution*, London: Hodder and Stoughton, 1894.

J. A. A., 'Witchell's "Evolution of Bird-Song". The Evolution of Bird-Song with Observations on the Influence of Mimicry and Imitation by Charles A. Witchell' (review), *The Auk* 13/3 (July 1896), 249–50.

Jacques, Edward, 'Modal Survivals in Folk-Song', *Journal of the Folk-Song Society* 1/1 (1899), 4–6 and 14–15.

'The Laws of Progress in Music', *Proceedings of the Musical Association* (1888–89), 109–31.

Jahoda, Gustav, *Images of Savages: Ancient Roots of Modern Prejudice in Western Culture*, London and New York: Routledge, 1999.

James, William, *The Principles of Psychology*, 2 vols., New York: Henry Holt and Co., 1890.

Jankélévitch, Vladimir, *Music and the Ineffable*, trans. Carolyn Abbate, Princeton, NJ: Princeton University Press, 2003.

Jann, Rosemary, *The Art and Science of Victorian History*, Columbus: Ohio State University Press, 1985.

Jirtle, James, 'Understanding Music's Theological Significance: A Kantian Approach', PhD diss., Durham University, 2010.

Johnson, Samuel, [no title], *Rambler* 154 (7 Sept. 1751), 237–38.

Johnston, J. Alfred, *The Art of Teaching Pianoforte Playing*, London: William Reeves, first ed., n.d.; 2nd ed., 1912.

Jones, Thomas Rymer, *The Animal Creation: A Popular Introduction to Zoology*, London: Society for Promoting Christian Knowledge, 1865.

Jones, Sir William, 'On the Arts, Commonly called Imitative', in Sir William Jones, *Poems, Consisting Chiefly of Translations from the Asiatick Languages, to which are added Two Essays, I. On the Poetry of the Eastern Nations [and] II. On the Arts, commonly called Imitative*, Oxford: Clarendon Press, 1772.

'On the Musical Modes of the Hindoos: Written in 1784, and Since Much Enlarged, by the President [of the Asiatic Society of Bengal]', reprinted in Sourindro Mohun Tagore, *Hindu Music*, Delhi: Low Price Publications, 1882/1994, 125–60.

Kant, Immanuel, *Kant's Critique of Judgement* (*Critik der Urtheilskraft*, Berlin and Libau: Lagarde und Friederich, 1790), trans. J. H. Bernard, London: Macmillan and Co., 1914.

Karpeles, Maud, *Cecil Sharp: His Life and Work*, London: Routledge and Kegan Paul, 1967.

Kassler, Jamie, 'The Royal Institution Music Lectures, 1800–1831: A Preliminary Study', *Royal Musical Association Research Chronicle* 29 (1983–85), 1–30.

Kay, James Phillips, *Minutes of the Committee of Council on Education: With Appendices. 1840–41*, London: William Clowes and Sons, 1841.

Report to the Secretary of State for the Home Department, From the Poor Law Commissioners, on the Training of Pauper Children; with Appendices, London: W. Clowes and Sons, 1841.

The Training of Pauper Children: A Report Published by the Poor Law Commissioners in Their Fourth Annual Report, London: William Clowes and Sons, 1839.

Keats, John, Letter to James Hessey (8 Oct. 1818), in John Keats, *Letters of John Keats*, ed. Edward Rollins Hyder, 2 vols., Vol. 1, Cambridge: Cambridge University Press, 1958.

Letters of John Keats, ed. Edward Rollins Hyder, 2 vols., Cambridge: Cambridge University Press, 1958.

Keble, John, 'On the Mysticism Attributed to the Early Fathers of the Church', *Tracts for the Times*, Vol. 5, London: J. G. F. and J. Rivington, 1840.

Kenna, John C., 'Sir Francis Galton's Contribution to Anthropology', *The Journal of the Royal Anthropological Institute of Great Britain and Ireland* 94/2 (July - Dec. 1964), 80–93.

Kenyon, John, *The History Men: The Historical Profession in England since the Renaissance*, London: Weidenfeld and Nicolson, 1983.

Ker, Ian, *John Henry Newman: A Biography*, Oxford: Oxford University Press, 1988/2010.

Kidson, Frank, 'English Folk-Song', *The Musical Times* 49/779 (1 Jan. 1908), 23–24.

Old English Country Dances, London: William Reeves, 1890.

Traditional Tunes: A Collection of Ballad Airs, Oxford: Charles Taphouse & Son, 1891.

Kiesewetter, Rafael Georg, *History of the Modern Music of Western Europe*, trans. Robert Müller, London: T. C. Newby, 1848, translated from Rafael Georg Kiesewetter, *Geschichte der europäisch-abendländischen oder unsrer heutigen Musik*, Leipzig: Breitkopf und Härtel, 1834.

Killeen, Jarlath, 'Emptying Time in Anthony Trollope's *The Warden*', in Trish Ferguson (ed.), *Victorian Time: Technologies, Standardizations, Catastrophes*, London: Palgrave Macmillan, 2013, 38–56.

Kindi, Vasso, and Theodore Arabatzis (eds.), *Kuhn's* The Structure of Scientific Revolutions *Revisited*, Chicago: University of Chicago Press, 2012.

Kivy, Peter, 'Charles Darwin on Music', *Journal of the American Musicological Society* 12/1 (Spring 1959), 42–48.

Klein, Julie Thompson, 'A Taxonomy of Interdisciplinarity', in Robert Frodeman, Julie Thompson Klein and Carl Mitcham (eds.), *The Oxford Handbook of Interdisciplinarity*, Oxford: Oxford University Press, 2012, 15–30.

Knox, Robert, *The Races of Men*, London: H. Renshaw, 1850.

Koeveler, Diane Long, and Jeffrey Cass (eds.), *Interrogating Orientalism: Contextual Approaches and Pedagogical Practices*, Columbus: Ohio State University Press, 2006.

Kraft, David, *Birdsong in the Music of Messiaen*, CreateSpace Independent Publishing Platform, 2013.

Kuhn, Thomas S. *The Structure of Scientific Revolutions*, Chicago and London: University of Chicago Press, 1962.

Kunst, Jaap, *Musicologica: A Study of the Nature of Ethno-musicology, its Problems, Methods and Representative Personalities*, Amsterdam: Indisch Instituut, 1950.

Kuukkanen, Jouni-Matti, 'Revolution as Evolution: The Concept of Evolution in Kuhn's Philosophy', in Vasso Kindi and Theodore Arabatzis (eds.), *Kuhn's* The Structure of Scientific Revolutions *Revisited*, Chicago: University of Chicago Press, 2012, 134–52.

Laavanyan, Ratnapalan, 'E. B. Tylor and the Problem of Primitive Culture', *History and Anthropology* 19/20 (June 2008), 131–42.

Lamarck, Jean-Baptiste, *Philosophie Zoologique*, Paris: Dentu, 1809.

Lang, Andrew, 'The Method of Folklore', in *Custom and Myth*, London: Longmans, Green and Co., 1884/1910.

Langley, Leanne, 'The English Music Journal in the Early Nineteenth Century', PhD diss., University of North Carolina, Chapel Hill, 1983.

'The Musical Press in Nineteenth-Century England', *Notes* 46/3 (1990), 583–92.

'Sainsbury's *Dictionary*, the Royal Academy of Music, and the Rhetoric of Patriotism', in Christina Bashford and Leanne Langley (eds.), *Music and British Culture, 1785–1914: Essays in Honour of Cyril Ehrlich*, Oxford: Oxford University Press, 2000, 65–97.

Large, David C., and William Weber, *Wagnerism in European Culture and Politics*, Ithaca, NY, and London: Cornell University Press, 1984.

Larsen, Timothy, 'E. B. Tylor, Religion and Anthropology', *The British Journal for the History of Science* 3 (September 2013), 467–85.

Latham, J. E., *Search for a New Eden: James Pierrepont Greaves (1777–1842)*, Fairleigh Dickinson University Press, 1999.

Lau, Catherine, *Cheesecake Seduction*, Singapore: Marshall Cavendish Cuisine, 2010.

Lawrence, A. H., *Sir Arthur Sullivan: Life-Story, Letters and Reminiscences*, London: J. Bowden, 1899.

Leach, Elizabeth Eva, *Sung Birds: Music, Nature, and Poetry in the Later Middle Ages*, Ithaca and London: Cornell University Press, 2007.

Lee, Sidney (ed.), *Dictionary of National Biography*, London: Smith, Elder, & Co., 1882–1901.

Leighton, Angela, 'Aesthetic Conditions: Returning to Pater', in Laurel Brake, Lesley Higgins and Carolyn Williams (eds.), *Walter Pater: Transparencies of Desire*, 1880–1920 British authors series 16, Greensboro, NC: ELT Press, 2002, 12–23.

Levine, George, *Darwin and the Novelists: Patterns of Science in Victorian Fiction*, Cambridge, MA: Harvard University Press, 1988.

Dying to Know: Scientific Epistemology and Narrative in Victorian England, Chicago and London: University of Chicago Press, 2002.

Levitas, Ruth, 'In eine bess're Welt entrückt: Reflections on Music and Utopia', *Utopian Studies* 21/2 (2010), 215–31.

Lewes, G. H., *Problems of Life and Mind*, London: Trübner & Co., 1874–79.

The Study of Philosophy, London: Trübner & Co., 1879.

Lewis, Elinor Spalding, 'Aboriginal Britain: Ethnological Folkoristics and the Origin of the Nation in the Work of Sir George Laurence Gomme, Sir John Rhŷs and Alfred Nutt', PhD diss., University of York, 2010.

Lightman, Bernard, 'The International Scientific Series and the Communication of Darwinism', *Journal of Cambridge Studies* 5 (2010), 27–38.

The Origins of Agnosticism: Victorian Unbelief and the Limits of Knowledge, Baltimore: Johns Hopkins University Press, 1987.

'The Popularization of Evolution and Victorian Culture', in Bernard Lightman and Bennett Zon (eds.), *Evolution and Victorian Culture*, Cambridge: Cambridge University Press, 2014, 286–311.

Victorian Popularizers of Science: Designing Nature for New Audiences, Chicago and London: University of Chicago Press, 2007.

(ed.), *Victorian Science in Context*, Chicago and London: University of Chicago Press, 1997.

Global Spencerism: The Communication and Appropriation of a British Evolutionist, Leiden: Brill, 2015.

Lightman, Bernard, and Bennett Zon (eds.), *Evolution and Victorian Culture*, Cambridge: Cambridge University Press, 2014.

Lippman, Edward, *A History of Western Musical Aesthetics*, Lincoln and London: University of Nebraska Press, 1992.

Littlejohn, S. W., 'An Overview of Contributions to Human Communication Theory from Other Disciplines', in F. E. X. Dance (ed.), *Human Communication Theory: Comparative Essays*, New York: Harper & Row, 1982, 243–85.

Lloyd, A. L., *Folk Song in England*, London: Panther Arts, 1967.

Lockhart, John Gibson, *Memoirs of the Life of Sir Walter Scott*, Edinburgh: Robert Cadell and London: John Murray, 1837–38.

Lovejoy, Arthur O., *The Great Chain of Being: A Study of The History of An Idea*, New York: Harper Torchbooks, 1936/1960.

'On the Discrimination of Romanticisms', *Proceedings of the Modern Language Association* 39 (1924), 229–53.

Lowenthal, David, *The Past is a Foreign Country*, Cambridge: Cambridge University Press, 1985.

Lubbock, John, *The Origin of Civilization and the Primitive Condition of Man*, New York: Appleton and Co., 1870/1879.

Prehistoric Times, as Illustrated by Ancient Remains, and the Manners and Customs of Modern Savages, 2nd ed., London: Williams and Norgate, 1869.

Lyell, Charles, *A Manual of Elementary Geology*, 3rd rev. ed., London: John Murray, 1851.

Principles of Geology: Being an Inquiry How the Former Changes of the Earth's Surface are Referable to Causes Now in Operation, 2 vols., Philadelphia: J. Kay, Jr, & Brother, Pittsburgh (1st American ed. from the 5th and last London ed.), 1837.

Maaß, Johann Gebhard Ehrenriech, 'Ueber die Instrumentalmusick', *Neue Bibliothek der schönen Wissenschaften* 48 (1792), 3–40.

Mabilat, Claire, *Orientalism and Representations of Music in Nineteenth-Century British Popular Arts*, Aldershot: Ashgate, 2008.

Macaulay, Thomas Babington, *History of England from the Accession of James II*, 5 vols., London, London: Longman, Green, Brown and Longmans, 1849.
Speeches of the Right Honorable T. B. Macaulay, M.P., 2 vols., Leipzig: Bernhard Tauchnitz, 1853.
The Works of Lord Macaulay Complete, ed. Lady Trevelyan, 8 vols., London: Longmans, Green, and Co., 1866.
Macfarlane, Robert, *Original Copy: Plagiarism and Originality in Nineteenth-Century Literature*, Oxford: Oxford University Press, 2007.
Macfarren, George Alexander, *Musical History*, Edinburgh: Adam and Charles Black, 1885.
Mackray, Rev. A. N., *Bird-Preachers: Friendly Talks with Little Folks about Bible-Birds*, London: The Religious Tract Society, n.d. [1880s?].
Macquarrie, John, *Jesus Christ in Modern Thought*, London: SCM Press, and Philadelphia: Trinity Press International, 1990.
Mactaggart, Peter and Ann Mactaggart (eds.), *Musical Instruments in the 1851 Exhibition: A Transcription of the Entries of Music Interest from the Official Illustrated Catalogue of the Great Exhibition of the Art and Industry of all Nations, with Additional Material from Contemporary Sources*, Welwyn: Mac & Me Ltd, 1986.
Magnum, Teresa, 'Animal Angst: Victorians Memorialize their Pets', in Deborah Denenholz Morse and Martina A. Danahay (eds.), *Victorian Animal Dreams: Representations of Animals in Victorian Literature and Culture*, Aldershot: Ashgate, 2007, 15–34.
'Dog Years, Human Fears', in Nigel Rothfels (ed.), *Representing Animals*, Bloomington and Indianapolis: Indiana University Press, 2002, 35–47.
Maienschein, Jane, and Michael Ruse (eds.), *Biology and the Foundation of Ethics*, Cambridge: Cambridge University Press, 1999.
Maine, Henry, *Ancient Law: Its Connection with the Early History of Society and Its Relation to Modern Ideas*, London: John Murray, 1861/1908.
Mainwaring, John, *Memoirs of the Life of the Late George Frederic Handel*, London: R. and J. Dodsley, 1760, translated into German by Johann Mattheson as *Georg Friedrich Händels Lebensbeschreibung* (1761), with an afterword by Walther Siegmund-Schultze, Leipzig: Deutscher Verlag für Musik, VEB, 1976.
Maitland, J. A. Fuller, *English Music in the XIXth Century*, New York: E. P. Dutton & Co., and London: Grant Richards, 1902.
Marcus, George E., and Michael M. J. Fischer, *Anthropology as Cultural Critique: An Experimental Moment in the Human Sciences*, Chicago and London: University of Chicago Press, 1986.
Massey, Reginald, and Jamila Massey, *The Music of India*, New Delhi: Abhinav Publications, 1993/1996.
Maurice, Frederick Denison, *The Kingdom of Christ; or, Hints to a Quaker Respecting the Principles, Constitution and Ordinances of the Catholic Church*, 2 vols., 3rd ed., London: Macmillan, 1883.

Mayo, Elizabeth, *Lessons on Objects as Given in a Pestalozzian School at Cheam, Surrey*, London: R. B. Seeley & W. Burnside, 1830.

Mayr, Ernst, 'Recapitulation Reinterpreted: 'The Somatic Program', *The Quarterly Review of Biology*, 69/2 (June 1994), 223–32.

McCarty, Willard, 'Humanities Computing as Interdiscipline', A Seminar in the Series, 'Is Humanities Computing an Academic Discipline?', held under the auspices of the Institute for Advanced Technology in the Humanities (IATH), at the University of Virginia, Guy Fawkes Day 1999 (Ver. 13 October 1999; rev. 14 October, 15 October, 22 October 1999), www.iath.virginia.edu/hcs/mccarty.html.

'Looking through an Unknown, Remembered Gate: Millennial Speculations on Humanities Computing', *Interdisciplinary Science Reviews* 26/3 (Autumn 2001), 173–82, www.mccarty.org.uk/essays/McCarty,%20Through%20an%20unknown%20remembered%20gate.pdf.

McCool, Gerald A., *Nineteenth-Century Scholasticism: The Search for a Unitary Method*, New York: Fordham University Press, 1999.

McGrath, Alister E., *Inventing the Universe: Why We Can't Stop Talking About Science, Faith and God*, London: Hodder & Stoughton Ltd, 2015.

Darwinism and the Divine: Evolutionary Thought and Natural Theology, The 2009 Hulsean Lectures, University of Cambridge, Malden, MA, and Oxford: Wiley Blackwell, 2011.

McGuire, Charles Edward, 'Elgar, Judas, and the Theology of Betrayal', *19th-Century Music* 23/3 (Spring 2000), 236–72.

Music and Victorian Philanthropy: The Tonic Sol-fa Movement, Cambridge: Cambridge University Press, 2009.

McNeill, W. H. (ed.), *Essays in the Liberal Interpretation of History by Lord Acton: Selected Papers*, Chicago: University of Chicago Press, 1967.

Meckel, J. F., *System der Vergleichenden Anatomie*, 7 Bde., Halle: Rengersche Buchhandlung, 1821.

Merriam, Alan P., 'Definitions of "Comparative Musicology" and "Ethnomusicology": An Historical-Theoretical Perspective', *Ethnomusicology* 21/2 (May 1977), 189–204.

Merrigan, Terrance, 'Revelation', in Ian Ker and Terrance Merrigan, *The Cambridge Companion to John Henry Newman*, Cambridge: Cambridge University Press, 2009, 47–72.

Michaelis, Emilie, and Henry Keatley Moore, 'Introduction', in Friedrich Froebel, *Fröbel's Letters on the Kindergarten*, ed., trans. and annotated by Emilie Michaelis and Henry Keatley Moore, London: W. Swan Sonnenschein & Co., 1891.

Michi, Bianca Beatric, 'The Principle of Plenitude, Utopia and Heterotopia', in Ligia Tomoiaga, Minodora Barbul and Ramona Demarcsek (eds.), *From Francis Bacon to William Golding; Utopias and Dystopias of Today and of Yore*, Newcastle: Cambridge Scholars Publishing, 2012, 92–108.

Middleton, Richard, 'Music Studies and the Idea of Culture', in Martin Clayton, Trevor Herbert and Richard Middleton (eds.), *The Cultural Study of Music: A Critical Introduction*, New York and London: Routledge, 2003, 251–62.

Mill, James, *Analysis of the Phenomena of the Human Mind*, 2 vols., 2nd ed., London: Baldwin & Cradick, 1829.

The History of British India, 4th ed., 8 vols., London: James Madden and Co., 1840–48.

'Section 1: Theory of the Human Mind. – Its Importance in the Doctrine of Education', Supplement to the Encyclopedia Britannica, London: J. Innes, 1825, http://oll.libertyfund.org/title/1764/92873.

Mill, John Stuart, 'Inaugural Address Delivered to the University of St Andrews, 1867'*, in The Collected Works of John Stuart Mill*, vol. 21 – Essays on Equality, Law, and Education, ed. John M. Robson, Toronto: University of Toronto Press, and London: Routledge and Kegan Paul, 1984, http://oll.libertyfund.org/title/255/21681.

'On Liberty' (1859), http://archive.org/stream/onliberty34901gut/pg34901.txt.

'Minutes on the Constructive Method of Teaching Reading, Writing, and Vocal Music', *Minutes of the Committee of Council on Education, 1840–41*, London: William Clowes and Sons.

Mitchell, D. W., *A Popular Guide to the Gardens of the Zoological Society of London*, London: Printed for the Author, 1852.

Mitchell, Rosemary, *Picture Theory: Essays on Verbal and Visual Representation*, Chicago and London: University of Chicago Press, 1994.

Picturing the Past: English History in Text and Image, 1830–1870, Oxford: Oxford University Press, 2000.

Mitchell, W. J. T., *Picture Theory: Essays on Verbal and Visual Representation,* Chicago and London: University of Chicago Press, 1994.

Mithen, Steven, *The Prehistory of the Mind: A Search for the Origins of Art, Religion and Science*, London: Phoenix, 1996/2003.

The Singing Neanderthals: The Origins of Music, Language, Mind and Body, London: Phoenix, 2005.

Mivart, St George, *The Groundwork of Science: A Study of Epistemology*, New York: G. P. Putnam's Sons, London: Bliss, Sands & Co., 1989.

On Truth: A Systematic Inquiry, London: Kegan Paul, Trench & Co., 1889, 134.

Monaghan, John, and Peter Just, *Social and Cultural Anthropology: A Very Short Introduction*, Oxford: Oxford University Press, 2000.

Moore, Henry Keatley, *Autobiography of Friedrich Froebel*, trans. and annoted by Emilie Michaelis and H. Keatley Moore, 2nd ed., London: Swan Sonnenschein, Lowry, 1888.

The Child's Pianoforte Book: Being a First Year's Course at the Pianoforte for the Home, the Kindergarten, and the School: With Upwards of Fifty Original Tunes and Songs, London: W. Swan Sonnenschein & Co., 1882.
The Child's Song and Game Book: A Collection of Original Songs and Games for Children, London: Swan Sonnenschein & Co., 1889.
Concert Pitch, London: Moore & Moore, 1901.
Friedrich Froebel's Education by Development, trans. Josephine Jarvis, New York: D. Appleton and Co., 1899.
Fröbel's Letters on the Kindergarten, ed. and annoted by Emilie Michaelis and H. Keatley Moore, London: C. W. Bardeen, 1891.
A Manual of Pianoforte Tuning, London: *Musical Opinion*, Chichester Chambers, Chancery Lane, W.C.2, 1919.
Music in the Kindergarten, London: W. Swan Sonnenschein, 1881.
The Nursery Song Book, London: G. Routledge & Sons, 1906.
The Post-Darwinian Controversies: A Study of the Protestant Struggle to Come to Terms with Darwin in Great Britain and America: 1870–1900, Cambridge: Cambridge University Press, 1979.
Moore, Thomas, *Life and Letters of Lord Byron* (1830), 2 vols., London: John Murray, 1830.
Interdisciplinarity, London and New York: Routledge, 2002/2010.
Moran, Joe, *Interdisciplinarity*, London and New York: Routledge, 2002/2011.
More, Henry, *Divine Dialogues, Containing Sundry Disquisitions and Instructions Concerning the Attributes of Goad and His Providence in the World*, London: James Flesher, 1668.
Morgan, C. Lloyd, *Animal Life and Intelligence*, Boston: Ginn and Co., 1891.
Morgan, Drew Phillip, 'Hermeneutical Aspects of John Henry Newman's *Essay on the Development of Christian Doctrine*, *Horizons* 16/2 (1989), 223–42.
Morris, Lydia T., *Famous Musical Composers being Biographies of Eminent Musicians*, London: T. Fisher Unwin, 1891.
Morse, Deborah Denenholz, and Martina A. Danahay (eds.), *Victorian Animal Dreams: Representations of Animals in Victorian Literature and Culture*, Aldershot: Ashgate, 2007.
Mossop, Rev. John, *Ornithologia Britannica: A Synopsis of British Land-Birds, in Verse and Prose*, London: Jackson and Walford, 1841.
Muir, Theresa, 'Wagner in England: Four Writers Before Shaw', PhD diss., City University of New York, 1997.
Mumford, Lewis, *Technics and Civilisation*, London: Routledge and Kegan Paul Ltd, 1934.
Mundy, Rachel, 'Birdsong and the Image of Evolution', *Society and Animals* 17 (2009), 206–223.

'Evolutionary Categories and Musical Style from Adler to America', *Journal of the American Musicological Society* 67/3 (Fall 2014), 735–768.
'Musical Evolution and the Making of Hierarchy', *The World of Music* 48/3 (2006), 13–27
'Nature's Music: Birds, Beasts, and Evolutionary Listening in the Twentieth Century', PhD diss., New York University, 2010.
Murray, Penelope (ed.), *Genius: The History of an Idea*, Oxford: Basil Blackwell, 1989.
Music in England, Ireland, Wales, and Scotland; from the Earliest Ages to the Present Time, London: Cradock and Co., 1845.
The Musical Guide for Singing and the Piano-Forte: in which A Clear Method of Explaining the Various Clefs, Scales, and Keys is Adopted; and A Variety of Information Necessary for the Musical Student is Given. With an Appendix on the Theory of Sound, London: Cradock & Co., 1845.
Myers, Charles Samuel, 'The Absurdity of any Mind–Body Relation' (1932), in Charles Samuel Myers, *In the Realm of Mind: Nine Chapters on the Application and Implications of Psychology*, Cambridge: Cambridge University Press, 1937, 189–216.
'The Beginnings of Music' (1913), in Charles Samuel Myers, *A Psychologist's Point of View: Twelve Semi-Popular Addresses on Various Subjects*, London: William Heinemann (Medical Books), 1933, 177–203.
'Hearing', in Alfred C. Haddon (ed.), *Reports of the Cambridge Anthropological Expedition to Torres Straits*, vol. 2: Physiology and Psychology, Cambridge: Cambridge University Press, 1903, 19–223.
'Human Improvability', in *A Psychologist's Point of View: Twelve Semi-Popular Addresses on Various Subjects*, London: William Heinemann (Medical Books), 1933, 1–16.
In the Realm of Mind: Nine Chapters on the Application and Implications of Psychology, Cambridge: Cambridge University Press, 1937.
'Instinct and Intelligence', *British Journal of Psychology* 3/3 (Oct. 1910), 209–18.
Letter to Sharp, Cecil Sharp House, no shelf mark.
'Music', in Alfred C. Haddon (ed.), *Reports of the Cambridge Anthropological Expedition to Torres Straits*, vol. 4: Arts and Crafts, Cambridge: Cambridge University Press, 1912, 238–69.
'On the Permanence of Racial Mental Differences', in G. Spiller (ed.), *Papers on the Inter-Racial Problems Communicated to the First Universal Races Congress held at The University of London July 26–29, 1911*, London: P. S. King & Son, 1911, 73–79.
'Principles of Development', in *A Psychologist's Point of View: Twelve Semi-Popular Addresses on Various Subjects*, London: William Heinemann (Medical Books), 1933, 17–27.
Review of *Some Conclusions*, *Folklore* 20/1 (30 March 1909), 97–98.

Myers, Helen (ed.), *Ethnomusicology: Historical and Regional Studies*, New York and London: W. W. Norton & Co., 1993.
Nadel, Ira Bruce (ed.), *Victorian Biography: A Collection of Essays from the Period*, New York and London: Garland Publishing, Inc., 1986.
Nägeli, Hans Georg, and Michael Traugott Pfieffer, *Gesangbildungslehre nach Pestalozzischen Grundsätzen, Pädagogisch Begründet von Michael Traugott Pfeiffer, Methodisch Bearbeitet von Hans Georg Nägeli*, Zurich: H. G. Nägeli, 1810.
Neale, John Mason, 'Around the Throne of God a Band', in J. M. Neale, *Hymns for Children*, www.hymnary.org/text/around_the_throne_of_god_a_band.
'Introduction', *Hymns of the Eastern Church*, London: J. T. Hayes, 1862.
Needham, John Turberville, *An Account of Some New Microscopical Discoveries*, London: F. Needham, 1745.
Nettl, Bruno, and Philip V. Bohlman (eds.), *Comparative Musicology and Anthropology of Music: Essays on the History of Ethnomusicology*, Chicago and London: University of Chicago Press, 1991.
Newman, John Henry, *An Essay on the Development of Christian Doctrine*, London: Basil Montegu Pickering, 1878.
Fifteen Sermons Preached Before the University of Oxford, 3rd ed., London, Oxford and Cambridge: Rivingtons, 1872.
The Idea of the University Defined and Illustrated, London, New York and Bombay: Longmans, Green, and Co., 1852/1905.
The Letters and Diaries of John Henry Newman, July 1850 to December 1851, vol. 12: Rome to Birmingham (January 1847–December 1848), ed. Charles Stephen Dessain, London: Thomas Nelson and Song Ltd, 1963.
The Letters and Diaries of John Henry Newman, July 1850 to December 1851, vol. 14: Papal Aggression (July 1850–December 1851), ed. Charles Stephen Dessain and Vincent Ferrer Blehl, London: Thomas Nelson and Song Ltd, 1963.
Discussions and Arguments on Various Subjects, London: Basil Montegu Pickering, 1872.
An Essay in Aid of a Grammar of Assent, London: Burns, Oates, & Co., 1870.
An Essay in Aid of a Grammar of Assent, London: Burns, Oates, & Co., 1874.
Newton, Alfred, *Zoology*, London: Society for Promoting Christian Knowledge, New York: Pott, Young & Co., 1874.
Nicholls, Angus, *Goethe's Concept of the Daemonic: After the Ancients*, Rochester, NY, and Woodbridge: Boydell & Brewer, 2006.
Nichols, Aidan, 'Catholic Theology of the Trinity in the Nineteenth Century', in Gilles Emery and Matthew Levering (eds.), *The Oxford Handbook of the Trinity*, Oxford: Oxford University Press, 2010, 281–313.

Redeeming Beauty: Soundings in Sacral Aesthetics, Aldershot: Ashgate, 2007.

Nicholson, Henry Alleyne, *A Manual of Zoology, vol. 1, Invertebrate Animals*, 1st ed., London, Robert Hardwicke, 1870.

Nisbet, Robert, *History of the Idea of Progress*, London: Heinemann, 1980.

'Obituary: Henry Keatley Moore', *The New Library World* 40/2 (1937), 40.

'Obituary: Henry Keatley Moore', *The Times*, 13 Aug 1937, issue 47762, col. D, 14.

Offer, John, *Herbert Spencer: Critical Assessments*, London and New York: Routledge, 2000.

O'Leary, Don, *Roman Catholicism and Modern Science: A History*, New York and London: Continuum, 2006.

'On the Gregorian Tones, and their Adaptation to English Psalmody', *The Parish Choir* 1/9 (Oct. 1846), 85–87.

Our Song Birds, London: The Religious Tract Society, [n. d., 1872?].

Palmer, H. R., *Palmer's Piano Primer: A Systematic and Clear Explanation of the Fundamental Principles of Piano-Forte Playing*, Philadelphia: Theodore Presser, 1885.

Palmer, John Williams, *Folk Song*, New York: Scribner, 1865.

Panchen, Alec L., *Classification, Evolution, and the Nature of Biology*, Cambridge: Cambridge University Press, 1992.

Pandora, Katherine, 'Knowledge Held in Common: Tales of Luther Burbank and Science in the American Vernacular', *Isis* 92/3, (Sept. 2001), 485–516.

Parker, Christopher, *The English Historical Tradition Since 1850*, Edinburgh: John Donald Publishers Ltd, 1990.

The Idea of History from Coleridge to Collingwood, Aldershot: Ashgate, 2000.

Parry, C. Hubert H., *The Evolution of the Art of Music*, ed. H. C. Colles, New York; Greenwood Press, 1893/1968.

'Inaugural Address', *Journal of the Folk-Song Society* 1/1 (1899), 1–3.

Johann Sebastian Bach: The Story of the Development of a Great Personality, New York and London: G. P. Putnam's Sons and The Knickerbocker Press, 1910.

'On Some Bearings of the Historical Method upon Music', *Proceedings of the Musical Association* 11, 1884–85, 1–17.

Studies of Great Composers, London: George Routledge and Sons, Ltd, and J. Curwen and Sons, Ltd, and New York: E. P. Dutton and Co., n.d. [1887].

Pater, Walter, *Appreciations, With an Essay on Style*, Rockville, MD: ARC Manor, 1888/2008.

'Coleridge', in Walter Pater, *Appreciations, With an Essay on Style*, London and New York: Macmillan and Co., 1895, 66–67.

'Coleridge's Writings', *Westminster Review*, New Series 29 (Jan. 1866), 106–32.

Plato and Platonism, London: Library Edition, 1893.

'The School of Giorgione', in Adam Phillips (ed.), *The Renaissance: Studies in Art and Poetry*, Oxford and New York: Oxford University Press, 1986.

Pauer, Ernest, *The Art of Piano-Forte Playing*, London: Novello, Ewer and Co., 1877.

Paul, Diane B., 'The Selection of "Survival of the Fittest"', *Journal of the History of Biology* 21/3 (Sept. 1988), 412–14.

Paulin, Roger, 'Adding Stones to the Edifice: Patterns of German Biography', in Peter France and William St Clair (eds.), *Mapping Lives: The Uses of Biography*, Oxford: Oxford University Press, 2002, 103–14.

Payne, Joseph, *Contributions to the Science of Education*, New York, Cincinnati and Chicago: American Book Co., 1886.

Penrhyn, Arthur Stanley, *The Life and Correspondence of Thomas Arnold*, London: Ward, Lock, 1844.

Pestalozzi, Johann Heinrich, *Enquiries concerning the Course of Nature in the Development of the Human Race* (1797), translation of *Meine Nachforschungen über den Gang der Natur in der Entwicklung des Menschengeschlechts*, ed. Dieter Jürgen Löwisch, WBG: Darmstadt, 2002.

How Gertrude Teaches Her Children: An Attempt to Help Mothers Teach their Children and an Account of the Method by Johann Heinrich Pestalozzi, trans. Lucy E Holland and Francis C. Turner, and ed. with introduction and notes by Ebenezer Cooke, London: Swann Sonnenschein, and Syracuse: C. W. Bardeen, 1894.

Leonard and Gertrude (Lienhard und Gertrud, 4 vols., 1781–87), translated and abridged by Eva Channing, Boston; D. C. Heath, 1898.

Pietsch, Theodore W., *Trees of Life: A Visual History of Evolution*, Baltimore, MD: Johns Hopkins University Press, 2012.

Pinker, Stephen, *How the Mind Works*, London: Penguin, 1997/1999.

Plotz, Judith, *Romanticism and the Vocation of Childhood*, Basingstoke: Palgrave, 2001.

Pole, William, *The Philosophy of Music, being the Substance of a Course of Lectures*, London, 1879.

Poliquin, Rachel, *The Breathless Zoo: Taxidermy and the Cultures of Longing*, University Park: Pennsylvania State University Press, 2012.

Poole, Reginald Lane, *Johann Sebastian Bach, 1685–1750*, London: Sampson Low, Marston & Co., Ltd, [n.d., 1880s].

Possing, Birgitte, 'Biography: Historical', www.possing.dk/pdf/historicalbio.pdf.

Potter, John, *Vocal Authority: Singing Style and Ideology*, Cambridge: Cambridge University Press, 1998.

Powell, Rev. Baden, *Essays on the Spirit of the Inductive Philosophy, the Unity of Worlds, and Philosophy of Creation*. London: Longman, Brown, Green, and Longmans, 1855.

Pugin, Augustus Welby Northmore, *An Earnest Appeal for the Revival of the Ancient Plain Song* (1850), London: R. & T. Washbourne, and New York, Cincinnati and Chicago: Benziger Bros., 1905.

The Collected Letters of A. W. N. Pugin 1851–1852, ed. Margaret Belcher, 5 vols., Oxford: Oxford University Press, 2001–15.

Pycraft, W. P., *The Story of Bird-Life*, London: George Newnes, Ltd, 1900.

Qureshi, Sadiah, 'Dramas of Development: Exhibitions and Evolution in Victorian Britain', in Bernard Lightman and Bennett Zon (eds.), *Evolution and Victorian Culture*, Cambridge: Cambridge University Press, 2014, 261–85.

Peoples on Parade: Exhibitions, Empire, and Anthropology in Nineteenth-Century Britain, 2011.

Radford, Andrew, and Mark Sandy (eds.), *Romantic Echoes in the Victorian Era*, Aldershot: Ashgate, 2008.

Ragan, Mark A., 'Trees and Networks Before and After Darwin', *Biology Direct* 4/43 (16 Nov. 2009), 1–38.

Rai, Lala Lajpat, *Lala Lajpat Rai: Writing and Speeches*, ed. V. J. Joshi, 2 vols., Delhi: University Publishers, 1966.

Rainbow, Bernarr, *The Land Without Music: Musical Education in England 1800–1860 and Its Continental Antecedents*, London: Novello and Co., Ltd, 1967.

Raitiere, Martin N., *The Complicity of Friends, How George Eliot, G. H. Lewes, John Hughlings-Jackson Encoded Herbert Spencer's Secret*, Lewisburg, PA: Bucknell University Press, 2012.

Raychaudhuri, Tapan, 'Europe in India's Xenology: The Nineteenth-Century Record', *Past and Present* 137: The Cultural and Political Construction of Europe (Nov. 1992), 156–82.

Europe Reconsidered: Perceptions of the West in Nineteenth-Century Bengal, 2nd ed., Oxford: Oxford University Press, 2002 (first published 1988).

Reed, Joseph W. Jr, *English Biography in the Early Nineteenth Century, 1801–1838*, New Haven and London: Yale University Press, 1966.

'Reed Organs in England: A Comprehensive Study of Reed Organs in England, Scotland and Wales', http://tardis.dl.ac.uk/FreeReed/English/organ_book/node14.html.

Reports of the Council and Auditors of the Zoological Society of London (30 April, 1855), London: Taylor and Francis, 1855.

Reventlow, Henning Graf, and William Farmer (eds.), *Biblical Studies and the Shifting of Paradigms, 1850–1914*, Sheffield: Sheffield Academic Press, 1995.

Review of Joseph Wolf, *Zoological Sketches, made for the Zoological Society of London, from Animals in their Vivarium in the Regent's Park, Edited, with Notes, by D. W. Mitchell B.A., F.L.S., Secretary to the Society, London, 1856*; D. W. Mitchell, *A Popular Guide to the Gardens of the Zoological Society of London, London, 1855*; and Philip Henry Gosse, *The Aquarium: An Unveiling of the Wonders of the Deep Sea, London, 1854, Quarterly Review* 98 (1855), 220–48.

Reynolds, Sir Joshua, *The Discourses of Sir Joshua Reynolds*, Discourse 6 (1774), London: T. Cadell, Jn., and W. Davies, 1801.

Rich, Paul B., *Race and Empire in British Politics*, 2nd ed., Cambridge: Cambridge University Press, 1990.
Richards, Graham, 'Getting a Result: The Expedition's Psychological Research 1890–1913', in Anita Herle and Sandra Rouse (eds.), *Cambridge and the Torres Strait: Centenary Essays on the 1898 Anthropological Expedition*, Cambridge: Cambridge University Press, 1998, 136–57.
'Race', Racism and Psychology: Towards a Reflexive History, London and New York: Routledge, 1997.
Richards, Robert J., *Darwin and the Emergence of Evolutionary Theories of Mind and Behavior*, Chicago and London: University of Chicago Press, 1987.
'Darwin's Romantic Biology: The Foundation of His Ethics', in Jane Maienschein and Michael Ruse (eds.), *Biology and the Foundation of Ethics*, Cambridge: Cambridge University Press, 1999, 113–53.
The Meaning of Evolution: The Morphological Construction and Ideological Reconstruction of Darwin's Theory, Chicago and London: University of Chicago Press, 1992.
The Romantic Conception of Life: Science and Philosophy in the Age of Goethe, Chicago and London: University of Chicago Press, 2002.
The Tragic Sense of Life: Ernst Haeckel and the Struggle over Evolutionary Thought, Chicago and London: University of Chicago Press, 2008.
Richardson, Angelique, *Love and Eugenics in the Late Nineteenth Century: Rational Reproduction and the New Woman*, Oxford: Oxford University Press, 2003.
Riley, Matthew, 'Heroic Melancholy: Elgar's Inflected Diatonicism', in J. P. E. Harper-Scott and Julian Rushton (eds.), *Elgar Studies*, Cambridge: Cambridge University Press, 2007, 284–307.
Ritter, Frederic Louis, *The History of Music, from the Christian Era to the Present Time*, London; William Reeves, 1870/1892.
Music in England, New York: Charles Scribner's Sons, 1883.
Ritvo, Harriet, *The Animal Estate: The English and Other Creatures in the Victorian Age*, Cambridge, MA, and London: Harvard University Press, 1987.
'The Order of Nature: Constructing the Collections of Victorian Zoos', in R. J. Hoage and William A. Deiss, *New Worlds, New Animals: From Menagerie to Zoological Park in the Nineteenth Century*, Baltimore and London: Johns Hopkins University Press, 1996, 43–50.
'Zoological Nomenclature and the Empire of Victorian Science', in Bernard Lightman (ed.), *Victorian Science in Context*, Chicago and London: University of Chicago Press, 1997, 334–53.
Roberts, Rev. Alexander, and James Donaldson (eds.), *The Writings of the Fathers down to A.D. 325*, Vol. 1, New York: Charles Scribner's Sons, 1913.
Robinson, Phil, *The Poets' Birds*, London: Chatto and Windus, 1883.

Rockstro, William, *A General History of Music from the Infancy of Greek Drama to the Present Period*, London: Sampson Low, Marston, Searle, & Rivington, 1886.
Rodwell, Anne, *The Juvenile Pianist, or A Mirror of Music for Infant Minds*, 2nd ed., London: Chapman and Hall, 1838.
Roemer, Kenneth M., 'Paradise Transformed: Varieties of Nineteenth-Century Utopias', in Gregory Claeys (ed.), *The Cambridge Companion to Utopian Literature*, Cambridge: Cambridge University Press, 2010, 79–106.
Rogers, James Allan, 'Darwinism and Social Darwinism', *Journal of the History of Ideas*, 33/2 (April 1972), 265–80.
Romanes, George John, *Animal Intelligence*, New York: Appleton and Co., 1888.
Mental Evolution in Man: Origin of Human Faculty, New York: D. Appleton and Co., 1889.
Ronge, Johannes, and Bertha Ronge, *A Practical Guide to the English Kinder Garten (Children's Garden)*, London: Hodson and Son, 1855.
Rothenberg, David, *Why Birds Sing: One Man's Quest to Solve An Everyday Mystery*, London: Allen Lane, 2005.
Rothfels, Nigel (ed.), *Representing Animals*, Bloomington and Indianapolis: Indiana University Press, 2002.
Rousseau, Jean-Jacques, *Dictionnaire de Musique*, Paris: Chez la Veuve Duchesne, 1768.
Rowbotham, John Fredrick, *A History of Music*, 3 vols., London: Trübner & Co., 1885–87.
The Private Life of the Great Composers, London: Isbister and Co., Ltd, 1892.
Rowlinson, Matthew, 'Interdisciplinarity, Estrangement, and Method in Victorian Studies', *Victorian Review* 33/1 (Spring 2007), 52–55.
Rudall, H. A., *Beethoven*, London: Sampson, Low, Marston & Co., n.d. [1880s].
Ruse, Michael, *The Darwinian Revolution: Science Red in Tooth and Claw*, Chicago: University of Chicago Press, 1979.
Darwinism and Its Discontents, Cambridge: Cambridge University Press, 2006.
Russell, Ian, *Popular Music in England 1840–1914*, Manchester: Manchester University Press, 1987.
Rylance, Rick, *Victorian Psychology and British Culture 1850–1880*, Oxford: Oxford University Press, 2000.
Saakwa-Mante, Norris, 'Western Medicine and Racial Constitutions: Surgeon John Atkins Theory of Polygenism and Sleepy Distemper in the 1730s', in Ernst Waltraud and Bernard Harris (eds.), *Race, Science and Medicine, 1700–1960*, London and New York: Routledge, 1999, 29–57.
Sainsbury, John S., *Dictionary of Musicians, from the Earliest Ages to the Present Time. Comprising The Most Important Biographical Contents of the Works of Gerber, Choron, and Fayolle, Count Orloff, Dr. Burney, Sir John Hawkins, &c. &c. Together with Upwards of a Hundred Original Memoirs of the Most*

Eminent Living Musicians. And A Summary of the History of Music, 2 vols., London: Printed for Sainsbury and Co., 1824.

Scaife, Nigel, 'British Music Criticism in a New Era: Studies in Critical Thought, 1894–1945', DPhil diss., University of Oxford 1994.

Schaffer, Simon, 'Genius in Romantic Natural Philosophy', in Andrew Cunningham, and Nicholas Jardine (eds.), *Romanticism and the Sciences*, Cambridge: Cambridge University Press, 1990, 82–98.

Schelling, Friedrich Wilhelm Joseph, *System of Transcendental Idealism* (1800), cited in Jacques Taminiaux, *Poetics, Speculation, and Judgment: The Shadow of the Work of Art from Kant to Phenomenology*, trans. Michael Gendre, Albany: State University of New York, 1993.

Scherren, Henry, *The Zoological Society of London: A Sketch of Its Foundation and Development and the Story of Its Farm, Museum, Gardens, Menagerie and Library*, London: Cassell and Co. Ltd, 1904.

Schiller, Friedrich, *On the Aesthetic Education of Man in a Series of Letters, Über die äesthetische Erziehung des Menschen*, trans. Elisabeth M. Wilkinson and L. A. Willoughby, Oxford: Oxford University Press, 1967.

Über die äesthetische Erziehung des Menschen (Letter 22), 1794 [1795?].

Schlegel, August Wilhelm, and Friedrich Schlegel, *Charackteristiken und Kritiken*, 2 vols., Königsberg: Friedrich Nicolovious, 1801.

Schleiermacher, Friedrich, *Ästhetik: Über den Begriff der Kunst*, ed. Thomas Lehnerer, Hamburg: Felix Meiner, 1831–32/1984.

The Christian Faith, ed. H. R. Mackintosh and J. S. Stewart (*Der Christliche Glaube*, 1st ed., 2 vols., Berlin: G. Reimer, 1821–22; 2nd ed., 2 vols., Berlin: G. Reimer, 1830–31), 2 vols. in 1 vol. London and New York: T & T Clark, 1999.

Schmitt, Cannon, 'Evolution and Victorian Fiction', in Bernard Lightman and Bennett Zon (eds.), *Evolution and Victorian Culture*, Cambridge: Cambridge University Press, 2014, 17–38.

Schmitz, Oskar Adolf Hermann, *Das Land ohne Musik*, München: Georg Müller, 1914.

Schneck, Daniel J., and Judith K. Schneck (eds.), *Music in Human Adaptation*, Blacksburg: Virginia Polytechnic Institute and State University, 1997.

Scholtz, Gunter, 'Schleiermacher', in Stefan Lorenz Sorgner and Oliver Fürbeth (eds.), *Music in German Philosophy: An Introduction*, trans. Susan H. Gillespie, Chicago and London: University of Chicago Press, 2010, 547–67.

Schubart, Christian Friederich Daniel, *Schubart's Leben und Gesinnungen*, Stuttgart: Gebrüdern Mantler, 1791–93.

Schumann, Robert, 'Englischer Komponist, kein Komponist', *Neue Zeitschrift für Musik* (24 Feb. 1837).

Schwarz, Hans, *Christology*, Grand Rapids, MI, and Cambridge, UK: William B. Eerdmans Publishing Company, 1998.

Secord, James A., *Victorian Sensation: The Extraordinary Publication, Reception and Secret Authorship of Vestiges of the Natural History of Creation*, Chicago and London: University of Chicago Press, 2000.

Seeley, John Robert, *The Expansion of England: Two Courses of Lectures*, London: Macmillan and Co., 1883.

Sen, Keshub Chunder, *Keshub Chunder Sen's Lectures in India*, London, Paris, New York and Melbourne: Cassell and Co., 1904 (first published [1882]).

'That Marvelous Mystery – The Trinity', in *Keshub Chunder Sen's Lectures in India*, London, Paris, New York and Melbourne: Cassell and Co., 1904 (first published [1882]), 1–48.

'Philosophy and Madness in Religion', in *Life and Works of Brahmananda Keshav*, ed. Prem Sundar Basu, Calcutta: Navavidhan Publication Committee, [1877]/1940, 284–85.

Sessa, Richard, *Richard Wagner and the English*, Rutherford, NJ: Fairleigh Dickinson University Press, 1979.

Shairp, John Campbell, *John Keble: An Essay on the Christian Year*, Edinburgh: Edmonston & Douglas, 1866.

Sharp, Cecil J., *English Folk-Songs of English Origin Collected from the Southern Appalachians*, London: Novello, 1912.

English Folk Songs for Schools, with S. Baring Gould, London: Curwen, 1905.

Folk-Dancing in Elementary and Secondary Schools, London: English Folk Dance Society, 1912.

Folk-Singing in Schools, London: English Folk Dance Society, 1912.

Folk-Songs from Dorset, with H. E. D. Hammond, London: Novello, 1908.

Folk Songs from Somerset, with Charles L. Marson, London: Simpkin, Marshall & Co., 1904–9.

Folk Songs from Various Counties, London: Novello, 1912.

'General Preface to the Folk-Song Series', *Folk-Songs from Various Counties*, London: Novello and Co., Ltd, n.d.

One Hundred English Folk Songs, Boston: Oliver Ditson Co., 1916.

English Folk-Songs: Some Conclusions, London: Simpkin & Co., Ltd, and Taunton: Barnicott & Pearce, Athenaeum Press, 1907.

Shattock, Joanne, 'Where Next in Victorian Literary Studies? – Revising the Canon, Extending Cultural Boundaries, and the Challenge of Interdisciplinarity', *Literature Compass* 4/4 (2007), 1280–91.

Shipman, Pat, *The Evolution of Racism: Human Differences and the Use and Abuse of Science*, Cambridge, MA: Harvard University Press, 1994/2002.

Shires, Linda, 'Victorian Studies and Cultural Studies: Interdisciplinarity, The Market, and a Call for Critical Realism', *Victorian Literature and Culture* 27/2 (1999), 481–86.

Shuttleworth, Sally, and Jenny Bourne Taylor, *Embodied Selves: An Anthology of Psychological Texts 1830–1890*, Oxford: Oxford University Press, 1998.

Simpson, J. Y., *Landmarks in the Struggle between Science and Religion*, London: Hodder & Stoughton, 1925.

Smiles, Aileen, *Samuel Smiles and His Surroundings*, London: Robert Hale Ltd, 1956.

Smiles, Samuel, 'The Education of the Working Classes: An Address given by D. Samuel Smiles to the Members and Friends of the Leeds Mutual Improvement Society', Leeds: Hobson, 1845.

'Self-Help, – National and Individual', *Self-Help; with Illustrations of Character and Conduct*, Boston: Ticknor and Fields, 1863, http://oll.libertyfund.org/title/297/36621/906086.

Smith, Charles, *Charles Darwin and Victorian Visual Culture*, Cambridge: Cambridge University Press, 2006.

Solie, Ruth A., 'The Living Work: Organicism and Musical Analysis', *19th-Century Music* 4/2 (Autumn 1980), 147–56.

'Melody and the Historiography of Music', *Journal of the History of Ideas* 43/2 (April–June 1982), 297–308.

Music in Other Words: Victorian Conversations, Berkeley, Los Angeles and London: University of California Press, 2004.

Solis, Gabriel, 'Thoughts on an Interdiscipline: Music Theory, Analysis, and Social Theory', *Ethnomusicology* 56/3 (Fall 2012), 530–554.

Sorgner, Stefan Lorenz, and Oliver Fürbeth (eds.), *Music in German Philosophy: An Introduction*, trans. Susan H. Gillespie, Chicago and London: University of Chicago Press, 2010.

Spencer, Herbert, 'Bain on The Emotions and The Will', *The Medico-Chirurgical Review* (January 1860), in *Illustrations of Universal Progress: A Series of Discussions*, New York: D. Appleton and Co., 1865, 300–324.

'The Development Hypothesis', *The Leader* (20 March 1852), in *Essays: Scientific, Political and Speculative*, London: Williams and Norgate, 1891, http://archive.org/stream/essaysscientific29869gut/pg29869.txt.

Education: Intellectual, Moral, and Physical, London: Williams and Norgate, 1861/1893.

Essays: Scientific, Political and Speculative, London: Williams and Norgate, 1891, http://archive.org/stream/essaysscientific29869gut/pg29869.txt.

Illustrations of Universal Progress: A Series of Discussions, New York: D. Appleton and Co., 1865.

First Principles, 4th ed., New York: D. Appleton and Co., 1896.

Literary Style and Music, London: Watts & Co., 1950.

'The Origin and Function of Music', *Fraser's Magazine* (Oct. 1857), in Herbert Spencer, *Literary Style and Music: Including Two Short Essays on Gracefulness and Beauty*, New York: Philosophical Library, 1951, 45–106.
'The Origin of Music', *Mind* 15/60 (Oct. 1890), 449–68.
'The Primitive Man – Emotional', *Principles of Sociology*, 3 vols., A System of Synthetic Philosophy, vols. 6–8, New York: D. Appleton and Co., 1883, vol. 1, 159–82.
'The Primitive Man – Intellectual, *Principles of Sociology*, 3 vols., A System of Synthetic Philosophy, vols. 6–8, New York: D. Appleton and Co., 1883, vol. 1, 83–104.
'The Primitive Man – Physical', *Principles of Sociology*, 3 vols., A System of Synthetic Philosophy, vols. 6–8, New York: D. Appleton and Co., 1883, vol. 1, 44–58.
Principles of Biology, 2 vols., n.p.: Williams and Norgate, 1864.
The Principles of Psychology, London: Longman, Brown, Green and Longmans, 1855, §206, http://oll.libertyfund.org/simple.php?id=1394.
Principles of Sociology, 3 vols. (1874–96), A System of Synthetic Philosophy 6–8, New York: D. Appleton and Co., vol. 1, 1883.
'Progress: Its Law and Cause', *Humboldt Library of Popular Science Literature* 17/1 (1881), 233–85.
Social Statics; Or the Conditions Essential to Human Happiness Specified and the First of Them Developed, London: J. Chapman, 1851.
The Study of Sociology, London and New York; D. Appleton, 1873.
'A Theory of Population, Deduced from the General Law of Animal Fertility', *Westminster Review* 57/112 (April 1852), 468–501.
Spiller, G. (ed.), *Papers on the Inter-Racial Problems Communicated to the First Universal Races Congress held at The University of London July 26–29, 1911*, London: P. S. King & Son, 1911.
Spotteswoode, William, 'Letter, Addressed by William Spottiswoode, Esq., M.A., F.R.S., to Some Leading Members of the Musical and Scientific World, Originated this Association (8 April 1874)', *Proceedings of the Musical Association* 1 (1874–75), iii.
Stafford, William C., *A History of Music*, Edinburgh: Constable and Co., and London: Hurst, Chance, and Co., 1830.
Stanfield, James Field, *An Essay on the Study and Composition of Biography*, Sunderland: George Garbutt, 1813.
Stanley, Matthew, *Huxley's Church and Maxwell's Demon: From Theistic Science to Naturalistic Science*, Chicago and London: University of Chicago Press, 2015.
Staudenmaier, Franz Anton, *Geist der göttlichen Offenbarung, oder Wissenschaft der Geschichtsprincipien des Christenthums*, Giessen: B. G. Ferber, 1837.
Die Lehre von der Idee. In Verbindung mit einer Entwicklungsgeschichte der Ideenlehre und der Lehre vom göttlichen Logos, Giessen: B. G. Ferber, 1840.

Steedman, Carolyn, *Strange Dislocations: Childhood And The Idea Of Human Interiority, 1780–1930*, Cambridge, MA: Harvard University Press, 1995.
Steege, Ben, *Helmholtz and the Modern Listener*, Cambridge: Cambridge University Press, 2012.
Steiner, George, *Grammars of Creation*, London: Faber and Faber, 2001.
Stephen, Leslie, *Principles of Biography*, Cambridge: Cambridge University Press, 1911.
Stewart, Dugald, *The Collected Works of Dugald Stewart*, ed. Sir William Hamilton, 2 vols., Vol. 2, Edinburgh: Thomas Constable, and Boston: Little, Brown and Co., 1854.
Stocking, George W. Jr., *Race, Culture, and Evolution: Essays in the History of Anthropology*, Free Press and University of Chicago Press, 1968.
After Tylor: British Social Anthropology 1888–1951, Madison: University of Wisconsin Press, 1995.
Victorian Anthropology, New York: The Free Press, 1987.
Stone, William Henry, 'On Extending the Compass and Increasing the Tone of Stringed Instruments, with Especial Reference to the Author's and Mr. Meeson's Elliptical Tension-Bars', *Proceedings of the Musical Association* 1 (1874–75), 1–3.
The Scientific Basis of Music, London: Novello, Ewer & Co., 1878.
Stott, John, *The Birds: Our Teachers: Essays in Orni-Theology*, Abingdon: Candle Books, 2007.
Stout, G. F., *Analytical Psychology*, 2 vols., London: Swan, Sonnenschein & Co., Ltd, and New York: Macmillan & Co., 1896.
Stow, David, *The Training System, Established in the Glasgow Normal Seminary, and Its model Schools*, Glasgow: Blackie and Son, 1840.
Straley, Jessica, 'Of Beasts and Boys: Kingsley, Spencer, and the Theory of Recapitulation', *Victorian Studies* 49/4 (2007), 583–609.
Strangways, A. H. Fox, *Cecil Sharp*, in collaboration with Maud Karpeles, Oxford: Oxford University Press, and London: Humphrey Milford, 1933.
Stratton, Stephen S., *Mendelssohn*, London: J. M. Dent & Sons, Ltd, 1901/1910.
Strauss, David Friedrich, *The Life of Jesus Critically Examined*, trans. Marian Evans [George Eliot], New York: Calvin Blanchard, 1856, translated from the fourth edition of David Friedrich Strauss, *Das Leben Jesu, Kritisch Bearbeitet,* Tübingen: C. F. Osiander, 1835–36/1840.
Stubbs, William, *The Constitutional History of England in its Origin and Development*, 3 vols., Oxford: Oxford University Press, 1874–78.
Stuchtey, Benedikt, and Peter Wende (eds.), *British and German Historiography 1750–1950: Traditions, Perceptions, and Transfers*, Oxford: Oxford University Press, 2000.
Sully, James, 'Animal Music', *The Cornhill Magazine* (Nov. 1879), 605–21.
Outlines of Psychology with Special Reference to The Theory of Education, London: Longmans, Green & Co., 1884.

Sutton, Emma, *Aubrey Beardsley and British Wagnerism in the 1890s*, Oxford: Oxford University Press, 2002.
Swainson, William, *A Preliminary Discourse on the Study of Natural History*, London: Longman, Ress, Orme, Brown, Green, and Longon, 1834.
Symcox, Geoffrey, 'The Wild Man's Return: The Enclosed Vision of Rousseau's Discourses', in Edward J. Dudley and Maximillian E. Novak (eds.), *The Wild Man Within: An Image in Western Thought from the Renaissance to Romanticism*, Pittsburgh: University of Pittsburgh Press, 1973, 223–248.
Tagore, Sourindro Mohun, *Hindu Music*, Delhi: Low Price Publications, 1882/1994.
'Hindu Music from the "Hindoo Patriot"', in *Hindu Music*, Delhi: Low Price Publications, 1882/1994, 339–97.
Six Principal Ragas with a Brief View of Hindu Music, Delhi: Low Price Publications, 1877/2006.
ṭhākura', śaurīndramohana, jātīyasaṅgītaviṣayakaprastāva [A Proposal Concerning National Music], Calcutta, 1870.
Universal History of Music Compiled from Divers Sources Together With Various Original Notes on Hindu Music, Delhi: Low Price Publications, 1896/1999.
Taneja, G. R., and Vinod Sena (eds.), *Literature: East and West*, Bombay: Allied Publishers Ltd, 1995.
Taylor, Edward, *The English Cathedral Service: Its Glory, Its Decline, and Its Designed Extinction*, London: Simpkin, Marshall and Co., 1845.
Taylor, Sedley, 'On a Suggested Simplification of the Established Pitch-Notation', *Proceedings of the Musical Association* 1 (1874–75), 18–40.
Sound and Music: A Non-Mathematical Treatise on the Physical Constitution of Musical Sounds and Harmony, London: Macmillan & Co., 1873.
Temperley, Nicholas, 'Xenophilia in British Musical History', in Bennett Zon (ed.), *Nineteenth-Century British Music Studies*, 3 vols., Aldershot: Ashgate, 1999, 3–19.
Temple, Frederick, 'Lecture IV: Apparent Conflict Between Religion and the Doctrine of Evolution', in *The Relations between Religion and Science: Eight Lectures Preached Before the University of Oxford in the Year 1884* (London, 1903), no publisher, Project Gutenbergwww.gutenberg.org/files/17194/17194.txt.
Thakur, A. P., *The Legacy of Social Darwinism*, New Delhi: Global Vision Publishing House, 2005.
Thomas, David Wayne, *Cultivating Victorians: Liberal Culture and the Aesthetic*, Philadelphia: University of Pennsylvania Press, 2004.
Thoms, William, 'Folk-Lore', *Athenaeum* 982 (22 Aug. 1846), 862–63.
'Folk-Lore', *Athenaeum* 983 (29 Aug. 1846), 886–87.
Tiedemann, Friedrich, *Zoologie, zu seinen Vorlesungen Entworfen*, 3 vols., Landshut: Weber, 1808–14.
Tipper, Henry, *The Growth and Influence of Music in Relation to Civilization*, London: Elliot Stock, 1898.

Tomlinson, Stephen, 'From Rousseau to Evolutionism: Herbert Spencer on the Science of Education', *History of Education* 25/3, 235–54.
Tomoiaga, Ligia, Minodora Barbul and Ramona Demarcsek (eds.), *From Francis Bacon to William Golding; Utopias and Dystopias of Today and of Yore*, Newcastle: Cambridge Scholars Publishing, 2012.
Torrence, John (ed.), *The Concept of Nature*, Oxford: Oxford University Press, 1992.
Towheed, Shafquat, 'The Creative Evolution of Scientific Paradigms: Vernon Lee and the Debate over the Hereditary Transmission of Acquired Characteristics', *Victorian Studies* 49/1 (Autumn 2006), 33–61.
Tracey, James M., 'The Power of Music Over Animals', *The Musical Standard* 47/32 (11 Aug. 1894), 106–107.
Trasoff, David, 'Tagore, Sir Sourindro Mohun', *Grove Music Online, Oxford Music Online*, www.oxfordmusiconline.com/subscriber/article/grove/music/27381.
Treutler, W. J., 'Music in Relation to Man and Animals', *Proceedings of the Musical Association* 25 (1998–99), 71–91.
Trevelyan, George Otto, *The Life and Letters of Lord Macaulay*, London: Longmans, Green, 1876.
Trollope, Anthony, *The Warden*, 1855.
Turner, Frank M., 'Newman', in David A. Fergusson (ed.), *The Blackwell Companion to Nineteenth-Century Theology*, Malden, MA, and Oxford: Wiley Blackwell, 2010, 119–38.
'John Henry Newman and the Challenge of a Culture of Science', *The European Legacy* 1/5 (1996), 1694–1704.
Turner, John, *Manual of Instruction in Vocal Music*, London, 1833.
Tylor, E. B., *Anahuac, or Mexico and the Mexicans, Ancient and Modern*, London: Longman, Green, Longman, and Roberts, 1861.
'On the Survival of Savage Thought in Modern Civilization', *Notices of the Proceedings at Meetings of the Members of the Royal Institution of Great Britain*, 5 (1869), 522–35.
'On Traces of the Early Mental Condition of Man', *Notices of the Proceedings at Meetings of the Members of the Royal Institution* 5, 1867, 83–93.
Primitive Culture: Researches into the Development of Mythology, Philosophy, Religion, Language, Art and Custom, 2 vols., New York: Henry Holt and Co., 1871/1889.
Researches into the Early History of Mankind and the Development of Civilization, London: John Murray, 1870.
Urry, James, *Before Social Anthropology: Essays on the History of British Anthropology*, Reading: Harwood Academic Publishers, 1993.
Voss, Julia, *Darwin's Pictures: Views of Evolutionary Theory, 1837–1874*, trans. Lori Lanz, New Haven and London: Yale University Press, 2007/2010.
Walker, Ernest, *A History of Music in England*, Oxford: Clarendon Press, 1907.

Wallace, William, *The Threshold of Music: An Inquiry into the Development of the Musical Sense*, London: Macmillan and Co., Ltd, 1908.
Wallaschek, Richard, *Primitive Music: Inquiry into the Origin and Development of Music, Songs, Instruments, Dances, and Pantomimes of Savage Races*, London and New York: Longmans, Green and Co., 1893.
Walls, Jerry L., *The Oxford Handbook of Eschatology*, Oxford: Oxford University Press, 2008.
Waltraud, Ernst, and Bernard Harris (eds.), *Race, Science and Medicine, 1700–1960*, London and New York: Routledge, 1999.
Warriner, John, *National Portrait Gallery of British Musicians*, London: Sampson, Low & Co., 1896.
Waterhouse, Celia, 'An Appraisal of the Norwich Sol-Fa Method and Materials for Choir Training and Music Teaching, Devised by Sarah Anna Glover in 19th Century Norwich', www.britishkodalyacademy.org/public_downloads/Sarah%20Glover's%20Music%20Education%20Scheme.pdf.
Watson, John Richard, 'Echoes of That Voice: Romantic Echoes in Victorian Poetic Birdsong', in Andrew Radford and Mark Sandy (eds.), *Romantic Echoes in the Victorian Era*, Aldershot: Ashgate, 2008, 155–74.
The English Hymn: A Critical and Historical Study, Oxford: Oxford University Press, 1999.
Watt, Emma, *The Cheesecake Bible*, Homemade Cookbook, 9, n.p.: Emma Watt, 2015.
Weber, Friedrich, *A Popular History of Music From the Earliest Times*, London: Simpkin, Marshall, Jamilton, Kent & Co., 1891.
Weliver, Phyllis, *Woman Musicians in Victorian Fiction: Representations of Music, Science and Gender in the Leisured Home*, Aldershot: Ashgate, 2000.
Weliver, Phyllis (ed.), *The Figure of Music in Nineteenth-Century British Poetry*, Aldershot: Ashgate, 2005.
Westell, W. Percival, *The Young Ornithologist: A Guide to the Haunts, Homes, and Habits of British Birds*, London: Methuen and Co. Ltd, 1911.
Westfall, Richard S., 'The Scientific Revolution of the Seventeenth Century: A New World View', in John Torrence (ed.), *The Concept of Nature*, Oxford: Oxford University Press, 1992, 63–93.
Wheeler, Michael, *Heaven, Hell, and the Victorians*, Cambridge: Cambridge University Press, 1994.
Whewell, William, 'The General Bearing of the Great Exhibition on the Progress of Art and Science', *Lectures on the Results of the Exhibition of 1851, Delivered Before The Society of Arts, Manufactures, and Commerce*, London: David Bogue, 1852, 3–34.
White, Andrew Dickson, *A History of the Warfare of Science with Theology in Christendom*, 2 vols., New York: D. Appleton & Co., 1896.
White, Hayden, *The Historical Imagination in Nineteenth-Century Europe*, Baltimore and London: Johns Hopkins University Press, 1973/1975.

Wilcox, Peter C., *John Henry Newman, Spiritual Director 1845–1890*, Eugene, OR: Pickwick Publications, 2013.

Wilderspin, Samuel, *Early Discipline Illustrated, or The Infant System Progressing and Successful*, London: Westley and Davis, 1832.

On the Importance of Educating the Infant Poor, London: T. Goyder, 1823.

Wiley, Christopher Mark, '"A Relic of an Age still Capable of a Romantic Outlook": Musical Biography and The Master Musicians Series, 1899–1906', *Comparative Criticism* 25 (2004), 161–202.

'Re-writing Composer's Lives: Critical Historiography and Musical Biography', PhD diss., Royal Holloway University of London, 2008.

Wilhem, G. L. Bocquillon, *Wilhem's Method of Teaching Singing: Adapted to English Use* (*Manuel Musical*, 1841), new ed., rev. and improved, London: John W. Parker, 1841.

Willard, N. Augustus, *A Treatise on the Music of Hindoostan: Comprising a Detail of the Ancient Theory and Modern Practice*, Calcutta: Baptist Mission Press, 1834, reprinted in Sourindro Mohun Tagore, *Hindu Music*, Delhi: Low Price Publications, 1882/1994, 1–122.

Willeby, Charles, *Masters of English Music*, London: J. R. Osgood, McIlvaine & Co., 1893.

Williams, C. F. Abdy, *Handel*, London: J. M. Dent & Co., 1901.

Williams, Ralph Vaughan, 'English Folk-Song', *The Musical Times* (1 Feb. 1911), 101–104.

National Music, London: Oxford University Press, 1934.

Wilson, Daniel J., 'Lovejoy's The Great Chain of Being after Fifty Years', *Journal of the History of Ideas* 48/2 (April–June 1987), 187–206.

Wilson, Flora, 'Hearing Things: Musical Objects at the 1851 Great Exhibition', in James Q. Davies and Ellen Lockhart (eds.), *Sound Knowledge: Music and Science in London, 1789–1851*, Chicago and London: Chicago University Press, 2016, 227–45.

Wilson, John, 'On the Analogy Between the Growth of Individual and National Genius', *Blackwood's Edinburgh Magazine* 6/34 (Jan. 1820), 375–81.

Witchell, Charles A., *Cries and Call-Notes of Wild Birds*, London: L. Upcott Gill, 1899.

The Cultivation of Man According to the Teachings of Common-Sense: Being a Treatise on the Application to Man of the Laws that He Applies to All Other Creatures, London: W. Steward & Co., 1904.

The Evolution of Bird-Song, London: Adam and Charles Black, 1896.

'The Evolution of Bird-Song', *The Zoologist*, Third Series, 14/163 (1890), 233–46.

The Natural History of Sin, Cheltenham, 1905.

Wolf, Joseph, *Zoological Sketches, made for the Zoological Society of London, from Animals in their Vivarium in the Regent's Park, Edited, with Notes, by D. W. Mitchell B.A., F.L.S., Secretary to the Society*, London, 1856.

Wolfsheim, S. M., *Kindergarten Child's Piano-Instructor*, New York: G. Schirmer, 1876.
Wormell, Deborah, *Sir John Seeley and the Uses of History*, Cambridge: Cambridge University Press, 1980.
Wylde, Henry, *Music in Its Art-Mysteries*, London: L. Booth, 1867.
Wyndham, Henry Saxe, *Arthur Sullivan*, London: G. Bell & Sons, 1903.
Yanni, Carla, 'Development and Display: Progressive Evolution in British Victorian Architecture and Architectural Theory', in Bernard Lightman and Bennett Zon (eds.), *Evolution and Victorian Culture*, Cambridge: Cambridge University Press, 2014, 227–60.
Nature's Museums: Victorian Science the Architecture of Display, New York: Princeton Architectural Press, 2005.
Yannis, Kanarakis, *Victorian Network* 2/1 (Summer 2010), 88–105.
Yardley, Mary (ed.), *Select Discourses from The Idea of a University*, Cambridge: Cambridge University Press, 1931.
Yi, Susan Won (ed.), *Music, Mind and Science*, Seoul: Seoul National University Press, 1999.
Young, Edward, *Conjectures on Original Composition*, London: A. Millar and R. and J. Dodesley, 1759.
Young, Paul, *Globalization and the Great Exhibition*, Houndsmill: Palgrave Macmillan, 2009.
'Mission Impossible: Globalization and the Great Exhibition', in Jeffrey A. Auerbach and Peter H. Hoffenberg (eds.), *Britain, the Empire, and the World at the Great Exhibition of 1851*, Aldershot: Ashgate, 2008, 3–25.
Young, Percy M., *A History of British Music*, London: Ernest Benn Ltd, 1967.
Young, Robert M., *Mind, Brain and Adaptation in the Nineteenth Century: Cerebral Localization and Its Biological Context from Gall to Ferrier*, Oxford: Oxford University Press, 1970, http://human-nature.com/mba/chap5.html.
Ziter, Edward, 'Teaching Nineteenth-Century Orientalist Entertainments', in Diane Long Koeveler and Jeffrey Cass (eds.), *Interrogating Orientalism: Contextual Approaches and Pedagogical Practices*, Columbus: Ohio State University Press, 2006, 224–244.
Zon, Bennett, *The English Plainchant Revival*, Oxford: Oxford University Press, 1999.
'From Great Man to Fittest Survivor: Reputation, Recapitulation and Survival in Victorian Concepts of Wagner's Genius', *Musicae Scientiae* (2009–10), 415–45.
'Histories of British Music and the Land Without Music: National Identity and the Idea of the Hero', in Emma Hornby and David Maw (eds.), *Essays on the History of English Music: Style, Performance, Historiography*, Woodbridge: Boydell & Brewer, 2010, 311–24.
'"Loathsome London": Ruskin, Morris, and Henry Davey's *History of English Music* (1895)', *Victorian Literature and Culture* 37 (2009), 359–75.

'"More than the Sum of the Life of Its Several Parts": Individual Differences in the Psychology and Ethnomusicology of C. S. Myers', *Nineteenth-Century Contexts* 29/4 (Dec. 2007), 373–88.

'Music', in *The Oxford Handbook of Nineteenth-Century Christian Thought*, Oxford: Oxford University Press, 2016.

Music and Metaphor in Nineteenth-Century British Musicology, Aldershot: Ashgate, 1999.

'The Non-Darwinian' Revolution and the Great Chain of Musical Being', in Bernard Lightman and Bennett Zon (eds.), *Evolution and Victorian Culture*, Cambridge: Cambridge University Press, 2014, 196–226.

Representing Non-Western Music in Nineteenth-Century Britain, Rochester, NY: University of Rochester Press, 2007.

Zon, Bennett (ed.), *Nineteenth-Century British Music Studies*, 3 vols., vol. 1, Aldershot: Ashgate, 1999.

Zuckerman, Lord (ed.), *The Zoological Society of London, 1826–1976 and Beyond*, London: Academic Press for the Zoological Society of London, 1976.

Index